Centrifugal Pump Handbook

Centrifugal Pump Handbook

Third edition

Sulzer Pumps Ltd
Winterthur, Switzerland

ELSEVIER

AMSTERDAM • BOSTON • HEIDELBERG • LONDON
NEW YORK • OXFORD • PARIS • SAN DIEGO
SAN FRANCISCO • SINGAPORE • SYDNEY • TOKYO
Butterworth-Heinemann is an imprint of Elsevier

Butterworth-Heinemann is an imprint of Elsevier
The Boulevard, Langford Lane, Kidlington, Oxford OX5 1GB, UK
30 Corporate Drive, Suite 400, Burlington, MA 01803, USA

Second edition 1998
Third edition 2010

British Library Cataloguing in Publication Data
A catalogue record for this book is available from the British Library

Library of Congress Cataloging-in-Publication Data
A catalog record for this book is availabe from the Library of Congress

ISBN–13: 978-0-75-068612-9

For information on all Butterworth-Heinemann publications
visit our web site at books.elsevier.com

Printed and bound in the UK

10 11 12 13 14 10 9 8 7 6 5 4 3 2 1

Contents

Preface

This new edition of the *Centrifugal Pump Handbook* takes full account of the recent progress that has been made in pump construction and technology.

- Close contact with the market has opened up new fields of application.
- Consistent evaluation of operating experience, backed by continuous collaboration with pump users, has led to improved operating performance and reliability.
- Intensified research and development using powerful computer modeling tools in areas such as cavitation, rotor dynamics and materials technology have led to steep rises in output.

The aim of the *Handbook* is to provide an overview of the current state of the art in pump construction. It is intended for planners and operating companies alike. To make the content accessible we have included the most important information but in some instances limited the degree of detail provided to that we believe is most commonly required.

All the experience gained by ourselves and others in the industry has been assembled and related to various common fields of application. Problem aspects such as cavitation, erosion, selection of materials, rotor vibration behavior, forces acting on pumps, operating performance in various systems, drives, acceptance testing and many more are dealt with in detail.

The *Handbook* is the product of collaboration by many members of the Sulzer Pumps team and we take this opportunity to thank all our personnel for their support in this project.

Chapter | one

Physical Principles

1.1 ENERGY CONVERSION IN CENTRIFUGAL PUMPS

In contrast to displacement pumps, which generate pressure hydrostatically, energy is converted in centrifugal pumps by hydrodynamic means. A one-dimensional representation of the complex flow patterns in the impeller allows the energy transfer in the impeller to be computed from the fluid flow momentum theorem (Euler equation) with the aid of vector diagrams as follows (Fig. 1.1):

The torque acting on the impeller is defined as:

$$T_{LA} = \rho Q_{LA}(c_{2u}R_2 - c_{ou}R_1) \tag{1}$$

With $u = R\omega$, the energy transferred to the fluid from the impeller is defined as:

$$P_{LA} = T_{LA}\omega = \rho Q_{LA}(c_{2u}u_2 - c_{ou}u_1) \tag{2}$$

The power transferred per unit mass flow to the fluid being pumped is defined as the specific work Y_{LA} done by the impeller. This is derived from equation (2) as:

$$Y_{LA} = \frac{P_{LA}}{\rho Q_{LA}} = c_{2u}u_2 - c_{ou}u_1 \tag{3}$$

The useful specific work Y delivered by the pump is less than that done by the impeller because of the losses in the intake, impeller and diffuser.

These losses are expressed in terms of hydraulic efficiency η_h:

$$Y = \eta_h Y_{LA} = \eta_h(c_{2u}u_2 - c_{ou}u_1) \tag{4}$$

The specific work done thus depends only on the size and shape of the hydraulic components of the pump, the flow rate and the peripheral velocity. It is independent of the medium being pumped and of gravitational acceleration. Therefore any given pump will transfer the same amount of energy to completely different media such as air, water or mercury.

Centrifugal Pump Handbook. DOI: 10.1016/B978-0-7506-8612-9.00001-2

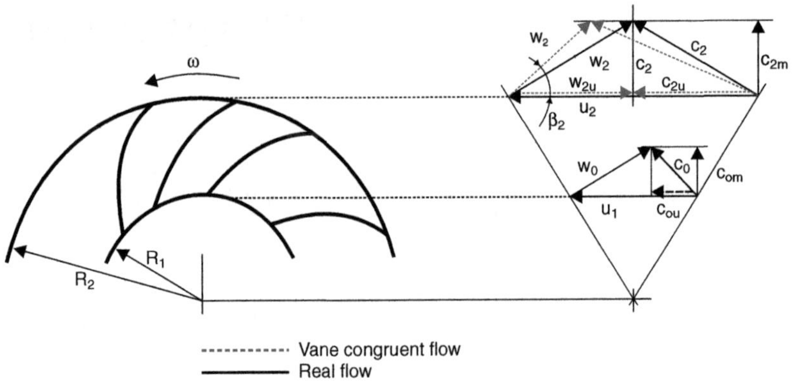

Figure 1.1 Vector diagrams

In order to use equation (4) to calculate the specific work done by the pump, the flow deflection characteristics of the impeller and all the flow losses must be known. However, these data can only be determined with sufficient precision by means of tests.

In all the above equations the actual velocities must be substituted.

If it were possible for the flow to follow the impeller vane contours precisely, a larger absolute tangential flow component $c_{2u\infty}$ would be obtained for a given impeller vane exit angle β_2 than with the actual flow c_{2u}, which is not vane-congruent (see Fig. 1.1). The difference between $c_{2u\infty}$ and c_{2u} is known as "slip". However, slip is not a loss that causes any increase in the pump's power consumption.

In order to examine the interrelationship between a pump and the pumping plant in which it is installed (Fig. 1.2), it is necessary to consider

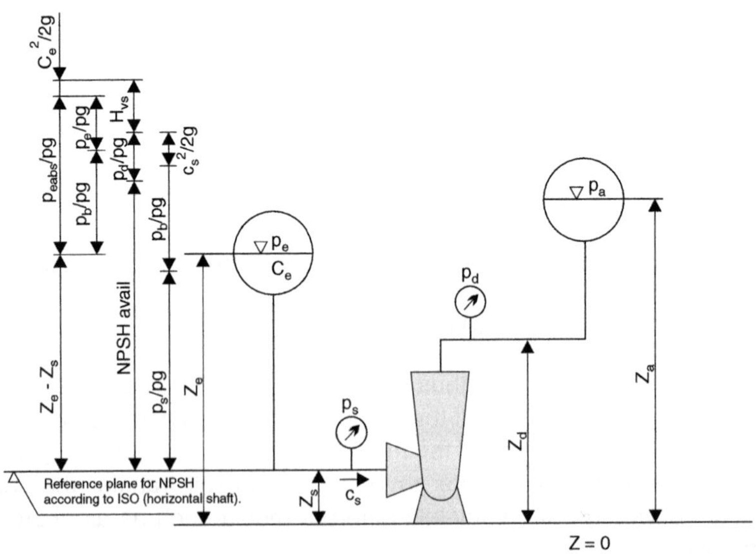

Figure 1.2 Pumping plant

the energy equation (Bernoulli equation). In terms of energy per unit mass of pumped fluid it can be written:

$$\underbrace{\underbrace{\frac{P_s}{Q} + Z_s \cdot g + \frac{C_s^2}{2}}_{\text{Pump suction nozzle}} + \underbrace{Y}_{\text{Pump}}}_{\text{Energy input}} = \underbrace{\frac{P_d}{Q} + Z_d \cdot g + \frac{C_d^2}{2}}_{\text{Energy output (pump discharge nozzle)}} \tag{5}$$

1.2 POWER, LOSSES AND EFFICIENCY

Equations (1) to (4) hold only as long as no part load recirculation occurs. The same assumption is implied below.

The impeller flow Q_{LA} generally comprises three components:

- the useful flow rate (at the pump discharge nozzle): Q;
- the leakage flow rate (through the impeller sealing rings): Q_L;
- the balancing flow rate (for balancing axial thrust): Q_E.

Taking into account the hydraulic losses in accordance with equation (4), the power transferred to the fluid by the impeller is defined as:

$$P_{LA} = \rho(Q + Q_L + Q_E)\frac{Y}{\eta_h} \tag{6}$$

The power input required at the pump drive shaft is larger than P_{LA} because the following losses also have to be taken into account:

- disc friction losses P_{RR} (impeller side discs, seals);
- mechanical losses P_m (bearings, seals);
- frictional losses P_{ER} in the balancing device (disc or piston).

The power input required at the pump drive shaft is calculated from:

$$P = \rho(Q + Q_L + Q_E)\frac{Y}{\eta_h} + P_{RR} + P_m + P_{ER} \tag{7}$$

If the volumetric efficiency η_v is:

$$\eta_v = \frac{Q}{Q + Q_L + Q_E} \tag{8}$$

the power input required by the pump can be written as:

$$P = \frac{\rho \cdot Q \cdot Y}{\eta_v \cdot \eta_h} + P_{RR} + P_m + P_{ER} \tag{9}$$

Pump efficiency is defined as the ratio of the useful hydraulic power $P_Q = \rho \cdot Q \cdot Y$ to the power input P at the pump drive shaft:

$$\eta = \frac{P_Q}{P} = \frac{\rho \cdot Q \cdot Y}{P} = \frac{\rho \cdot Q \cdot g \cdot H}{P} \tag{10}$$

Pump efficiency can also be expressed in the form of individual efficiencies:

$$\eta = \eta_h \cdot \eta_v \left(\eta_m - \frac{P_{RR} + P_{ER}}{P} \right) \tag{10a}$$

If the hydraulic losses and disc friction losses are combined to:

$$\eta_{hR} = \eta_h \cdot \left(1 - \frac{P_{RR} + P_{ER}}{P \cdot \eta_m} \right) \tag{10b}$$

efficiency may also be expressed as:

$$\eta = \eta_{hR} \cdot \eta_v \cdot \eta_m \tag{10c}$$

In the above equation η_m represents the mechanical efficiency:

$$\eta_m = 1 - \frac{P_m}{P} \tag{10d}$$

The internal efficiency incorporates all losses leading to heating of the pumping medium, such as:

- hydraulic losses;
- disc friction losses;
- leakage losses including balancing flow, if the latter is returned to the pump intake as is normally the case

$$\eta_i = \frac{\rho \cdot Q \cdot \rho \cdot H}{P - P_m} \tag{10e}$$

The power balance of the pump expressed as equation (7) is illustrated in Fig. 1.2a, where:

P_{vh} = hydraulic power losses = $\rho \cdot Q \cdot g \cdot H \cdot \left(\frac{1}{\eta_h} - 1 \right)$

$$P_L = (Q_L + Q_E) \cdot \rho \cdot g \cdot \frac{H}{\eta_h}$$

P_{ER} = frictional losses in the balancing device

P_{Rec} = hydraulic losses created by part load recirculation at the impeller inlet and/or outlet.

The pump efficiency depends on various factors such as the type and size of pump, rotational speed, viscosity (Reynolds number), hydraulic

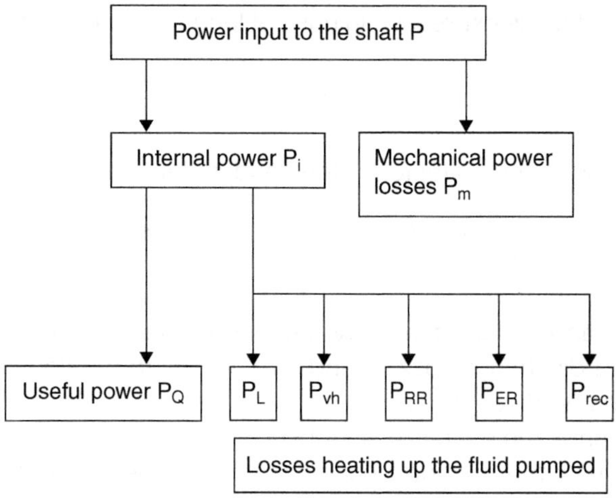

Figure 1.2a Power balance of a pump

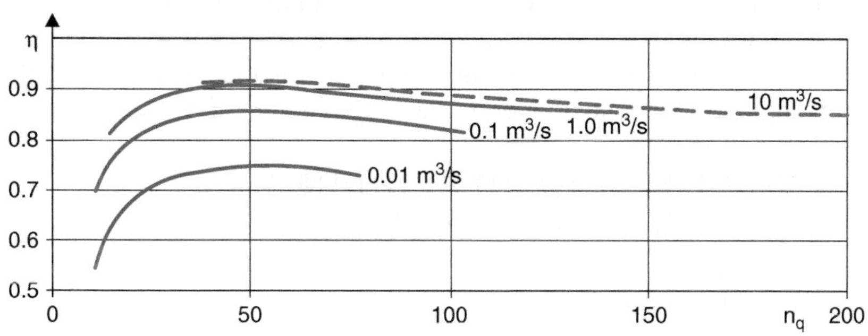

Figure 1.3 Efficiencies at the best efficiency point as a function of specific speed and flow rate

layout, surface finish and specific speed. Figure 1.3 shows the range of typical efficiencies. Maximum efficiencies are obtained in the range $n_q \cong 40$ to 60. At higher specific speeds efficiency starts to fall off due to an increasing proportion of hydraulic losses. With reduced specific speed there is above all a considerable increase in the proportion of disc friction losses, leakage and balancing flow losses.

For given values of density, flow rate and head the power input required by the pump can be calculated from:

$$P = \frac{\rho \cdot Q \cdot g \cdot H}{\eta}$$

1.3 TOTAL HEAD OF THE PUMP

In pump design it is normal practice to define the useful mechanical energy transferred to the pumped medium in terms of unit weight under gravitational acceleration, instead of unit mass as in equation (3).

This value, which is expressed in units of length, is known as the head H:

$$H = \frac{Y}{g} \tag{11}$$

By solving equation (5) for Y, the head is defined as:

$$H = \frac{Y}{g} = \frac{P_d - P_s}{\rho \cdot g} + Z_d - Z_s + \frac{c_d^2 - c_s^2}{2g} \tag{12}$$

The head can be determined by measuring the static pressure in the suction and discharge nozzles and by computing the flow velocities in the suction and discharge nozzles.

In summary, the head is a unit of energy and corresponds to the total head between the suction and discharge nozzles as defined by the Bernoulli equation. It is independent of the properties of the pumped medium, and for a given pump depends only on the flow rate and the peripheral speed of the impeller. In contrast to this, the pressure difference created by the pump and power consumption are proportional to fluid density. However, for extremely viscous fluids (such as oils) the pumping characteristics depend to a certain extent on viscosity.

1.4 THE TOTAL HEAD REQUIRED BY A PUMPING PLANT

In order to ensure that the correct pump is selected for a given application, the total head H_A required by the plant at design flow must be determined. This is again derived from the Bernoulli equation including all relevant head losses in the circuit (Fig. 1.2). Generally the pump draws from a reservoir whose surface is at a pressure p_e and delivers to a second reservoir at a pressure p_a (p_a and/or p_e can be at atmospheric pressure). Using the definitions shown in Fig. 1.2, the plant head requirement is then defined as:

$$H_A = \frac{p_a - p_e}{\rho \cdot g} + Z_a - Z_e + \frac{c_d^2 - c_s^2}{2g} + H_{vs} + H_{vd} \tag{13}$$

where:

H_{vs} = total head loss in the intake line

H_{vd} = total head loss in the discharge line

The head of the pump H must be selected to match the near H_A required by the plant: $H_A = H$ in accordance with equations (12) and (13).

Normally it is not advisable to apply large safety factors when determining the head requirement, since this leads to oversizing of the pump. Oversizing not only increases investment cost but is unfavorable because the pump then operates at part load, causing unnecessarily high energy consumption and premature wear.

1.5 CAVITATION AND SUCTION BEHAVIOR

1.5.1 Physical principles

The flow path over the impeller vane leading edge, as with an aerofoil, is subject to local excess velocities and zones in which the static pressure is lower than that in the suction pipe. As soon as the pressure at any point falls below the saturation pressure corresponding to the temperature of the liquid, vapor bubbles are formed. If the liquid contains dissolved gases, these are separated out to an increasing extent with falling pressure. The vapor bubbles are entrained in the flow (although the bubble zone often appears stationary to the observer) and implode abruptly at the point where the local pressure again rises above the saturation pressure. These implosions can cause damage to the material (cavitation erosion), noise and vibrations.

The criteria used for determining the extent of cavitation can best be illustrated by a model test at constant speed. For this purpose the suction head is reduced in steps at constant flow rate while measuring the head and observing the impeller eye with a stroboscope. Figure 1.4 shows such a test, with the suction head (NPSH) represented by a dimensionless cavitation coefficient. Points 1 to 5 correspond to the following situations:

1. Cavitation inception, $NPSH_i$: at this suction head the first bubbles can be observed on the impeller vanes (at higher suction heads no more cavitation occurs).
2. As the suction head is reduced, the bubble zone covers an increasing length of the impeller vane.
3. Start of head drop, $NPSH_0$ (0% head drop): if the suction head is reduced below the value at point 3, the head will start to fall off.
4. 3% head drop, $NPSH_3$, is a widely used cavitation criterion. This point is much easier to measure than the normally gradual onset of fall-off in the head, and is less subject to manufacturing tolerances.
5. Full cavitation, $NPSH_{FC}$: at a certain suction head, the head falls off very steeply ("choking").

Figure 1.4 also shows the distribution of cavitation bubbles over the impeller vanes and the amount of material removed by cavitation erosion, as a function of cavitation coefficient.

Figure 1.4a shows typical curves for the visual cavitation inception $NPSH_{inc}$, the NPSH required for 0%, 3% head drop and full cavitation (choking) as a function of flow. At shockless flow – where inlet flow angles match blade angles – $NPSH_{inc}$ shows a minimum. At flow rates above shockless, cavitation occurs on the pressure side of the blades, below shockless on the suction side. $NPSH_{inc}$ and $NPSH_{0\%}$ often show a maximum at part load where the effects of inlet recirculation become effective. When the rotation of the liquid upstream of the impeller, induced by the recirculation, is prevented by ribs or similar structures this maximum

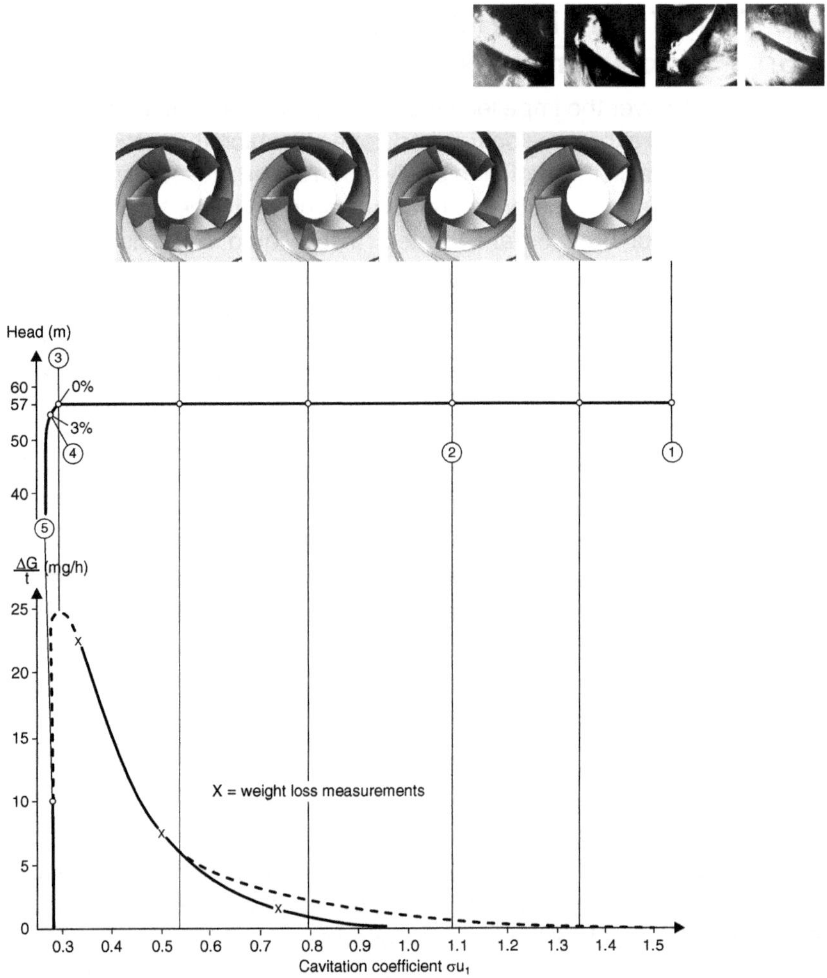

Figure 1.4 Cavitation bubble distribution and weight loss per unit time as a function of cavitation coefficient at constant head

can be suppressed and the NPSH will continue to increase with reduction in flow. $NPSH_{3\%}$ and full cavitation usually do not exhibit a maximum at part load; if they do, this in an indication of an oversized impeller eye which may be required in some applications where operation at flow rates much above BEP is requested.

Figure 1.5 shows the cavitation bubble distribution in the entire impeller eye.

1.5.2 Net positive suction head (NPSH)

The difference between total head (static head plus dynamic head) and the vapor pressure head at the pump inlet is defined by DIN 24260 as the

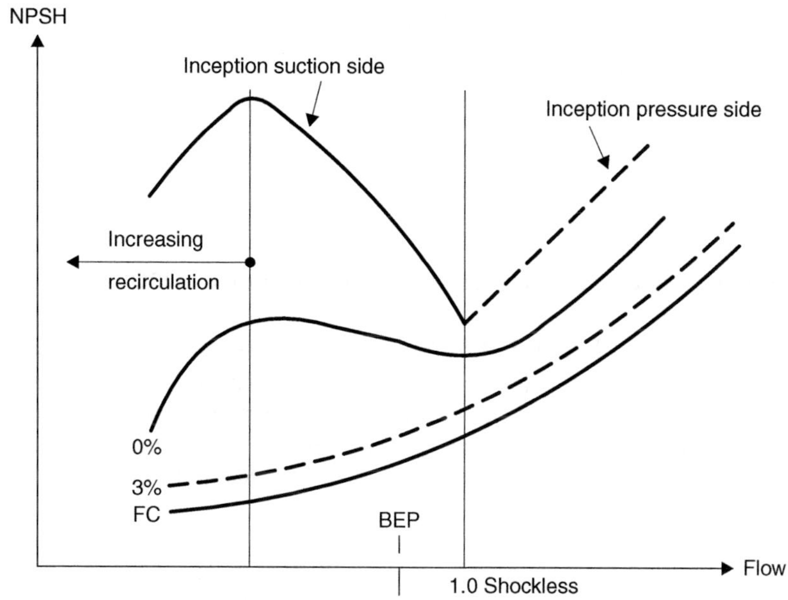

Figure 1.4a Typical NPSH curves

Figure 1.5 Cavitation at the impeller inlet

total suction head H_H:

$$H_H = \frac{p_{ges} - p_D}{\rho \cdot g} \tag{14}$$

If the suction head is referred to the centerline of a horizontal pump, it is identical to the actual NPSH value (net positive suction head). For vertical shaft or slanting pumps the reference plane as defined by the

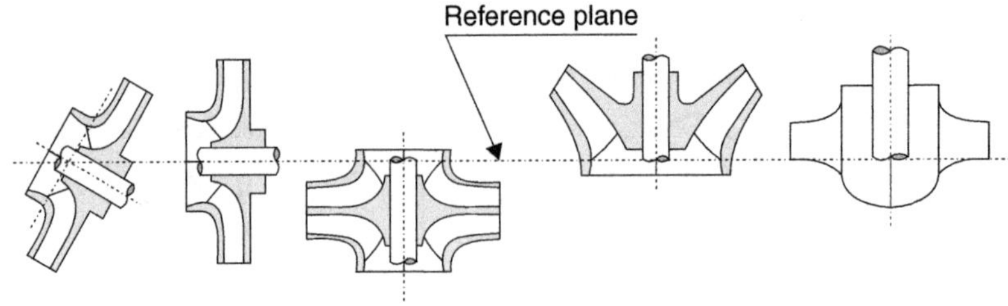

Figure 1.6 Reference plane for NPSH value according to ISO

ISO is shown in Fig. 1.6. The NPSH value determines the cavitation condition of the pump (see Fig. 1.4). Data on NPSH, suction head or cavitation coefficient of a pump must therefore always be related to the corresponding cavitation criterion (e.g. onset of cavitation bubbles, inception of head drop, 3% drop in head). The NPSH value for a pump is defined as:

$$\mathrm{NPSH} = \frac{p_{\mathrm{s_{abs}}} - p_{\mathrm{D}}}{\rho \cdot g} + \frac{c_{\mathrm{s}}^2}{2g} = \frac{p_{\mathrm{b}} + p_{\mathrm{s}} - p_{\mathrm{D}}}{\rho \cdot g} + \frac{c_{\mathrm{s}}^2}{2g} \tag{15}$$

where:

$p_{\mathrm{s_{abs}}}$ = absolute pressure in the suction pipe, related to the pump centerline $p_{\mathrm{s_{abs}}} = p_{\mathrm{b}} + p_{\mathrm{s}}$

p_{s} = gauge pressure (negative figures for pressures below atmosphere)

p_{b} = atmospheric pressure

p_{D} = absolute vapor pressure at the fluid temperature

c_{s} = flow velocity in the suction nozzle

The required $\mathrm{NPSH_R}$ of the pump defines the total pressure at the suction nozzle necessary for the pump to operate under *defined cavitation conditions* (e.g. with 1% head drop). Without specifying the cavitation criterion considered (e.g. $\mathrm{NPSH_3}$, $\mathrm{NPSH_i}$, etc.) any quantitative statement about the $\mathrm{NPSH_R}$ of a pump is (in principle) meaningless. However, in case the cavitation criterion is not specifically mentioned it is assumed that $\mathrm{NPSH_3}$ has been meant.

The NPSH value for a particular pump depends on the flow rate for a given speed and fluid.

Typical NPSH value for 3% head drop can be determined from Fig. 1.7a for pumps with overhung impellers, and from Fig. 1.7b for pumps with impellers between bearings.

The NPSH values given apply to the point of optimum efficiency.

In order to compare the suction performance of different pumps the suction-specific speed N_{SS} is a useful criterion:

$$N_{\mathrm{SS}} = n \frac{\sqrt{Q}}{\mathrm{NPSH}^{0.75}} \tag{15a}$$

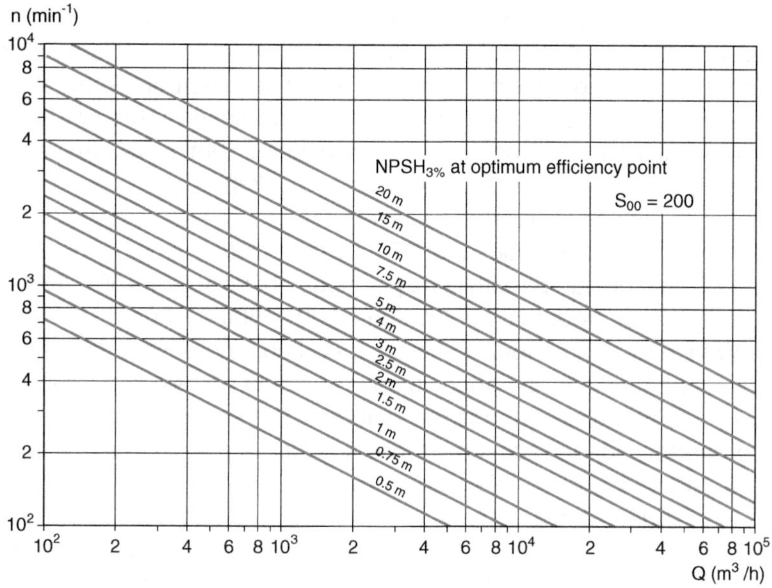

a)

Figure 1.7a NPSH at 3% head drop for pumps with overhung impeller, calculated for N_{SS} = 200

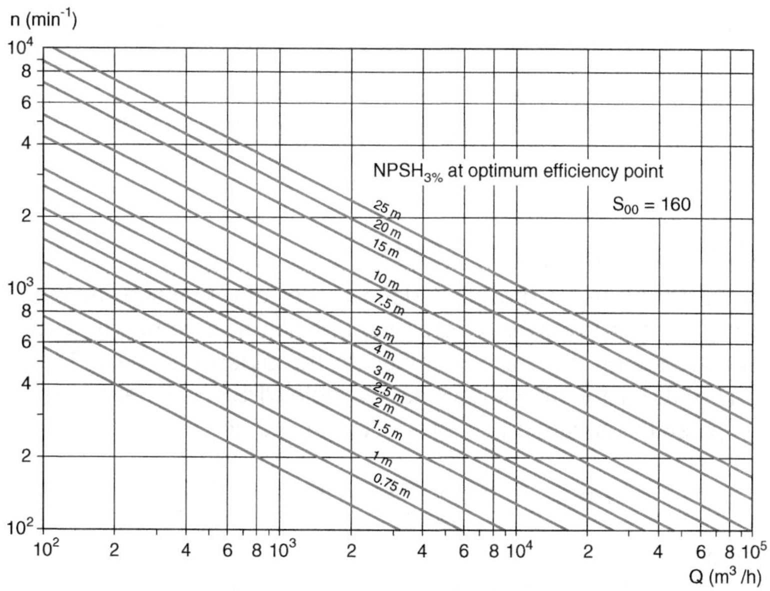

b)

Figure 1.7b NPSH at 3% head drop for pumps with impellers between bearings, calculated for N_{SS} = 200

It is defined with the flow per impeller eye for the best efficiency point and usually with the 3% head drop cavitation criterion. Depending on the impeller design, pump inlet (and to some extent on pump size) the suction-specific speed for a given type of pump can vary in a wide range. Typical values for N_{SS} as defined above (units: rpm, m^3/s, m) are:

- End suction pumps with axial inlet: 190 ÷ **230** ÷ 270
- Pumps with shaft through the eye:
  - moderate shaft blockage: 170 ÷ **200** ÷ 240
  - multistage pumps with head per stage above 500 m: 150 ÷ **180** ÷ 220
- Inducers for industrial applications: 350 ÷ **500** ÷ 700

With increasing impeller eye tip speed and increasing risk of cavitation erosion the suction-specific speed should be selected towards the lower end of the N_{SS} ranges given above. The suction capability can be improved by special suction impellers, which represent the upper range of N_{SS} figures, and double entry impellers.

In the 1980s it was suggested that the suction-specific speed should be limited for various pump types to some defined levels in order to reduce the risk of premature wear or damage to the pump. Such recommendations were based on statistical evaluations of pump failures in the petroleum industry which showed an increase of pump failures with increasing N_{SS}. The reasoning was that high N_{SS} required large impeller eye diameters which led to excessive and damaging part load recirculation. More recent investigations have shown that there is no unique relationship between onset of recirculation and N_{SS} and that correctly designed impellers with high N_{SS} give trouble-free long-term operation, if applied properly. It has also been recognized that pump failures attributed to large N_{SS} were most likely caused by impellers designed with excessive incidence, combined with an oversized inlet diameter. Therefore it cannot be recommended to attempt the prediction of suction recirculation from the N_{SS}, or to use a specific limit as a criterion for pump selection.

1.5.3 Available plant NPSH

The NPSH value of the plant defines the total pressure at the pump suction nozzle for given fluid characteristics at a certain flow rate (the dependency on flow rate is limited to the pipeline losses). The NPSH value of the plant, also known as available NPSH, is independent of the pump itself; it is defined according to Fig. 1.2 as:

$$\mathrm{NPSH_A} = \frac{p_{\mathrm{e\ abs}} - p_D}{\rho \cdot G} + Z_e - Z_s + \frac{c_e^2}{2g} - H_{vs} \tag{16}$$

where:

$p_{\mathrm{e\ abs}}$ = absolute pressure on the fluid surface at the intake

c_e = mean flow velocity in the intake vessel (or generally in the suction tank, usually very small).

When the pump is operating on the suction lift (water level below the impeller inlet), Z_e is negative. The term $(Z_e + Z_s)$ is then defined as the geodetic suction head.

If a boiling fluid is being pumped, the pressure above the water level is the vapor pressure ($p_{e\,abs} = p_D$); the pump can then only be operated with positive suction head (i.e. $Z_e > Z_s$).

NPSH figures stated for a pump without defining the corresponding cavitation criterion *usually refer to the suction head at 3% head drop.* For the majority of applications the available plant NPSH for continuous operation must exceed that at 3% head drop by an adequate safety margin in order to avoid a loss of performance, noise and vibrations or even cavitation erosion.

The required safety margin:

- increases with increasing peripheral speed at the impeller eye;
- is reduced if cavitation-resistant materials are used;
- increases with increasingly corrosive media;
- depends on the operating conditions of the pump and the type and temperature of the medium being pumped.

For suction impellers with high head per stage, stroboscopic observations of cavitation zones are often employed to establish the available plant NPSH necessary to avoid cavitation erosion. In such cases the necessary positive suction head is usually provided by a booster pump.

Figure 1.8 shows approximate values for selecting plant NPSH as a function of the NPSH at 3% head drop. Depending on the application,

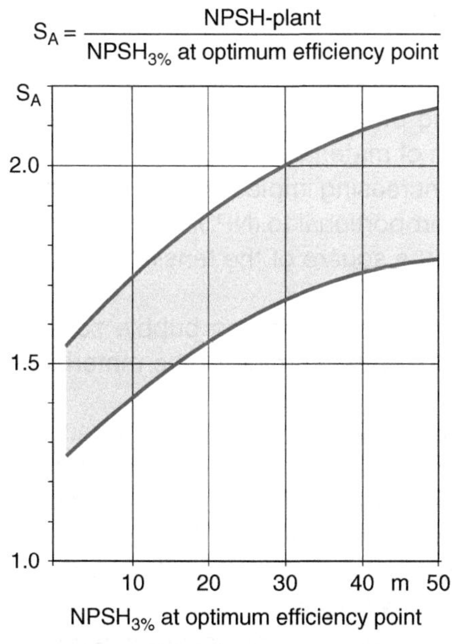

Figure 1.8 Approximate values for selecting the available $NPSH_A$ in the plant

deviations can be made from this curve; with seawater, for example, slightly higher margins are employed, and for hydrocarbons lower values can be used. If Fig. 1.8 leads to values less than 0.6 m, the absolute minimum to be maintained in the plant is $NPSH_{A,min} = 0.6$ m.

$$S_A = \frac{NPSH_A}{NPSH_{3\%} \text{ at optimum efficiency point}}$$

1.5.4 Methods of improving suction performance

In a great number of cases the highest permissible speed – and thus the size and cost – of a pump is determined by its suction performance or the required or available NPSH. For given values of flow, head and available plant NPSH the permissible speed can be increased by the following measures:

- using special suction impellers;
- installing a double-entry first stage impeller;
- installing an inducer;
- using a booster pump.

Inducers and suction impellers must be designed so that no pulsation (which could endanger the plant) occurs during part load operation. Suction performance increases slightly with increasing water temperature.

1.5.5 Cavitation erosion

When cavitation bubbles implode on the impeller or other pump components very high local pressures are generated which may exceed the fatigue strength, yield point or ultimate strength of the material. If this takes place, pitting of the material, known as cavitation erosion, may result. The erosion of material:

- increases with increasing implosion energy; the intensity of erosion is approximately proportional to $(NPSH)^3$ or the 6th power of the speed;
- decreases with the square of the tensile strength (or hardness of the material);
- increases with growing cavitation bubble zones, assuming that the bubbles implode on the surface of the material and not in the liquid (see Fig. 1.4);
- decreases with increasing water temperature;
- decreases with increasing gas content in the fluid, since non-condensable gases reduce implosion pressures;
- increases with more corrosive media;
- is usually considerably greater at part load than when pumping in the region of optimum efficiency.

Cavitation erosion hardly ever occurs when pumping hydrocarbon media.

1.6 MODEL LAWS AND SIMILARITY COEFFICIENTS

If all velocities shown in the vector triangles of Fig. 1.1 are divided by u_2, the resultant dimensionless vector triangles are valid for geometrically similar impellers of all sizes at the same relative inflow angle. By transforming equation (3), we obtain:

$$Y_{LA} = u_2^2\left(\frac{c_{2u}}{u_2} - \frac{c_{ou}}{u_2} \cdot \frac{D_1}{D_2}\right) \tag{17}$$

and it is apparent that the specific pumping work done by a given impeller at a given flow depends only upon u_2.

The flow rate can be expressed as:

$$Q = c_{2m} \cdot \pi \cdot D_2 \cdot B_2 \cdot = \left(\frac{c_{2m}}{u_2}\right) u_2 \cdot \pi D_2^2 \left(\frac{B_2}{D_2}\right) \tag{18}$$

Since u_2 is proportional to $n \cdot D_2$ these considerations lead to the following model laws for geometrically similar pumps. The subscript v refers to a given pump for which test results may, for example, be available:

$$\frac{Y}{Y_v} = \frac{H}{H_v} = \left(\frac{n}{n_v}\right) \cdot \left(\frac{D_2}{D_{2v}}\right)^2 \tag{19}$$

$$\frac{Q}{Q_v} = \frac{n}{n_v} \cdot \left(\frac{D_2}{D_{2v}}\right)^3 \tag{20}$$

$$\frac{P}{P_v} = \left(\frac{n}{n_v}\right)^3 \cdot \left(\frac{D_2}{D_{2v}}\right)^5 \tag{21}$$

$$\frac{c}{c_v} = \frac{n}{n_v} \cdot \frac{D_2}{D_{2v}} \tag{22}$$

$$\frac{\text{NPSH}}{\text{NPSH}_v} = \left(\frac{n}{n_v}\right)^2 \cdot \left(\frac{D_2}{D_{2v}}\right)^2 \tag{23}$$

The above equations also indicate the powers of D_2 and n required for converting the performance data into dimensionless coefficients. Of the various definitions of dimensionless coefficients, only the following will be given here:

$$\text{Head coefficient}: \psi = \frac{2gH}{u_2^2} = \frac{2Y}{u_2^2} \tag{24}$$

$$\text{Flow coefficient}: \varphi = \frac{c_{2m}}{u_2} = \frac{Q}{\pi \cdot D_2 \cdot B_2 \cdot u_2} \tag{25}$$

$$\text{Cavitation coefficient}: \sigma_{u1} = \frac{2g\,\text{NPSH}}{u_1^2} \tag{26}$$

Any pump characteristic represented in terms of these dimensionless coefficients is valid for all geometrically similar pumps, independently of their size and rotational speed.

Strictly speaking, the model laws are only applicable at constant hydraulic efficiency. This means that the relative surface roughness and Reynolds number of the model and full-scale pump must also be identical, which is an extremely difficult condition to fulfill. However, the resultant deviations are negligible in the majority of cases. The different types of impeller can best be compared by means of the specific speed (characteristic of the impeller shape):

Head coefficient $\psi = f(\text{specific speed } n_q)$

Impeller shape:

$$n_q = n \, . \, \frac{Q^{1/2}}{H_{st}^{3/4}}; \; \psi = \frac{2gH_{st}}{u_2^2}; \left(u_2 = \frac{D_2 \, . \, \pi \, . \, n}{60} \right)$$

where:

H_{st} = stage head in m
Q = flow rate in m^3/s
n = pump speed in rpm
D_2 = impeller diameter in m
g = acceleration due to gravity, normally = 9.81 m/s^2

$$n_q = n \frac{\sqrt{Q}}{H^{3/4}} \tag{27}$$

where n is measured in revolutions per minute, Q in m^3/s and H in m.

For multistage pumps the head per stage must be used in this equation, and for double-entry impellers the volumetric flow rate per impeller eye.

Figure 1.9 shows typical head coefficients per equation (24) as a function of specific speed, and in addition the way in which the impeller shape changes as a function of specific speed.

The shape of the head capacity and efficiency characteristics varies with the specific speed (Fig. 1.10): with increasing specific speed the characteristic becomes steeper and the efficiency curve more peaked. The NPSH characteristic is shown in Fig. 1.11. All the curves shown are to be considered only approximate characteristics. Depending on the type of pump and its hydraulic design, considerable differences can occur in the characteristics.

1.7 AXIAL THRUST

A pump rotor is subject to the following axial forces, which must be known in order to determine the capacity of the axial thrust bearing:

- hydraulic forces acting on the impellers, obtained by integrating the pressure distribution on the impeller shrouds (F_{SS}, F_{DS} in Fig. 1.12);

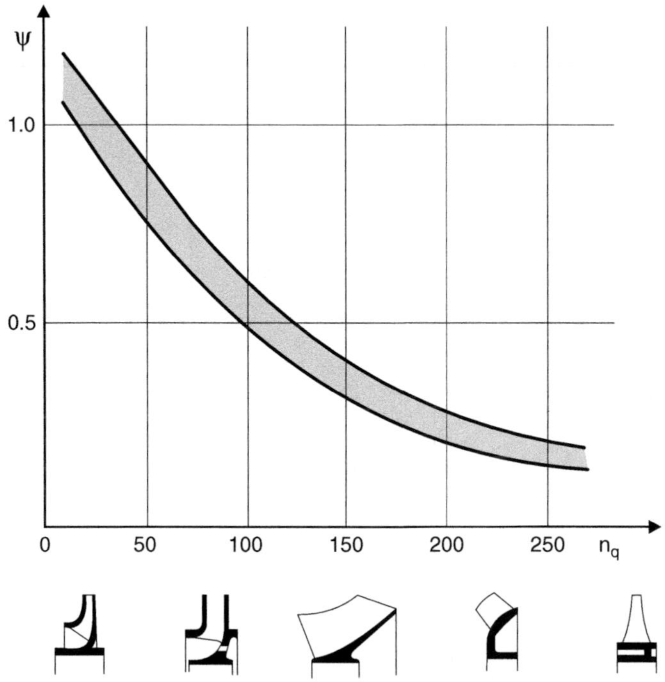

Figure 1.9 Impeller shapes and head coefficients

- momentum $F_1 = \rho Q c_{om}$ (for radial impeller outlets);
- rotor weight for vertical pumps;
- unbalanced shaft thrusts;
- forces exerted by any balancing device.

According to Fig. 1.12 the hydraulic axial force on the impeller is defined as:

$$F_{HY} = F_{DS} - F_{SS} - F_1 \tag{28}$$

In order to compute the forces acting on hub and shroud of the impeller, the pressure distribution must be known. On the hub this is given by:

$$p_{DS^{(R)}} = p_{DS} - K_{DS}^2 \frac{\rho \cdot \omega^2}{2} \left(R_2^2 - R^2\right) \tag{29}$$

The force acting on the hub is therefore:

$$F_{DS} = 2\pi \int_{R_{DS}}^{R_2} P_{DS}(R)\, R\, dR \tag{30}$$

Similar relationships apply to the shroud.

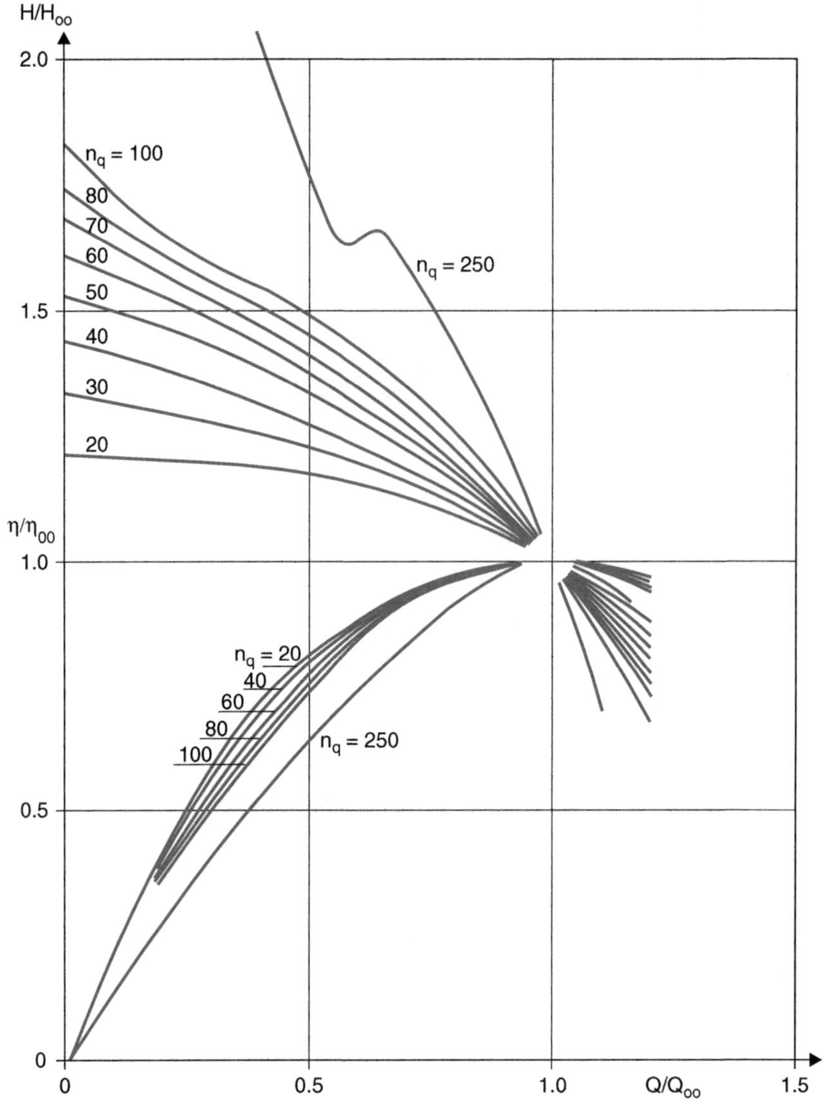

Figure 1.10 Approximate head and efficiency evolution as a function of flow rate ratio for various specific speeds

In general the increase in static pressure p_{DS} on the hub side may differ from the increase in pressure p_{SS} on the shroud side, which can be responsible for a portion of the axial thrust. The pressure distribution $p^{(R)}$ depends on the rotation of the water in the space between the impeller and the casing. This is expressed as $K = \beta/\omega$ (β = angular velocity of the fluid, ω = angular velocity of the impeller). The factor K depends on the size and direction of the leakage through the space between the impeller

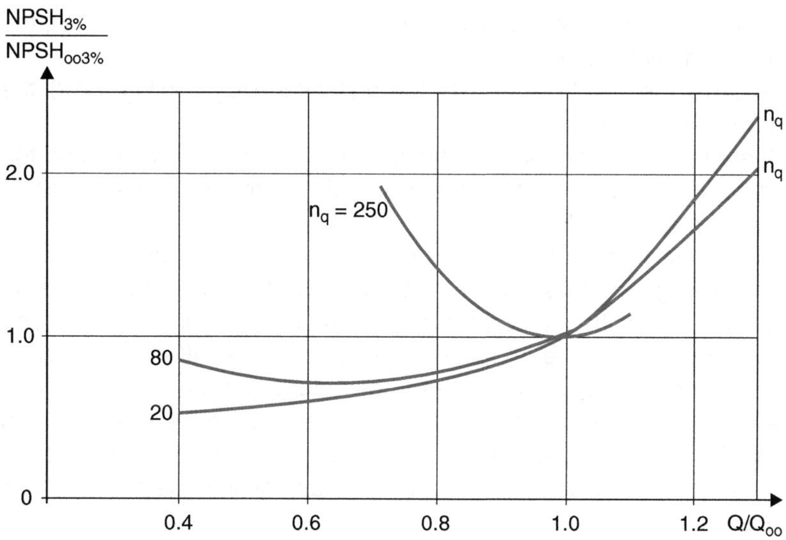

Figure 1.11 Approximate NPSH characteristics as a function of flow rate ratio for various specific speeds

and the casing (Q_{LSS} and Q_{LDS}; Fig. 1.12), and on the geometrical shape and surface roughness of that space:

- For narrow gaps between impeller and casing without leakage $K = 0.5$ is a good approximation.
- The larger the surface area and/or surface roughness of the pump casing compared to the impeller shroud, the greater the reduction of K below 0.5, assuming no leakage.
- By fitting ribs (rudimentary vanes) to the impeller shroud, K can be increased to a value approaching unity.

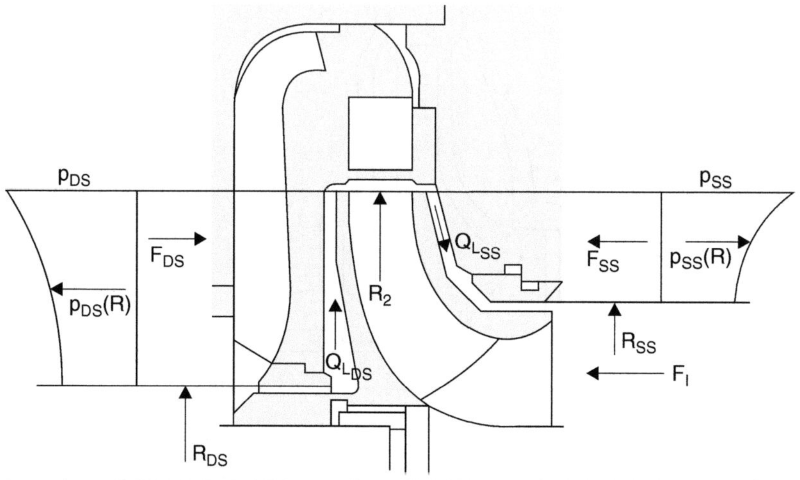

Figure 1.12 Axial forces acting on the impeller

- The larger the radially inwards flowing leakage (Q_{LSS} in Fig. 1.12), the greater the value of *K*.
- The larger the radially outwards flowing leakage (Q_{LDS} in Fig. 1.12), the smaller the value of *K*.
- If there is no leakage through the space between the impeller and the casing, *K* remains constant over the radius. If there is any leakage, *K* varies along the radius.

From these flow characteristics in the clearance between the impeller and the casing, it follows that for multistage, single-entry pumps the axial thrust increases with seal wear during operation. For pumps of this type the axial position of the impellers in relation to the diffusers, which is affected by manufacturing and assembly tolerances, can have an influence on the axial thrust. Since the forces acting on the impeller side plates (F_{DS}, F_{SS}) are large in comparison with the residual thrust, even small tolerances in p_{DS}, p_{SS}, K_{DS} and K_{SS} can cause large alterations in residual thrust.

The axial thrust can be reduced or partially compensated by:

- balancing holes and seals on the hub (see Fig. 1.13);
- rudimentary vanes on the hub (see Fig. 1.14);
- balancing pistons or seals (see Figs 1.15a and 1.15b);
- balancing disc (see Fig. 1.16);
- opposed arrangement of the impellers in multistage pumps ("back-to-back" design) (see Fig. 1.17).

All these measures (including the back-to-back arrangement of the impellers, in view of the overflow losses) affect pump efficiency.

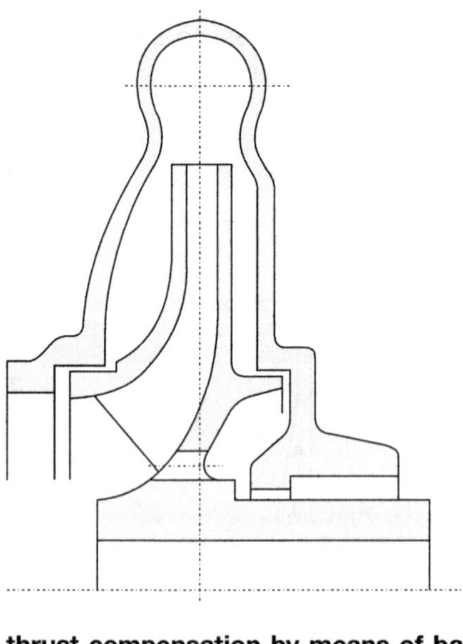

Figure 1.13 Axial thrust compensation by means of balancing holes and a seal

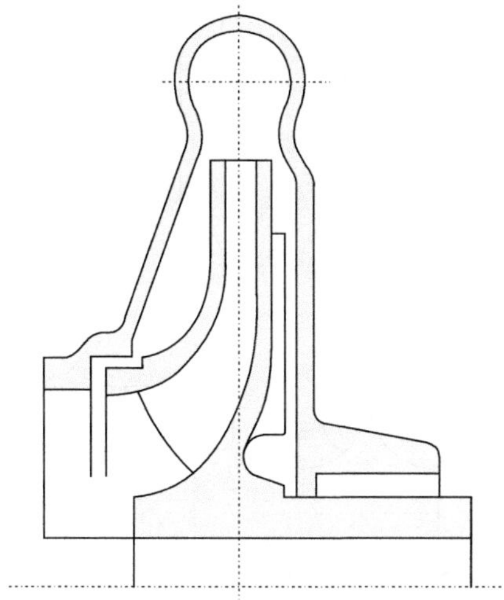

Figure 1.14 Impeller with rudimentary vanes

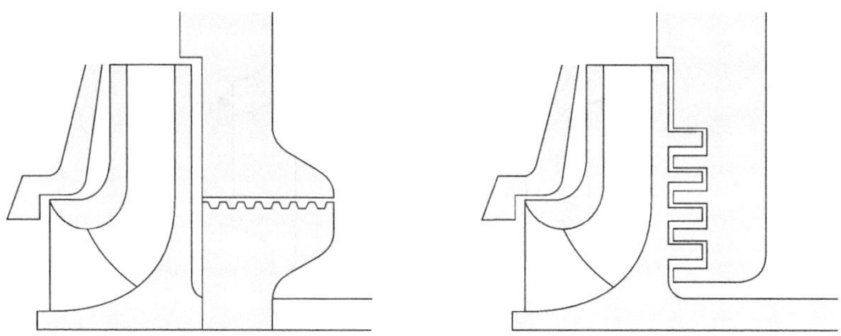

a) b)

Figure 1.15a Design with balancing piston

Figure 1.15b Design with balancing seals

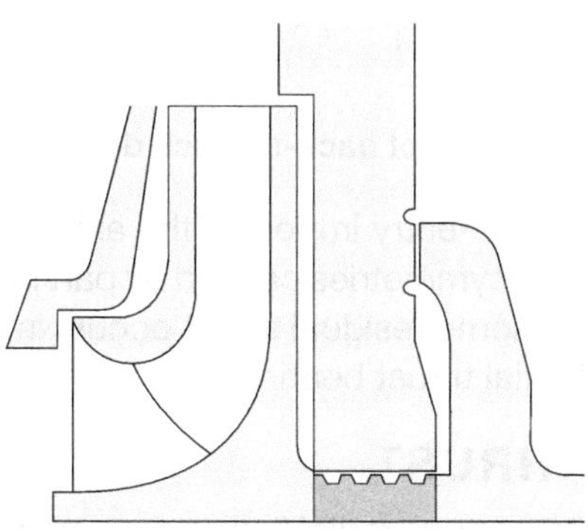

Figure 1.16 Design with balancing disc

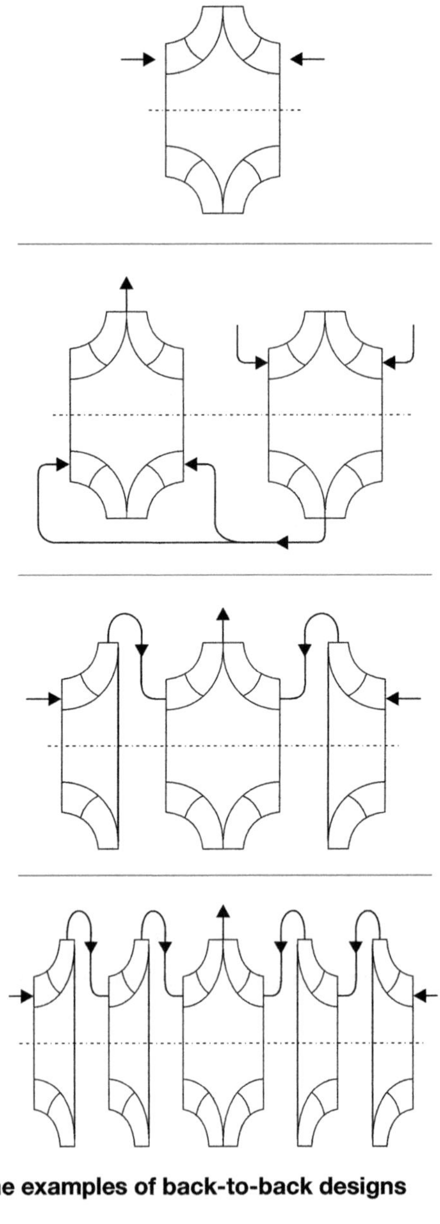

Figure 1.17 Some examples of back-to-back designs

In the case of double-entry impellers the axial thrust is theoretically balanced; due to flow asymmetries caused by part load recirculation and tolerances, however, some residual thrust occurs in practice and has to be absorbed by an axial thrust bearing.

1.8 RADIAL THRUST

In pumps with volute casings radial forces caused by non-uniform pressure distribution over the impeller circumference act on the impeller at off-design conditions. The radial thrust applicable with volute casings can be estimated from the following equation:

$$F_R = k_R \cdot \rho \cdot g \cdot H \cdot D_2 \cdot B_2^{*} \left[1 - \left(\frac{Q}{Q_{00}}\right)^2\right] \tag{31}$$

where B_2^{*} = impeller outlet width including shrouds.

The radial thrust factor K_R depends on the specific speed; for $n_q = 20$ approx., $K_R = 0.17$, increasing to $K_R = 0.38$ at $n_q = 60$.

For double volute casings with a common discharge branch, the radial thrust at 180° wrap angle between the two cutwaters is about 25% of the thrust for single volutes. With reducing wrap angle between the cutwaters, the radial thrust factor increases rapidly.

These radial thrust factors are based upon measurements carried out on pumps with annular seals. This type of seal exerts restoring forces on the impeller when it runs eccentrically to the pump casing. Since this is the case under the influence of hydraulic radial thrust, the radial thrust computed from equation (31) represents a combination of the hydraulic forces acting on the impeller and the forces acting in the seal. With increasing clearances, the resultant radial force thus becomes larger, and for radial flow seals higher radial thrusts can be expected than those obtained from equation (31). Wide impeller side spaces facilitate pressure equalization over the periphery and consequently can reduce the radial thrust.

Pumps with diffusers are subject to lower radial thrusts than pumps with volute casings. The steady component of the radial thrust becomes greater with increasing eccentricity of the impeller in the casing, while the unsteady component is greatest during concentric running.

1.9 ROTOR DYNAMICS

Rotor dynamics is concerned with the vibrational behavior of shafts. This includes both lateral and torsional vibrations. In the case of lateral vibrations, the critical speed is an important factor. This is defined as the speed at which the rotation frequency coincides with one of the natural bending frequencies of the rotor. In general, critical speeds should be excluded from the operational ranges of pumps. This is important especially when rotor damping is low. In order to ensure operational reliability of high-head pumps, it is important to make a careful analysis of the vibrational behavior of the rotor. For carrying out this analysis, a number of sophisticated computer programs are available today, and apart from analyzing complex rotor shapes, they can also take account of the bearings and seals. These programs deliver data not only on the natural frequencies associated with each vibration mode, but also on the corresponding damping values for each frequency. The forces occurring in the seals and any balancing pistons have a considerable influence on the vibrational behavior of the rotor. The restoring force acting in opposition to rotor deflection has the effect of increasing the natural frequency,

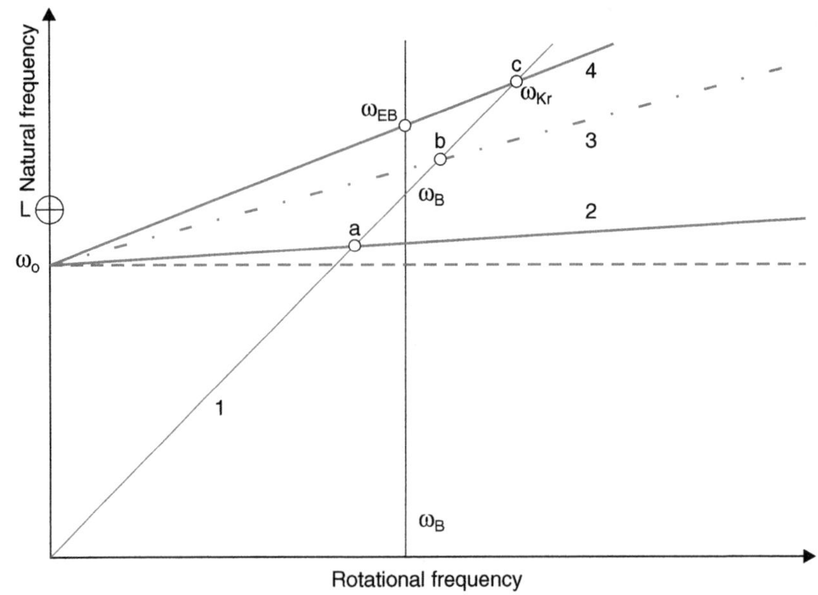

Figure 1.18 Natural frequencies of a multistage pump shaft as a function of the operating speed in air and water and of variable seal clearances

while the transverse force acting in the tangential direction can lead to instabilities in the form of self-excited vibrations.

In Fig. 1.18, the natural frequencies of a rotor are plotted against rotational frequency (speed). Such a plot is called a Campbell diagram. Point L corresponds to the natural frequency of the rotor in air. The stationary rotor has a slightly lower natural frequency in water, because the water has the effect of an additional mass (point ω_o). With increasing rotational speed, the pressure differential across the seals and balancing piston increases, as a result of which the above-mentioned restoring forces are built up and the natural frequency, which then depends on the speed, is raised. Curve 4 corresponds to the seal clearance in the new condition, curve 3 represents increased clearance, and curve 2 would be obtained if clearances were infinitely large. Curve 2 is not parallel to the abscissa, because the bearing stiffness is a function of the rotational speed. The intersection points a to c of this curve with the line of synchronous excitation ("1×" line) (line 1) represent the critical speeds of the rotor, which become lower with increasing seal wear. Calculation of the natural frequencies at operating speed (i.e. the frequency at which the rotating rotor vibrates when it is excited into bending vibration by a blow) shows that these are nearer to the operating speed than the critical speeds:

$$\frac{\omega_{EB}}{\omega_B} < \frac{\omega_{kr}}{\omega_B}$$

(see Fig. 1.18)

Computer programs for determining natural bending frequencies and damping rates supply several natural frequencies with various vibration modes for complex rotors of multistage pumps.

Apart from programs for computing natural frequencies, there are also programs for computing forced vibrations. These allow the calculation of the amplitudes of shaft vibrations caused by imbalance forces, for example. Vibrations excited by imbalance forces occur at the rotation frequency (synchronous vibrations) or at higher harmonic frequencies. Subsynchronous vibrations become noticeable in the case of self-excited vibrations or in the case of rotating stall or broad band excitation at part load. Self-excited vibrations can occur, for example, in boiler feed pumps with heavily worn seals, since with increasing seal clearance, not only the natural frequencies but also the rotor damping rates are reduced.

One of the main tasks of rotor dynamics, especially in the case of high performance pumps, is to optimize the rotor design so that self-excited vibrations and resonances are avoided. The natural frequency, vibration mode configuration and damping depend upon a large number of factors:

• Rotor geometry:	bearing span shaft diameter mass distribution overhung shaft sections
• Radial bearings:	type relative clearance loading width to diameter ratio oil viscosity (temperature)
• Axial bearings:	mass overhang stiffness damping loading
• Shaft coupling:	type mass overhang alignment errors
• Annular seals, balance pistons:	seal clearance length diameter clearance surface geometry pressure differential Reynolds number, axial Reynolds number, tangential inlet and outlet geometry of the gap balance piston/sleeve misalignment pre-rotation

• Hydraulic excitation forces due to hydraulic unbalance interaction:	Forces due to hydraulic unbalance interaction (unequal blade loading) and vane passing restoring and lateral forces caused by impeller/diffuser interaction dynamic axial thrusts flow separation phenomena (hydraulic instabilities).

Torsional vibrations are often given less attention in pump units although they can be of importance as well. Because torsional vibrations are very difficult to detect and monitor, the first sign of a torsional vibration problem is typically a failure, e.g. of a coupling, or even a shaft rupture.

Torsional excitation from the driver is important in the following cases (list not exhaustive):

- Startup of asynchronous electric motors (switching to the grid):
  - Excitation frequency
  - Grid frequency
- Short circuit at motor terminal. Excitation frequencies:
  - Grid frequency
  - Twice grid frequency
- Normal operation of variable speed electric motors
- Excitation frequencies depend on motor control. Examples:
  - Harmonics of rotational frequency
  - Slip-dependent frequencies
- Normal operation of reciprocating engines (diesel)
- Excitation frequencies. Harmonics of rotational frequency.

Some torsional excitation originates in speed increasing/reducing gears due to pitch circle runout. The magnitude of excitation depends on the quality of the gear. Excitation frequencies: the two rotational frequencies. Gear tooth mesh frequencies are of less importance.

Torsional excitation from pump impellers: typically very small, provided the vane numbers of impeller and stator have been selected properly.

This can be of some importance for waste water pumps with a small number of thick impeller vanes.

In cases considered critical (e.g. large size, high reliability required) a torsional natural frequency analysis shall be performed. Sophisticated computer programs are available today to perform this task. It is important to note that such an analysis has always to be performed for the entire shaft train including driver, couplings, gear and pump. Resonance with important excitation frequencies shall be avoided whenever possible; torsional damping in shaft trains is very low.

Often a forced torsional vibration analysis is also performed, especially when a resonance situation cannot be avoided. As a possible measure to reduce the fluctuating stress level in such a case the installation of damper couplings may be considered.

Chapter | two

Behavior of Centrifugal Pumps in Operation

2.1 CHARACTERISTICS OF CENTRIFUGAL PUMPS

The characteristic curves indicate the behavior of a pump under changing operating conditions (Fig. 2.1). The head H, power input P and efficiency η at constant speed n are plotted against the flow rate Q (Fig. 2.2).

Figure 2.3 shows the head-capacity curves (Q–H curves) at various speeds and the iso-efficiency curves.

Dependent on the specific speed, the slope of the head characteristic varies from flat (low specific speed) to steep (high specific speed); see Fig. 2.4.

Radial impellers cover the specific speed range up to about $n_q = 120$ ($NS_{USA} = 6200$).

Mixed-flow impellers are feasible from $n_q = 40$ to about $n_q = 200$ ($NS_{USA} = 2000-10\,000$).

Axial-flow impellers with very steep (but often unstable) characteristics are built with specific speeds $n_q = 160$ to about $n_q = 350$ ($NS_{USA} = 8000$ to $18\,000$) (see Fig. 2.1).

A head-capacity curve of a centrifugal pump is said to be *stable* when the head curve rises steadily from high flow towards shut-off, i.e. there is always a negative slope dH/dQ. For any given head there is a unique intersection with the Q–H curve. Pumps with low specific speeds are prone to flat Q–H curves or even a slight instability. Axial-flow pumps (high specific speed) usually have a flow range with a distinct instability, within which it is not permissible to operate (see Fig. 2.1).

On radial pumps, stable characteristics can be obtained by the following measures:

- small impeller outlet angle;
- extended return vanes on multistage pumps;
- relatively large impeller outlet width;
- small number of vanes.

Centrifugal Pump Handbook. DOI: 10.1016/B978-0-7506-8612-9.00002-4

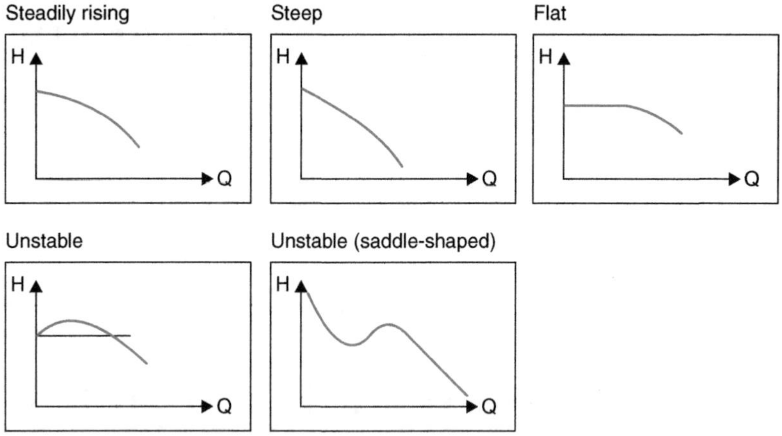

Figure 2.1 Typical shapes of the characteristics of centrifugal pumps

Figure 2.5 shows an unstable characteristic, the *Q–H* curves being intersected twice by the pump characteristic when, for example, the grid frequency drops. The pump may revert to minimum or zero flow with a risk of overheating because the power input to the pump is then converted into heat.

Pumps with unstable characteristics may interact with the system in closed circuits, leading to flow oscillations (surge) and possibly pipe vibrations. Stable characteristics are a fundamental requirement for the automatic control of centrifugal pumps.

2.2 CONTROL OF CENTRIFUGAL PUMPS

2.2.1 Piping system characteristic

The head to be overcome consists of a static component H_{geo} which is independent of the flow rate, and a head loss H_{dyn} increasing with the square of the flow and depending on pipework layout, pipe diameter and length. The sum $H_A = H_{geo} + H_{dyn}$ represents the head required by the system at any given flow rate. The resulting curve $H_A = f(Q)$ is termed the "system curve" (Fig. 2.6). Centrifugal pumps adjust themselves automatically to the intersection of the system and head characteristics. The operating point is given by the intersection of the *Q–H* curve with the system characteristic.

In many applications the static head is zero, and the system head is entirely due to hydraulic losses.

2.2.2 Pump control

Pump output may be controlled by the following methods:

1. Throttling

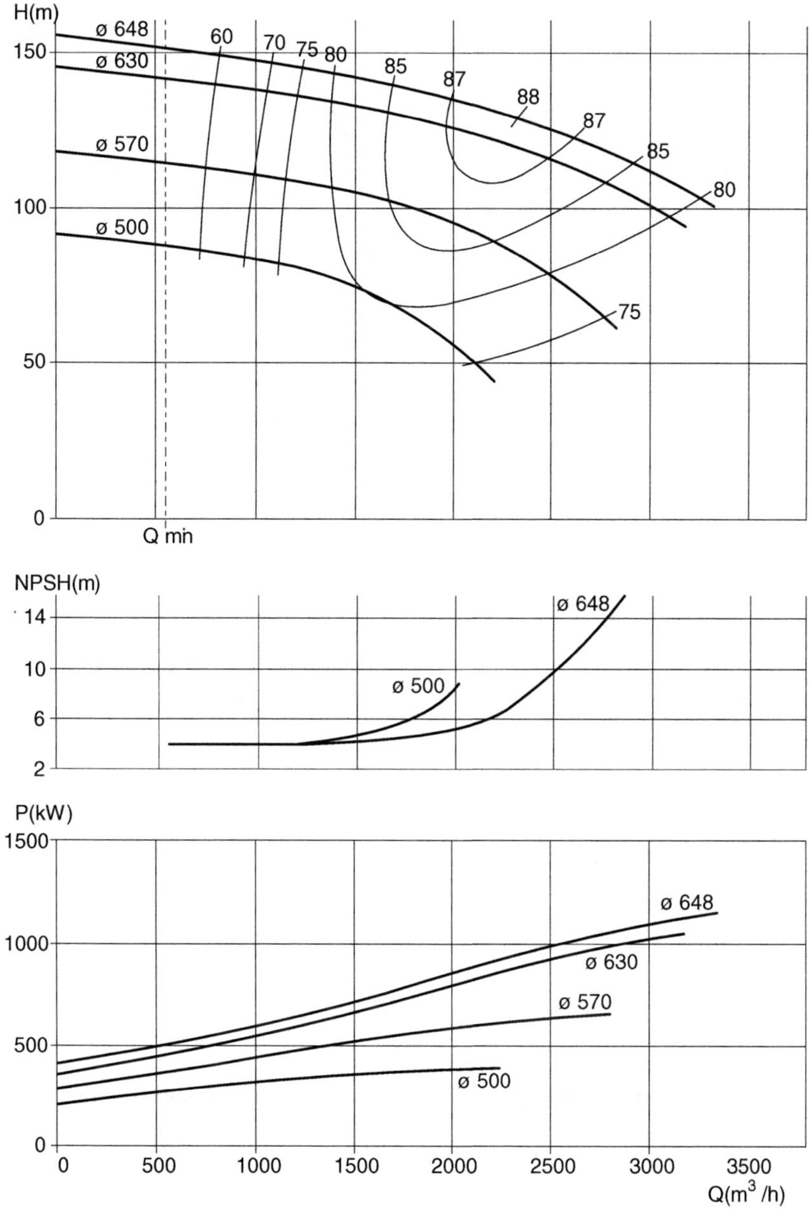

Figure 2.2 Typical pump characteristics with various impeller diameters and constant speed

2. Switching pumps on or off
   a. with pumps operating in parallel
   b. or with pumps operating in series
3. Bypass control
4. Speed control

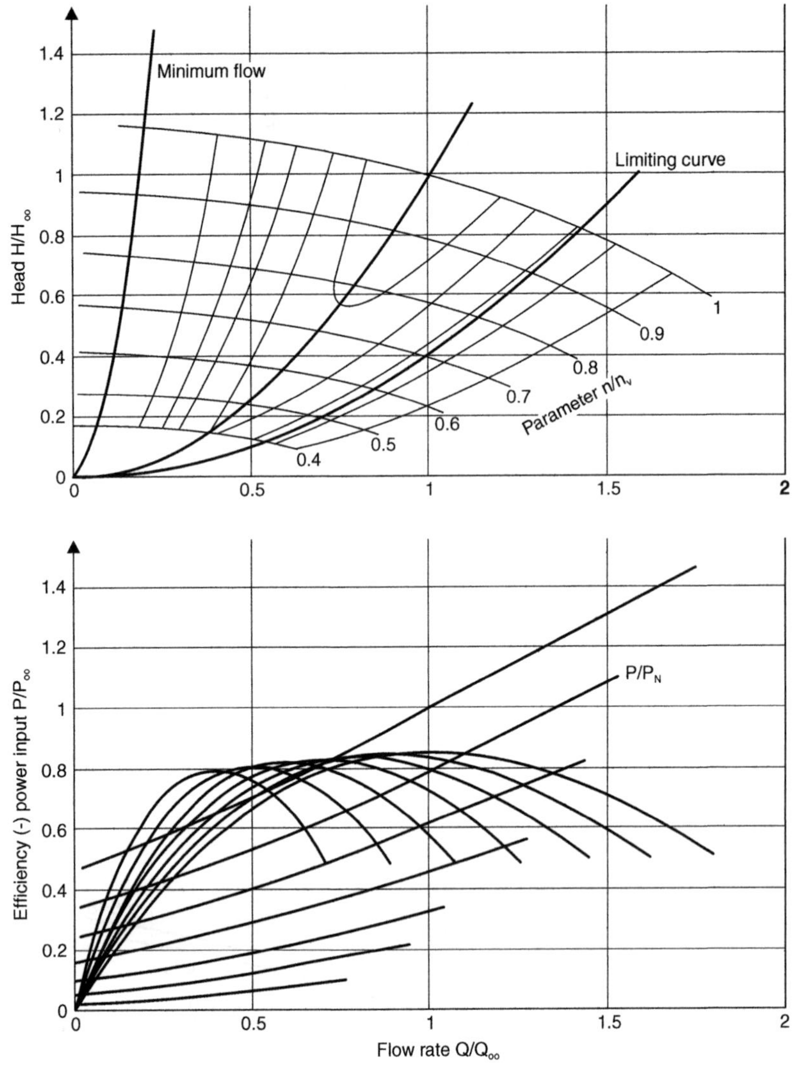

Figure 2.3 Typical pump characteristics for constant impeller diameter and variable speed

5. Impeller vane adjustment
6. Pre-rotation control
7. Cavitation control.

2.2.2.1 THROTTLING

Closing a throttle valve in the discharge line increases the flow resistance in the system. The system characteristic becomes steeper and intersects the pump characteristic at a lower flow rate. To avoid a reduction of the $NPSH_A$, throttling is not admissible in the suction pipe.

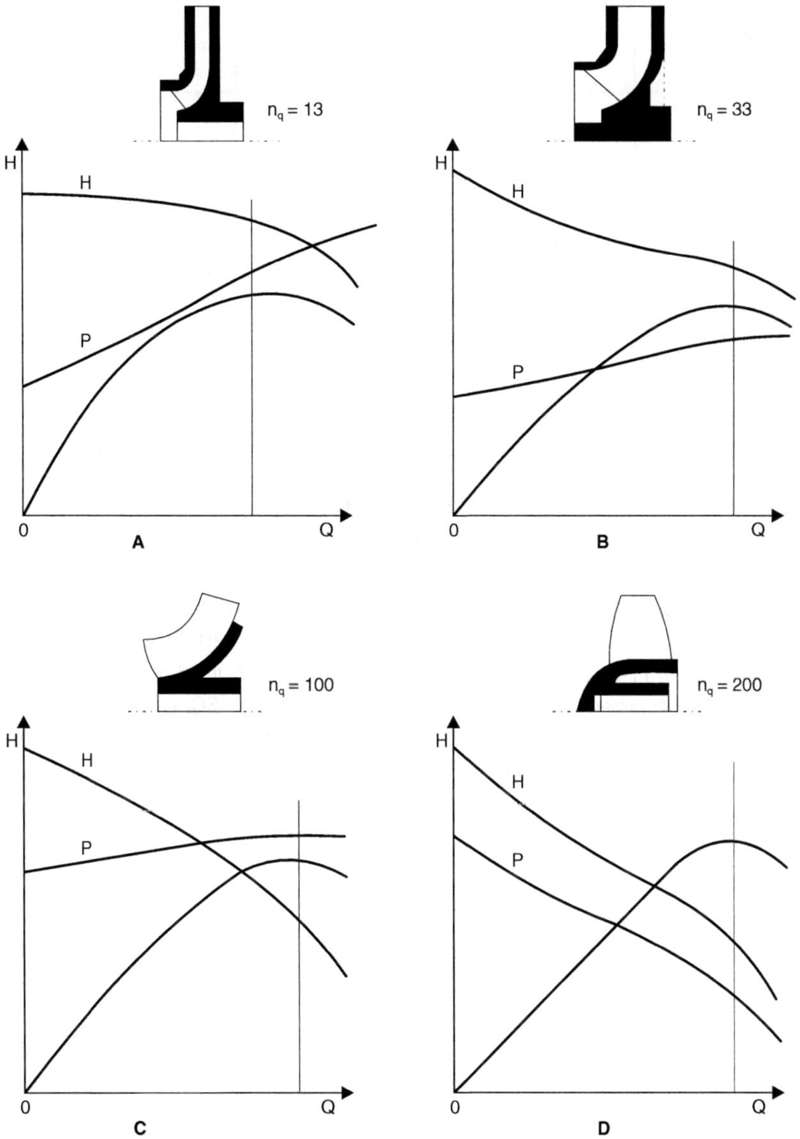

Figure 2.4 Influence of specific speed on the shape of the characteristics

Throttling control (Fig. 2.7) is generally employed where the required flow rate deviates from the nominal flow for short periods of operation.

The power input P' in kWh per m^3 liquid pumped increases with falling flow rate:

$$P' = \frac{\rho \cdot H}{367 \cdot \eta}\left[\frac{\text{kWh}}{\text{m}^3}\right] = const. \frac{H}{\eta} \qquad \begin{matrix}\rho \text{ in kg/dm}^3 \\ H \text{ in m}\end{matrix}$$

i.e. P' depends only on the ratio H/η.

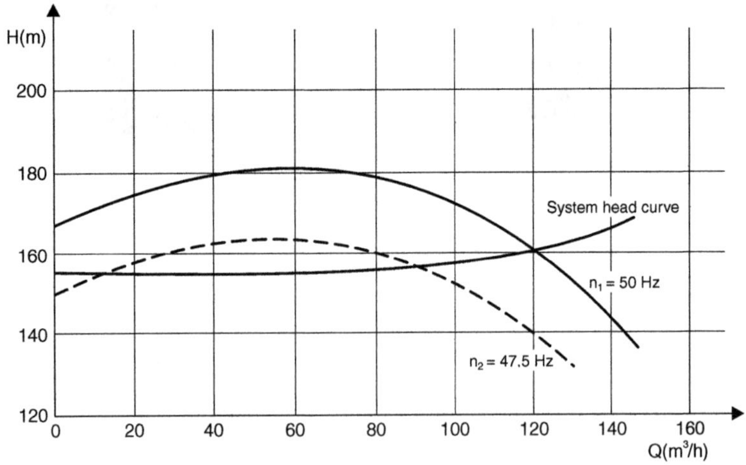

Figure 2.5 Influence of an unstable characteristic on the operating behavior of a pump under frequency variation

This reasoning demonstrates that throttling control ought to be employed mainly on radial pumps with $n_q \leq \sim 40$ ($NS_{USA} \leq 2000$), because these have a flatter *H*-to-*Q* characteristic and are better suited for such control.

With high specific speeds ($n_q > 100$, $NS_{USA} > 5000$) this control is wasteful of energy because the power input to the pump rises when throttling is effected. Even the driver may be overloaded that way. Moreover, the flow conditions in the pump deteriorate at low flow rates, causing uneven running.

2.2.2.2 STARTING AND STOPPING PUMPS

Parallel operation

Where the pumping requirement varies, it may be desirable to employ several small pumps instead of a few large units. When the required flow

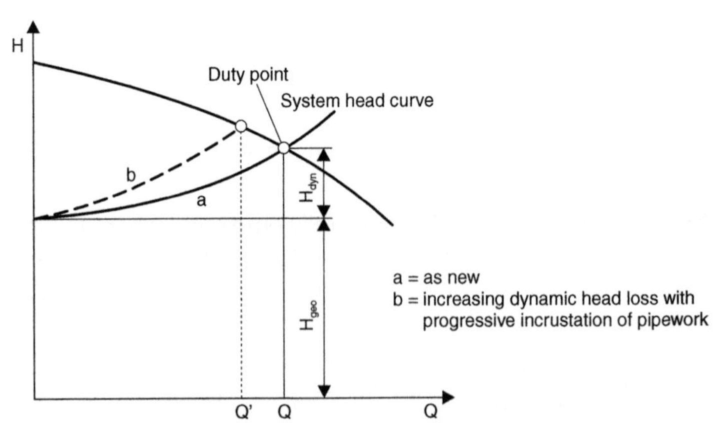

Figure 2.6 System characteristic

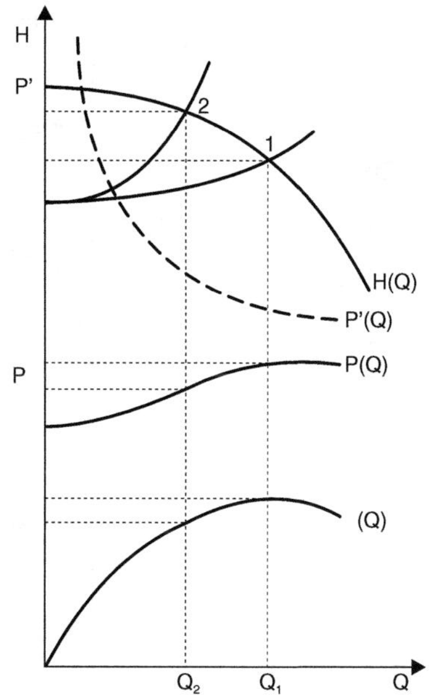

Figure 2.7 Throttling control and power input at constant rpm

rate drops, one or more pumps are stopped. The remainder then operate closer to their best efficiency point.

With a given suction head and flow rate, a greater number of pumps may sometimes yield benefits because higher pump speed can then be adopted enabling pump costs to be lowered. The optimum number of pumps must be chosen in the light of the constraints of the system (space requirements, costs of auxiliary equipment and erection).

Parallel operation is conditional upon a steadily rising pump head curve.

With $H_{geo} \geq H_{dyn}$, operating pumps in series is not advisable (Fig. 2.8a). If one pump shuts down, consideration must be given to how far the remaining pump can run out to the right and whether the NPSH available covers the NPSH required (Fig. 2.8).

At the same time the driver power must be sufficient to ensure pump runout with adequate margin.

Positive displacement pumps can also be run in parallel with centrifugal pumps. A positive displacement pump has an *H–Q* characteristic which is almost vertical. Here again *Q* values for the same head are added, giving the combined characteristic plotted in Fig. 2.9. The intersection of the system head curve with the combined pump characteristic is the duty point. By combining centrifugal and positive displacement pumps, one disadvantage of the latter (no throttle control) can be compensated partially.

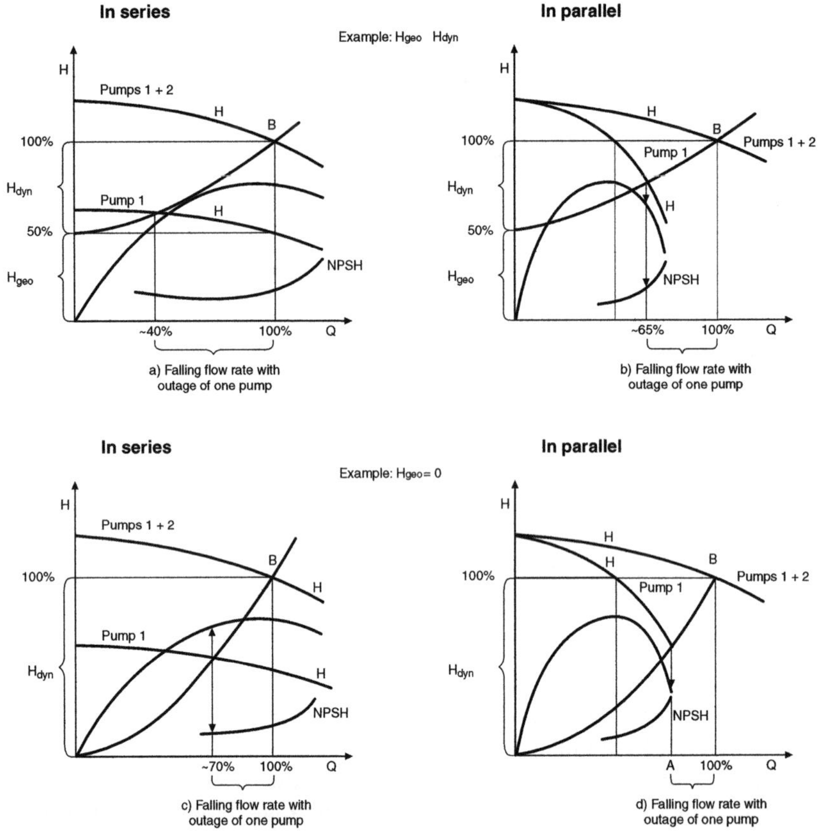

Figure 2.8 Series and parallel operation

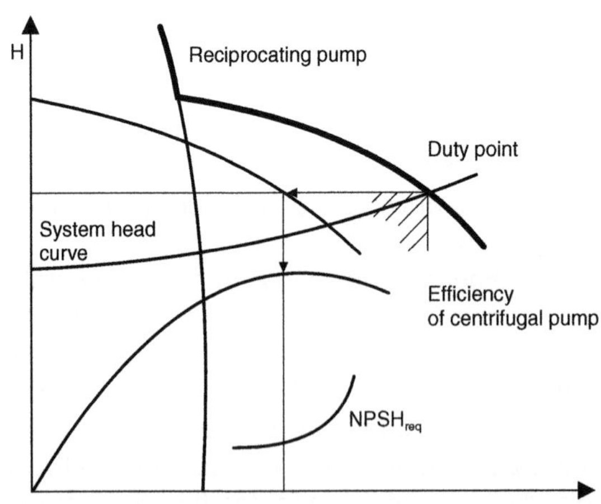

Figure 2.9 Reciprocating and centrifugal pumps operating in parallel

Series operation

When flow-dependent losses are dominating, i.e. $H_{geo} \leq H_{dyn}$, pumps may be operated in series. The reason for this will be clear from Fig. 2.8c. With predominantly dynamic losses, series operation is advisable because it is still possible to deliver about 70% of the original flow if one of 2 pumps is shut down. Moreover the pump operates with better NPSH and good efficiency.

If, however, $H_{geo} \geq H_{dyn}$ (i.e. a mainly static system) series operation is not recommended. When one pump fails in series operation, the other runs at lower flow and lower efficiency in a range where uneven performance must be expected (Fig. 2.8a). In this case, pumps in parallel operate much better and give greater flow at a higher efficiency.

Where pumps operate in series, it must be borne in mind that the seals and casings on the downstream pump are under higher pressure. The shaft seal may have to be balanced to the suction pressure of the upstream pump. Recirculation losses result from this, which must be taken into account in the overall efficiency calculation.

2.2.2.3 BYPASS CONTROL

In order to vary flow, centrifugal pumps are sometimes controlled by means of a bypass, i.e. a certain capacity is piped back to the suction nozzle or – if large capacities are involved – to a vessel at the suction end (see also section 2.8: Minimum flow rate). The pump then operates to the right of its optimum point (depending on the system head curve), and its cavitation behavior must be reviewed. Owing to recirculation losses, the overall efficiency drops markedly.

This method is used only rarely to control centrifugal pumps, and judged by the criterion of overall efficiency, it is the least satisfactory system compared to speed control and throttling (Figs 2.10, 2.11 and 2.12). Bypass control is, however, employed more often with pumps with

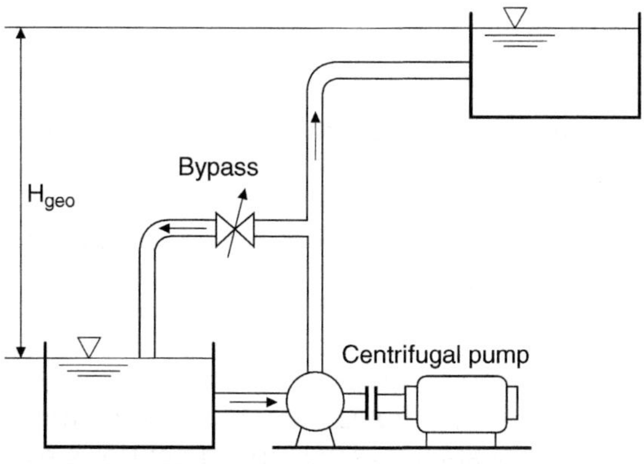

Figure 2.10 Typical bypass control system

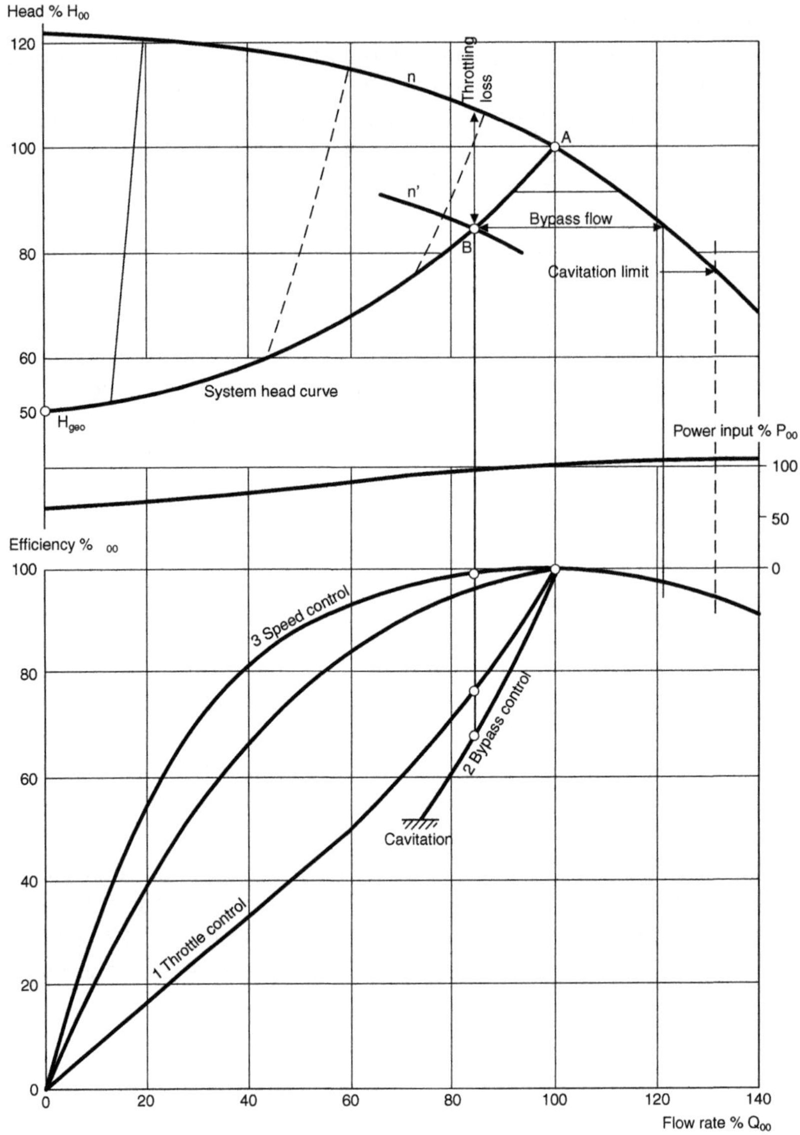

Figure 2.11 Various control systems with *steep* system head curve

high specific speed, such as axial-flow pumps, because the power input drops with increasing flow.

2.2.2.4 SPEED CONTROL

Where the head required by the plant is made up wholly or mainly from hydraulic losses, speed control is best, because pump efficiency then remains practically constant. Though speed control is expensive, wear on the pump and valves is less, so that the lifespan of both is generally

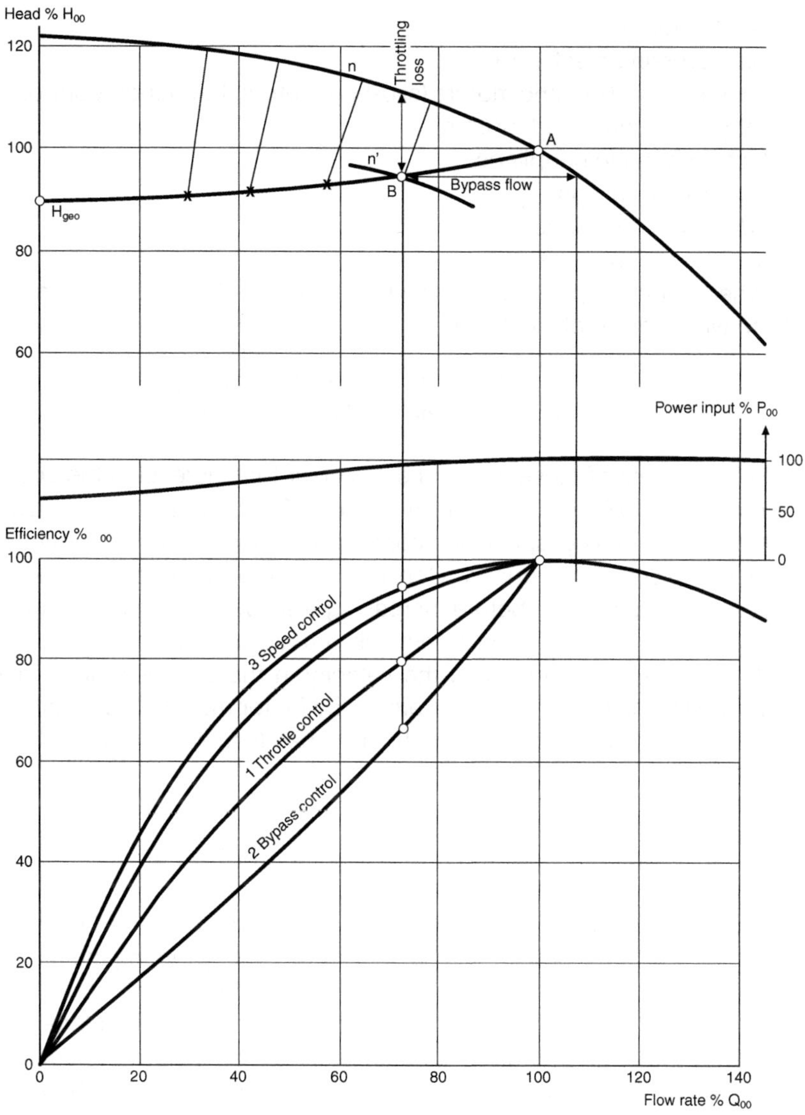

Figure 2.12 Various control systems with *flat* system head curve

increased. Furthermore, operating costs are reduced, because higher pump efficiency means less power input (Figs 2.11 and 2.12).

Unlike throttling, speed control leaves the piping system characteristic unchanged, while the pump characteristic moves up or down in accordance with the speed change. Two or more pumps serving a common discharge line are usually installed.

Speed control is achieved by employing the following driving machinery:

- steam turbines

- diesel engines
- pole-reversing electric motors
- variable-speed electric motors (slip-ring motors, frequency-controlled or thyristor-controlled motors)
- electromagnetic couplings
- hydraulic couplings
- hydrostatic couplings
- variable-speed gears.

When varying speed, the following equations hold good in accordance with the law of similarity:

$$\frac{Q_1}{Q_2}=\frac{n_1}{n_2} \quad \frac{H_1}{H_2}=\left(\frac{n_1}{n_2}\right)^2 \quad \frac{\text{NPSH}_1}{\text{NPSH}_2}=\left(\frac{n_1}{n_2}\right)^2$$

From the last of these equations it is clear that increasing the speed increases the NPSH_{req} as well. A higher NPSH_{av} must be provided. Moreover, increasing speed increases power input approximately by the third power of the speed. Consequently the strength of the shaft must always be verified before increasing the speed. With small speed changes (up to 10%) efficiency remains virtually unchanged.

With larger speed changes the velocity in the channels alters and with it the Reynolds number. The efficiency factor must be downgraded at lower speeds and upgraded at higher ones (see section 3.2.4: Conversion of test results).

2.2.2.5 IMPELLER BLADE ADJUSTMENT

For axial pumps (and even semi-axial pumps of high specific speed) impellers with adjustable blades can provide a most economical control for duties involving a widely varying flow rate. If the demanded flow rate increases, the impeller blade angles are increased by turning the blades around an axis essentially perpendicular to the hub.

Figure 2.13 shows a family of characteristics for various blade settings with the setting angle β as parameter at a constant speed. The higher the head, the greater is the cavitation coefficient and therefore the required NPSH. At low flow rates the characteristics of axial pumps are usually unstable; operation in this range must be avoided.

Impeller blade adjustment is used on water circulating pumps. With constant water level, these pumps can be controlled economically on the system characteristic.

2.2.2.6 PRE-ROTATION CONTROL

For pre-rotation control, the impeller is preceded by controllable inlet guide vanes. The pump characteristic is altered solely by varying inflow to the impeller. The setting of the pre-rotation guide imparts a circumferential component to the inflowing medium, thus altering the energy conversion process in the pump and with it the characteristic.

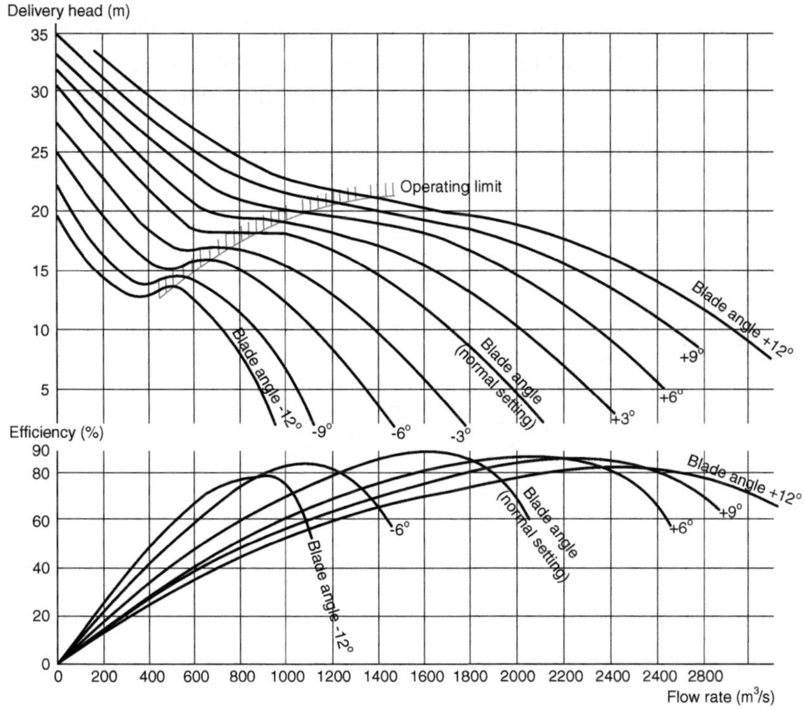

Figure 2.13 Characteristic of a mixed-flow pump with impeller blade adjustment

The inlet guide can be used on any pump, but its effect on the characteristic increases with specific speed. Accordingly, inlet guides are employed chiefly with mixed-flow or axial-flow pumps. With pre-rotation control, efficiency drops much faster in the part load and overload ranges than with impeller blade control.

Moreover, pre-rotation control is more susceptible to cavitation (Fig. 2.14). Consequently, the control range is narrower with

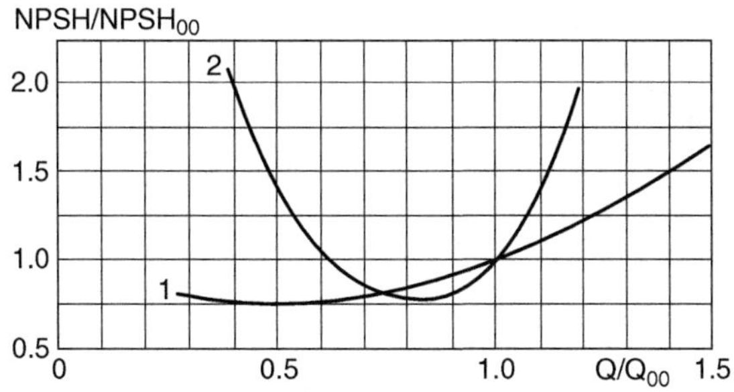

Figure 2.14 Dimensionless NPSH characteristic for (1) impeller blade adjustment and (2) pre-rotation control

this method than with impeller blade adjustment. Depending on the shape of the system head curve, the head can be controlled down to about 70 to 50% of its value at the best efficiency point (Fig. 2.15).

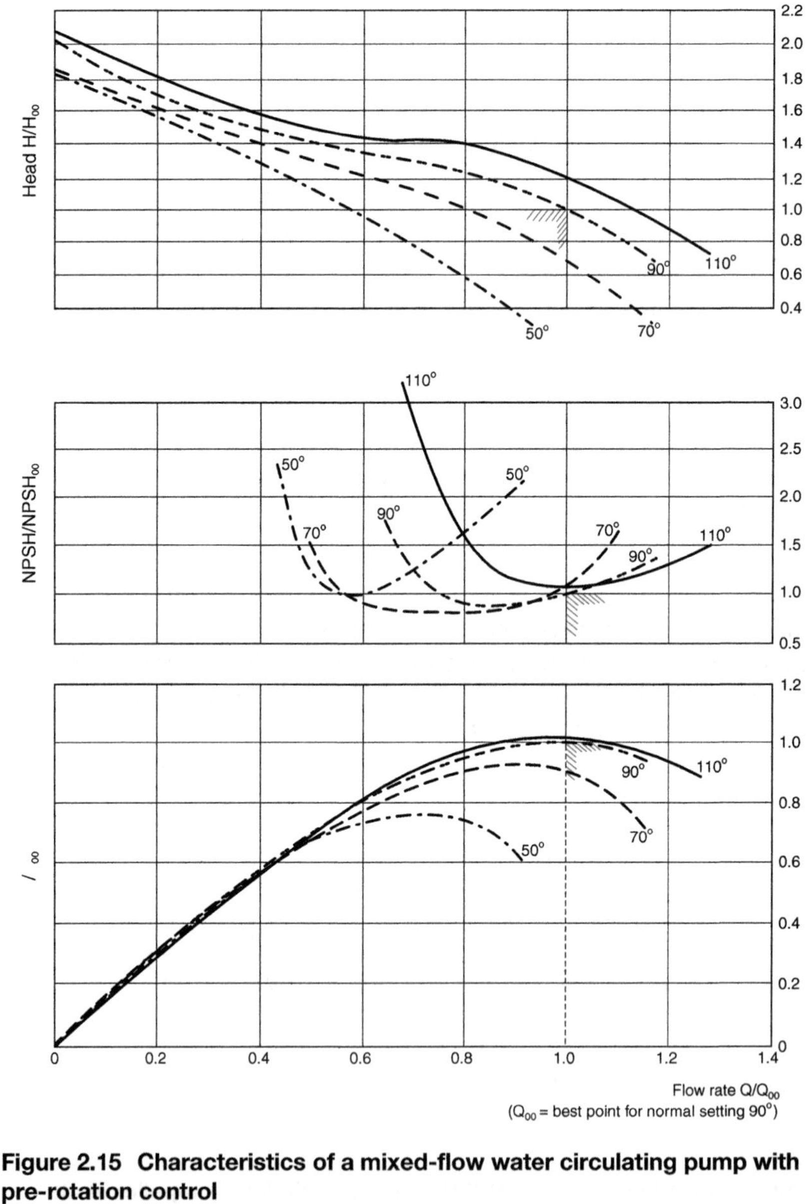

Figure 2.15 Characteristics of a mixed-flow water circulating pump with pre-rotation control

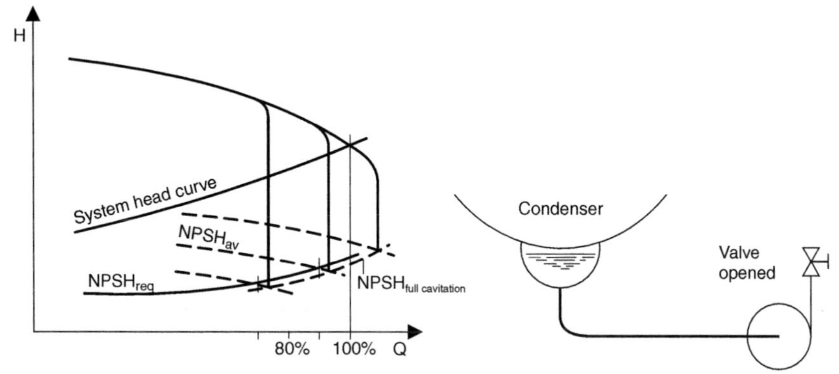

Figure 2.16 Cavitation control

2.2.2.7 CAVITATION CONTROL

Cavitation control is often used to vary the flow rate on smaller condensate pumps. The head per stage is generally limited to 50 m, otherwise the pump will run too unevenly and sustain excessive cavitation wear on the vanes. If condensate flow in the hotwell is below operating flow, the level will drop and with it the available NPSH, and the pump cavitates. The break-off characteristic now intersects elsewhere with the system head curve, at a lower flow rate.

If the flow of condensate into the hotwell increases above the momentary flow rate of the pump, the liquid level rises, and with it the available NPSH. This alters the cavitation status, and pump flow increases (Fig. 2.16). The intersection of variable $NPSH_{av}$ with pump $NPSH_{req}$ at full cavitation determines the flow rate. Simplicity is the argument for this control system.

2.3 MATCHING CHARACTERISTICS TO SERVICE DATA

2.3.1 Reduction of impeller diameter (impeller trimming)

If the flow required by the system is permanently less than the pump flow at best efficiency point, and speed control is not applicable, the impeller can be matched to the new conditions by reducing its diameter (see Fig. 2.17). On diffuser pumps, only the impeller blades are cut (Fig. 2.17b), while on volute casing pumps, usually the impeller blades with the shrouds are turned down (Fig. 2.17a and e).

With diffuser or double-entry pumps, it is often advisable to trim the impeller obliquely as in Fig. 2.17c or d, because such corrections usually yield a more stable curve.

The bigger the correction is and the higher the specific speed, the more the pump efficiency suffers from adaptation.

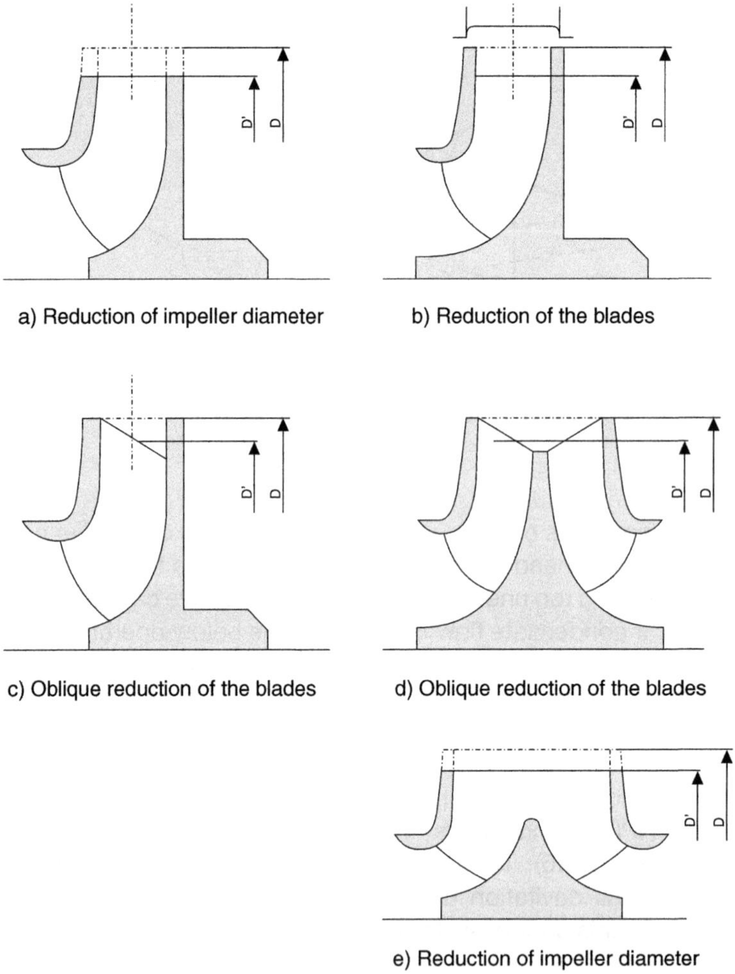

Figure 2.17 Impeller corrections

Bigger corrections increase pump NPSH in the overload range, because specific vane loading is raised by the reduced vane length, affecting velocity distribution at the impeller inlet.

If only a small correction ($\leq$5%) to impeller outside diameter is necessary, it may be assumed that the required NPSH will increase only slightly.

When pump impeller diameters are reduced, outlet width, blade angle and blade length are altered. The effect of this depends very much on the design of the impeller (i.e. on its specific speed). Consequently, only approximate statements can be made about the effects of reducing the impeller diameter on characteristics.

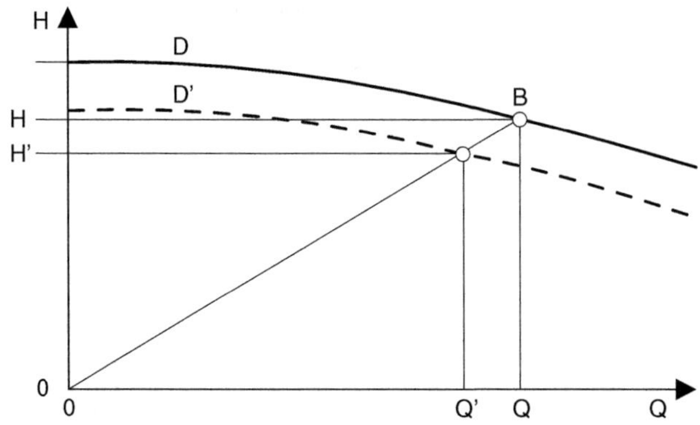

Figure 2.18 Alteration of the pump characteristic after reduction of the impeller diameter

When radial impellers are corrected, the pump characteristic alters according to Fig. 2.18 as follows:

$$Q' = Q\left(\frac{D'}{D}\right)^m \quad H' = H\left(\frac{D'}{D}\right)^m$$

where $m = 2$ to 3 may be assumed: $m = 2$ for corrections $\geq 6\%$; $m = 3$ for corrections $\leq 1\%$

It should be noted that the best efficiency point lies between Q' and Q, depending on n_q.

In accordance with Fig. 2.18, operating point B is joined to point $Q = 0$, $H = 0$ by a straight line. The new operating point is found by plotting Q' on the Q axis or H' on the H axis. In this way the corrected impeller diameter can be determined.

If an impeller with higher specific speed ($n_q > 50$, $NS_{USA} > 2500$) is to be matched, or the correction amounts to more than 3%, the matching to service data should be effected in stages, i.e. first turning-down to a larger diameter than the one calculated and carrying out a trial run. Only after this is the impeller corrected to its final diameter.

2.3.2 Sharpening of impeller blade trailing edges

By sharpening the impeller blades on the suction side (also called underfiling), the outlet angle can be enlarged to obtain up to 5% more head near the best efficiency point, depending on the outlet angle. At least 2 mm must be left (Fig. 2.19). In this way efficiency can be improved slightly (Fig. 2.20).

Where stage pressures are high, underfiling must be carried out with great care on account of the high static and dynamic stresses involved.

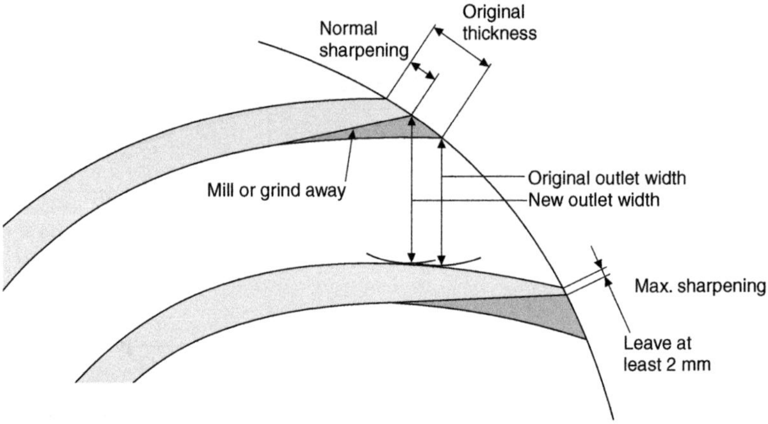

Figure 2.19 Sharpening of impeller blades

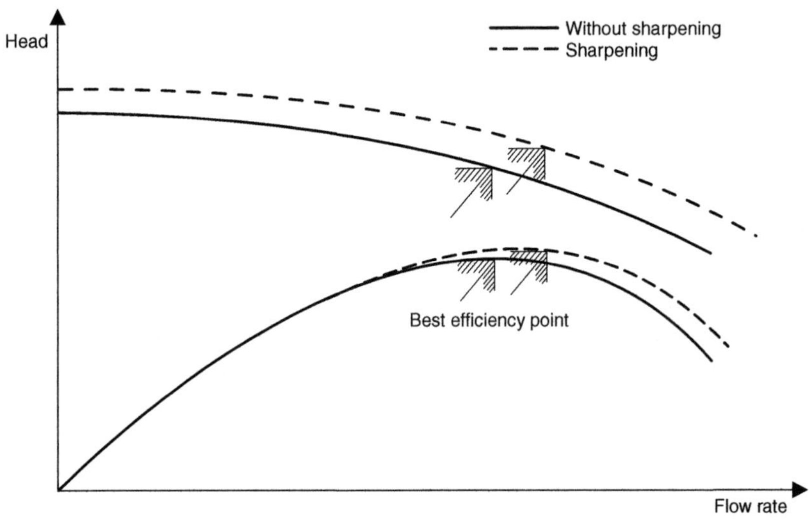

Figure 2.20 Influence of sharpening of impeller blades on pump characteristic

2.4 TORQUE/SPEED CURVES

2.4.1 Starting torque of centrifugal pumps

Normally, the centrifugal pump starting torque is so low that it does not require special consideration. Drivers like steam, gas and water turbines have very high driving torques. The difference between driver and pump torque is sufficient to accelerate a centrifugal pump to its operating speed in a very short time.

Where a pump is driven by an internal combustion engine, the starter must be sufficiently large to bring both the engine and pump up to operating speed. However, a clutch is usually interposed to separate the pump from the engine during startup.

Depending upon their type and size, electric motors have various torque characteristics (see section 7.1.4). With high power inputs in particular, startup must therefore be investigated thoroughly at the planning stage. At around 80% of the speed the torque of an electric motor may drop to the pump torque. Acceleration torque is then nil, and the group becomes locked into this speed. As a result, the motor takes up a high current, with the risk of destroying the windings (Fig. 2.21). Pump torque T is calculated as follows:

$$T = 9549\,(P/n)\ (\mathrm{Nm})$$
$$P = \text{pump power input (kW)} \quad n = \text{rpm}$$

In accordance with the affinity law, the torque varies as the square of the speed:

$$T_2 = T_1(n_2/n_1)^2$$

To achieve initial pump breakaway, a starting torque between 10 and 25% of pump torque at best efficiency point is normally required

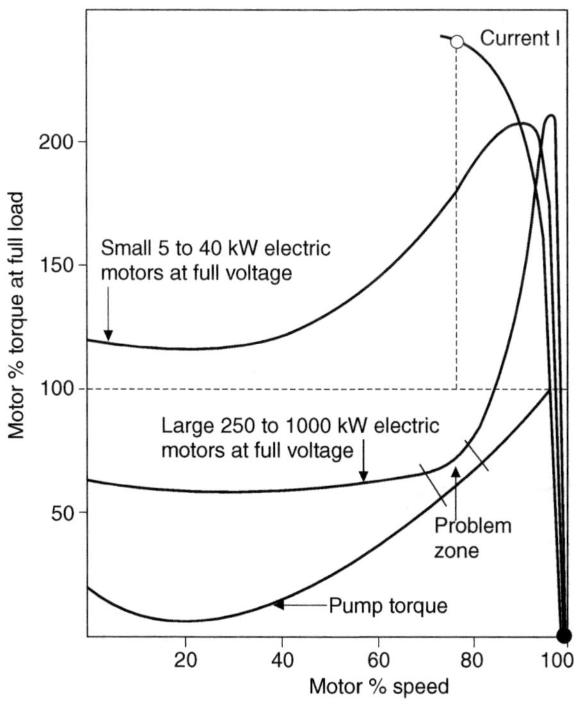

Figure 2.21 Starting torque curve for electric motor and pump

because of the static friction of the moving parts involved. For vertical pumps with long rising mains or pumps with very high inlet pressures, initial breakaway torque must be calculated specially (number of bearings, thrust bearing friction, stuffing box friction, etc.).

The torque curve depends very much on specific speed, as may be seen from Fig. 2.4. Pumps with lower specific speed have a torque characteristic that rises with flow rate, whereas pumps with high specific speed have a characteristic that falls, as flow increases.

2.4.2 Startup (excluding pressure surge)

Centrifugal pumps may be started in four different ways.

2.4.2.1 STARTUP AGAINST CLOSED VALVE

Figure 2.22 plots the startup of a centrifugal pump against a closed valve. With comparatively high ratings, the pump must start with a minimum flow valve opened, due to the risk that liquid will evaporate inside the pump. As soon as operating speed is reached, the discharge valve is opened. This method is possible only on pumps with low specific speed, where the power input and torque with closed valve are less than at the duty point.

2.4.2.2 STARTUP AGAINST CLOSED NON-RETURN VALVE WITH THE DISCHARGE VALVE OPEN (FIG. 2.23)

When starting, the non-return valve opens at point A, when the back pressure acting on it is reached. As the speed increases, flow rate and head follow the system curve.

2.4.2.3 STARTUP WITH DISCHARGE VALVE OPEN BUT NO STATIC HEAD (FIG. 2.24)

Here, the torque curve is a parabola through duty point B, provided that the pipe is short and that the time needed to accelerate the water column coincides with the pump startup time.

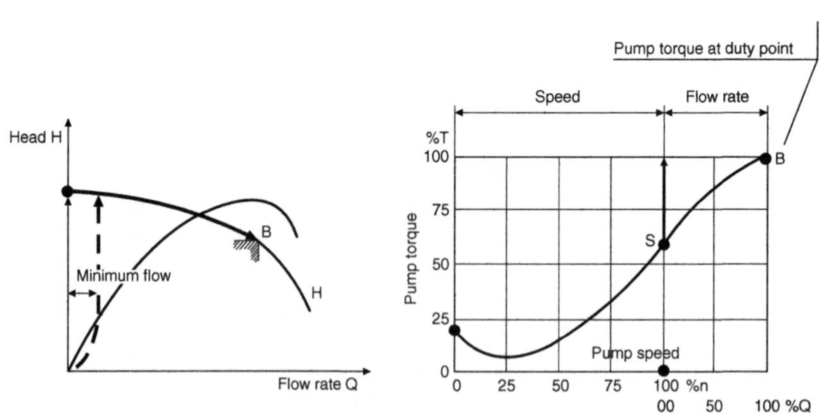

Figure 2.22 Starting against closed valve

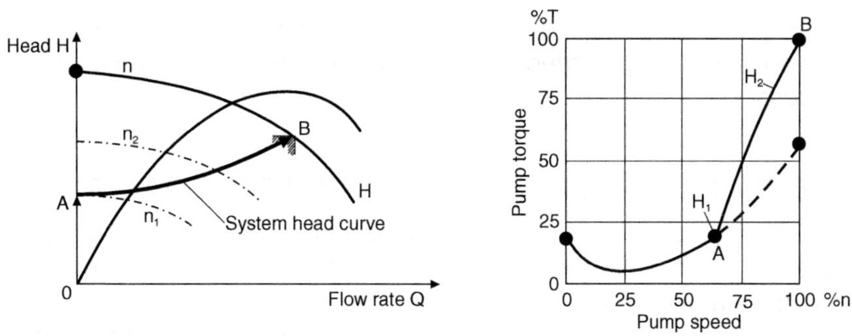

Figure 2.23 Startup against closed non-return valve with discharge valve open

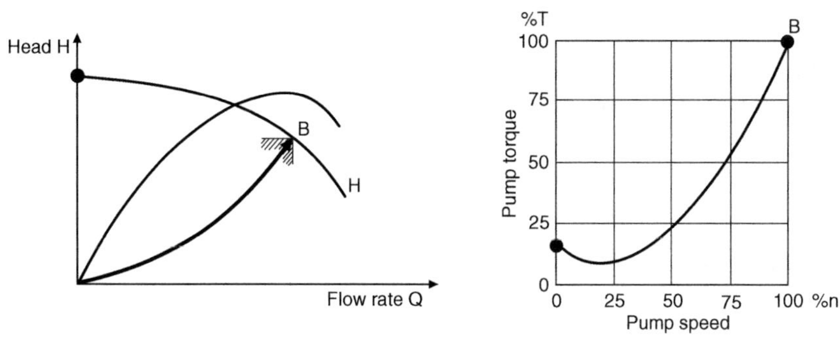

Figure 2.24 Startup with discharge valve open but no geodetic head

2.4.2.4 STARTUP WITH DISCHARGE VALVE OPEN AND PRESSURE PIPELINE DRAINED (FIG. 2.25)

Pumps with high specific speed are mostly started with the pipeline drained. A backpressure ought to be established as quickly as possible, if necessary by throttling with the discharge valve, to prevent high pump runout flows where there is danger of cavitation.

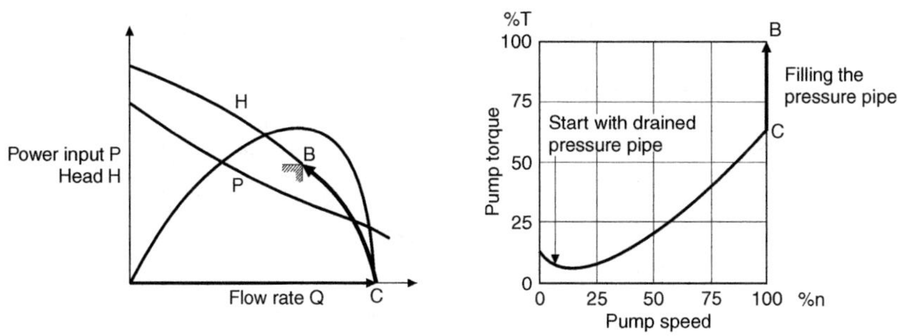

Figure 2.25 Startup with discharge valve open and pressure pipeline drained

Starting up pumps of this type with the pressure pipeline filled is also possible, provided considerable quantities are diverted via a bypass.

2.5 STARTUP TIME FOR A CENTRIFUGAL PUMP

The startup time for a centrifugal pump depends on the accelerating torque, i.e. on the difference between the torque of the driving unit T_A and the starting torque T_p required by the pump.

The accelerating torque varies with the speed (Fig. 2.26). The startup time must therefore be calculated for individual speed increments and then totaled:

$$T_B = T_A - T_p = J\varepsilon \text{ (Nm)}$$

where J = mass moment of inertia for all revolving parts including liquid, related to driving speed:

$$J = \frac{mD^2}{4} \text{ (kg m}^2\text{)}$$

where D = inertia diameter:

$$J_2 - J_1(n_1/n_2)^2 \text{ (kg m}^2\text{)}$$

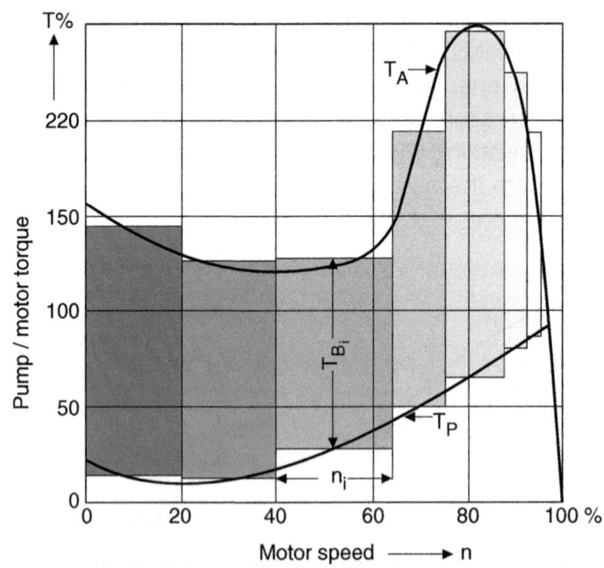

Figure 2.26 Calculating the startup time for a centrifugal pump

$$\varepsilon = d\omega/dt = \frac{\pi}{30} \cdot \frac{dn}{dt} = \text{angular acceleration of drive shaft} \left(\frac{1}{s^2}\right)$$

From the above equations, startup time is calculated as follows:

$$t_A = \Sigma t_i = \frac{\pi \cdot J}{30} \left(\frac{\Delta n_1}{T_{B1}} + \frac{\Delta n_2}{T_{B2}} + \cdots + \frac{\Delta n_i}{T_{Bi}} \right) \text{ (s)}$$

where J (kg m^2) n (min^{-1}) T_B (Nm).

Depending on whether the pump starts with the discharge valve open or closed, its starting torque (T_p) and hence its startup time will also vary.

2.6 RUNDOWN TIME FOR A CENTRIFUGAL PUMP (DISREGARDING THE PRESSURE SURGE)

When the driving unit is shut down or fails, the drive torque drops to zero. The following then holds good:

$$-T_p = J\varepsilon \text{ (Nm)}$$

T_p = runout torque = pump starting torque (Nm)

J = mass moment of inertia for all revolving parts related to drive speed (pump including liquid)

$$\varepsilon = d\omega/dt = \frac{\pi}{30} \cdot \frac{dn}{dt} = \text{angular acceleration of drive shaft} \left(\frac{1}{s^2}\right)$$

From the above equations, the runout time is calculated as follows:

$$t_H = \Sigma t_i = \frac{\pi \cdot J}{30} \left(\frac{\Delta n_1}{T_{P1}} + \frac{\Delta n_2}{T_{P2}} + \cdots + \frac{\Delta n_i}{T_{Pi}} \right) \text{ (s)}$$

with J (kg m^2) n (min^{-1}) T_p (Nm).

Depending on whether the pump is shut down with the discharge valve open or closed, its runout torque (T_p) varies and hence its runout time also.

Because the value $\Delta n_1/T_{P1}$ is very small immediately after shutdown, the speed drops very rapidly at first (Fig. 2.27). However, the effective rundown time is obtained from the pressure surge calculation (see section 4.3).

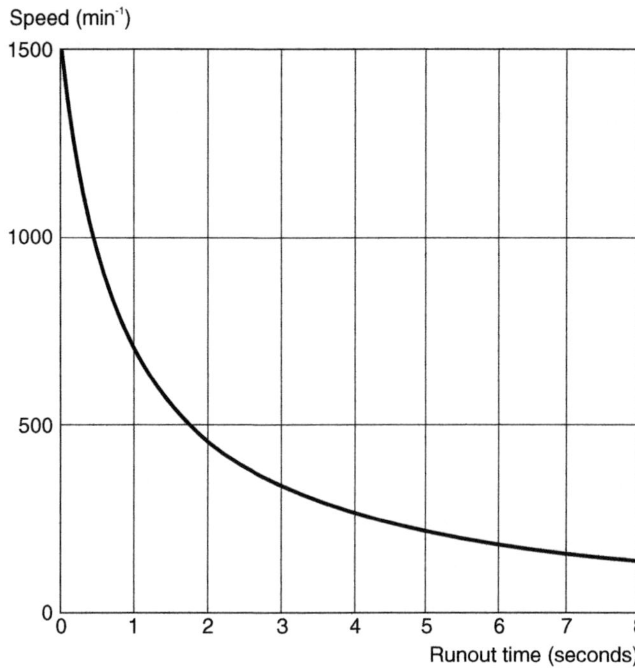

Figure 2.27 Runout time for a condensate pump

2.7 PUMPING SPECIAL LIQUIDS

2.7.1 Viscous liquids

2.7.1.1 VISCOSITY DEFINITIONS

If a plate is moved at constant speed c_o over a viscous medium (Fig. 2.28), the medium adheres to the underside of the plate and is entrained at the speed c_o. The internal friction of the liquid causes the speed to vary at right angles to the direction of flow between the fixed and moving plates:

$$F = \eta A \cdot dc/dy$$

related to the friction surface of the underside:

$$\tau = \eta \, dc/dy$$

The proportionality factor η is called *dynamic viscosity*, with the dimension mass/(length $\times$ time):

$$F = \eta \cdot A \cdot \frac{dc}{dy} \rightarrow \tau = \eta \frac{dc}{dy}$$

where:
F = frictional force
η = dynamic viscosity

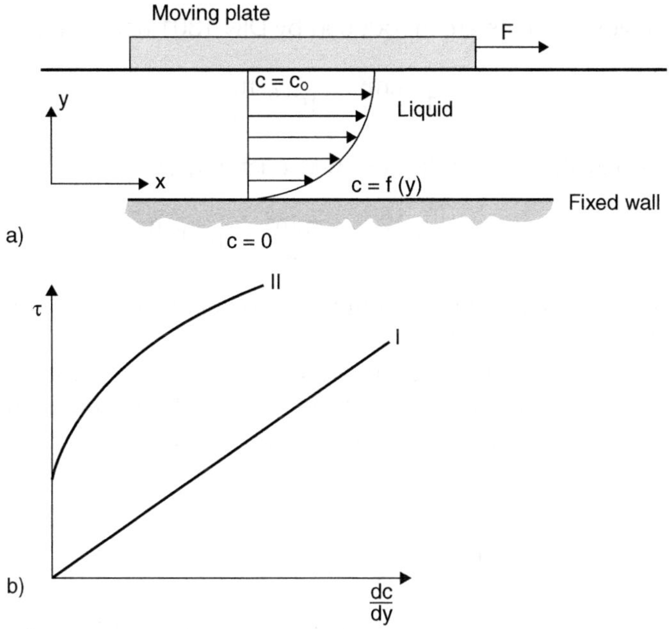

Figure 2.28 (I) Newtonian fluid. (II) Non-Newtonian fluid (e.g. carbon black in boiled oil, effluent sludge)

A = friction surface
τ = shear stress

The international units recommended according to DIN 1301 and 1342 are:

$$1\ \mathrm{Ps\,s} = 1\frac{\mathrm{kg}}{\mathrm{s\,m}}$$

Previously the unit "poise" (P) was used as viscosity unit:

$$1\ \mathrm{cP} = 10^{-2}\cdot \mathrm{P} = 10^{-3}\frac{\mathrm{kg}}{\mathrm{ms}} = 10^{-3}\ \mathrm{Pa\,s} = 10^{-2}\frac{\mathrm{g}}{\mathrm{cm}\cdot\mathrm{s}}$$

Liquids whose frictional behavior conforms to the above equation are called Newtonian fluids. Viscosity depends on temperature and pressure, though the influence of the latter is negligible compared with that of temperature. The viscosity of liquids diminishes with increasing temperature.

For hydrodynamic purposes it is convenient to relate dynamic viscosity η to density ρ. This gives us the *kinematic viscosity*:

$$v = \frac{\eta}{\rho} \text{ with the dimension length}^2/\text{time}$$

The following units are introduced by DIN 1301 and 1342:

$$1\ \frac{\text{mm}^2}{\text{s}} = 10^{-6}\ \frac{\text{m}^2}{\text{s}}$$

Previously the unit “stoke” (St) was customary:

$$1\ \text{cSt} = 10^{-2}\ \text{St} = \frac{\text{mm}^2}{\text{s}} = 10^{-6}\ \frac{\text{m}^2}{\text{s}}$$

Besides the official units set out above, other conventional units are in use for kinematic viscosity:

1. *Degrees Engler (E)*: The degree Engler expresses the ratio between the efflux time of 200 cm^3 distilled water at 20°C and of the same amount of test liquid at 20°C (DIN 51560). The following equation is used for conversion:

$$v =^{\circ} \text{E} \cdot 7.6^{\left(1^{-1}/^{\circ}E^{3}\right)} \left(10^{-6} \cdot \frac{\text{m}^2}{\text{s}}\right)$$

2. *Saybolt viscosity*: Saybolt seconds universal (SSU). This unit is customary in the USA and defines the efflux time for a given quantity of liquid from a defined viscosity meter.

Viscosity also depends to a considerable extent on the amount of gas dissolved in the liquid. Crude oil containing a lot of gas is less viscous than the same oil with little gas.

2.7.1.2 PUMPING VISCOUS LIQUIDS

By virtue of their favorable properties (little pulsation, no safety valve needed as with positive displacement pumps, simple flow control) the use of centrifugal pumps is being increasingly extended in the chemical industry and refineries to high-viscosity media (up to about $1000 \cdot 10^{-6}$ m^2/s), although an efficiency loss must be taken into account compared with positive displacement pumps with the same service data.

The economical application limit for centrifugal pumps is about (150 to 500) 10^{-6} m^2/s; this limit very much depends on the pump size and application. The use of a centrifugal pump is possible up to about 1500×10^{-6} m^2/s and even above. Tests are available with viscosities up to 3000×10^{-6} m^2/s. Reciprocating or rotary positive displacement pumps are employed for still higher viscosities (Fig. 2.29).

Due to increased losses, higher NPSH must be made available when pumping liquids with viscosities appreciably higher than cold water.

The pressure losses in the impeller and diffuser channels of centrifugal pumps, the impeller friction and internal leakage losses, depend to a large extent on the viscosity of the pumped liquid. Consequently, the characteristics ascertained for water lose their validity when pumping liquids of higher viscosity, such as oil.

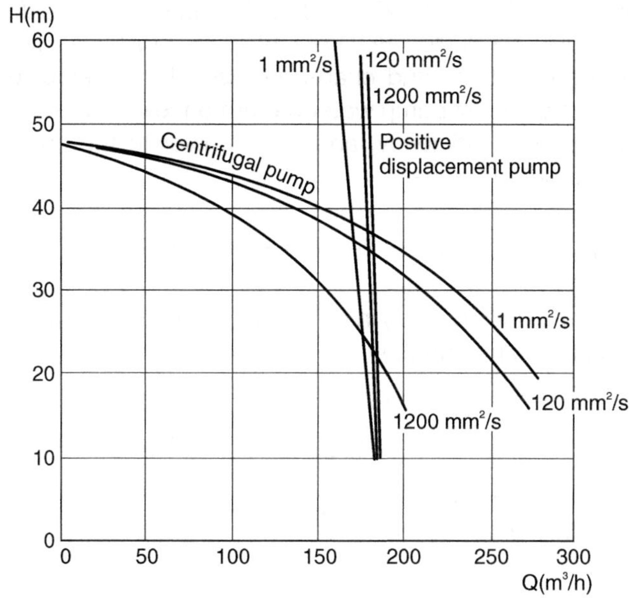

Figure 2.29 Comparison of characteristics for reciprocating and centrifugal pumps at constant speed and different viscosities

The higher the viscosity of a liquid compared with water, the greater is the loss of flow capacity and head for a given pump at a given speed. Consequently, the best efficiency point of the pump moves towards lower flow rates and efficiency. The power input shifts up parallel to that for water. The head against closed valve is generally retained.

For converting the water characteristic to a liquid with a defined viscosity, the correction factors laid down in ISO/TR 17766 2005 are employed internationally. This guideline is based on: "Effects of Liquid Viscosity on Rotodynamic Pump Performance", ANSI/HI 9.6.7-2004.

The correction factors are valid only as follows:

1. for Newtonian fluids, excluding solid mixtures, paper stock or other non-homogeneous liquids;
2. for open or closed radial hydraulics (not for mixed-flow or axial-flow hydraulics or side-channel pumps);
3. where there is adequate NPSHA;
4. within the limits: viscosity $< 3000 \times 10^{-6}$ m^2/s; $B < 40$; $n_q < 60$.

It should be noted that with multistage pumps only the stage head is taken, and with double-entry pumps only half the total flow rate.

If the characteristic for pumping water is known, that for the viscous liquid can be determined by multiplying the water values by the appropriate correction factors for flow, head and efficiency.

2.7.1.3 SIMPLIFIED INSTRUCTIONS FOR DETERMINING PUMP PERFORMANCE ON A VISCOUS LIQUID WHEN PERFORMANCE ON WATER IS KNOWN

The following equations and charts are used for developing the correction factors to adjust pump water performance characteristics of rate of flow, total head, efficiency, and input power to the corresponding viscous liquid performance.

Step 1

Calculate parameter B based on the water performance best efficiency flow (Q_{BEP-W}).

Given metric units Q_{BEP-W} in m^3/h, H_{BEP-W} in m, N in rpm and V_{vis} in cSt, use the following equation:

$$B = 16.5 \times \frac{(V_{vis})^{0.50} \times (H_{BEP-W})^{0.0625}}{(Q_{BEP-W})^{0.375} \times N^{0.25}}$$

Given USCS units of Q_{BEP-W} in gpm, H_{BEP-W} in ft, N in rpm and V_{vis} in cSt, use the following equation:

$$B = 26.6 \times \frac{(V_{vis})^{0.50} \times (H_{BEP-W})^{0.0625}}{(Q_{BEP-W})^{0.375} \times N^{0.25}}$$

If $1.0 < B < 40$, go to Step 2.

If $B \geq 40$, the correction factors derived using the equations above are highly uncertain and should be avoided. Instead a detailed loss analysis method may be warranted.

If $B \leq 1.0$, set $C_H = 1.0$ and $C_Q = 1.0$, and then skip to Step 4.

Step 2

Read the correction factor for flow (C_Q) (which is also equal to the correction factor for head at BEP [C_{BEP-H}]) corresponding to the water performance best efficiency flow (Q_{BEP-W}) using the chart in Fig. 2.30. Correct the other water performance flows (Q_W) to viscous flows (Q_{vis}).

Step 3

Read the head correction factors (C_H) using the chart in Fig. 2.30, and then the corresponding values of viscous head (H_{vis}) for flows (Q_W) greater than or less than the water best efficiency flow (Q_{BEP-W}).

Step 4

Determine the value for C_η from the chart in Fig. 2.31.

Step 5

Calculate the values for viscous pump shaft input power (P_{vis}). The following equations are valid for all rates of flow greater than zero.

For flow in m^3/h, head in m, shaft power in kW, and efficiency in percent, use the equation below:

$$P_{vis} \frac{Q_{vis} \times H_{vis-tot} \times s}{367 \times \eta_{vis}}$$

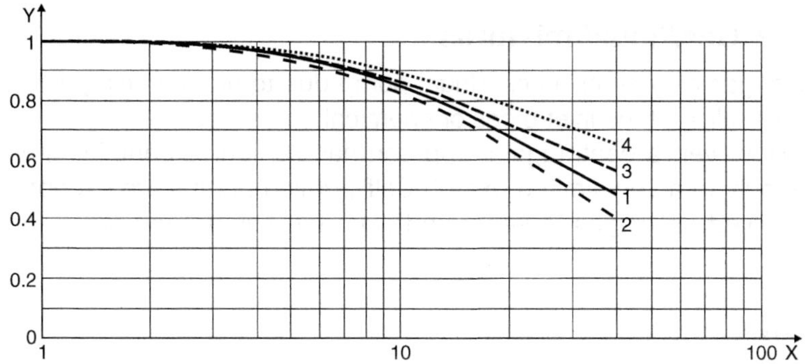

X parameter B
Y correction factors, CH and CC

Key

1. C_H and C_Q versus B at Q_{BEP-W}
2. C_H versus B at 1.2 × Q_{BEP-W}
3. C_H versus B at 0.8 × Q_{BEP-W}
4. C_H versus B at 0.6 × Q_{BEP-W}

Figure 2.30 Chart of correction factors for C_Q and C_H based on flow and head correction factors versus parameter *B*

For flow in gpm, head in ft, and shaft power in hp, use the equation below:

$$P_{vis} \frac{Q_{vis} \times H_{vis-tot} \times s}{3960 \times \eta_{vis}}$$

where s = specific gravity.

X parameter *B*
Y correction factors, C_η

Figure 2.31 Chart of correction factors for C_η based on the efficiency correction factor versus parameter *B*

2.7.2 Gas/liquid mixtures

Air or gas may enter a centrifugal pump due to air-entraining vortices, poor sealing of the inlet pipes or chemical processes.

However, a centrifugal pump can handle liquids containing air or gas only to a very limited extent. Self-priming pumps are exceptions. Dissolved air or gas is separated from the liquid upon reduction of

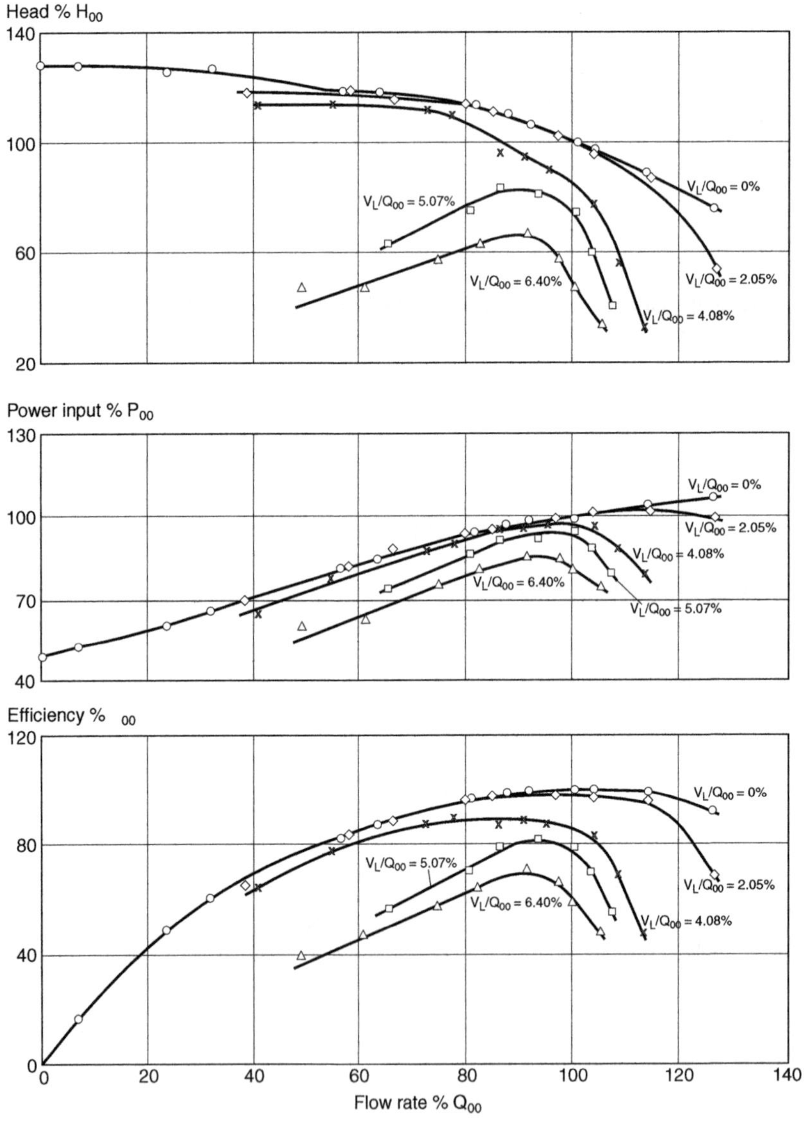

Figure 2.32 Influence of air on the characteristic of a single-stage centrifugal pump (n_q = 26) with inlet pressure 2.5 bar absolute

the local pressure (Henry's law). The air or gas can gather at the impeller inlet or anywhere in the flow passages. When these are sufficiently blocked by gas accumulations the flow may collapse.

Figure 2.32 shows the influence of air on the characteristics of a single-stage centrifugal pump with specific speed $n_q = 26$ ($NS_{USA} = 1350$). This influence depends on the suction pressure. Here, with low inlet pressure, the characteristic already drops at $V_L/Q_{00} = 2\%$ (volume percentage at inlet pressure 2.5 bar absolute), especially at high and low flow rates. As the air content rises, head, flow rate, efficiency and power input decline. The pump cannot be operated down to closed discharge valve. If there is more than 5 to 7% gas, the flow may collapse altogether even at best efficiency point.

For multistage pumps, the gas content limit is determined by the first stage. Susceptibility to air influence is, however, less with regard to the shape of the characteristic, since at each subsequent stage, the gas bubbles pass on to a higher pressure level and thus exert volume-wise less influence on the characteristic. With increasing air content the NPSH required also increases as shown in Fig. 2.33.

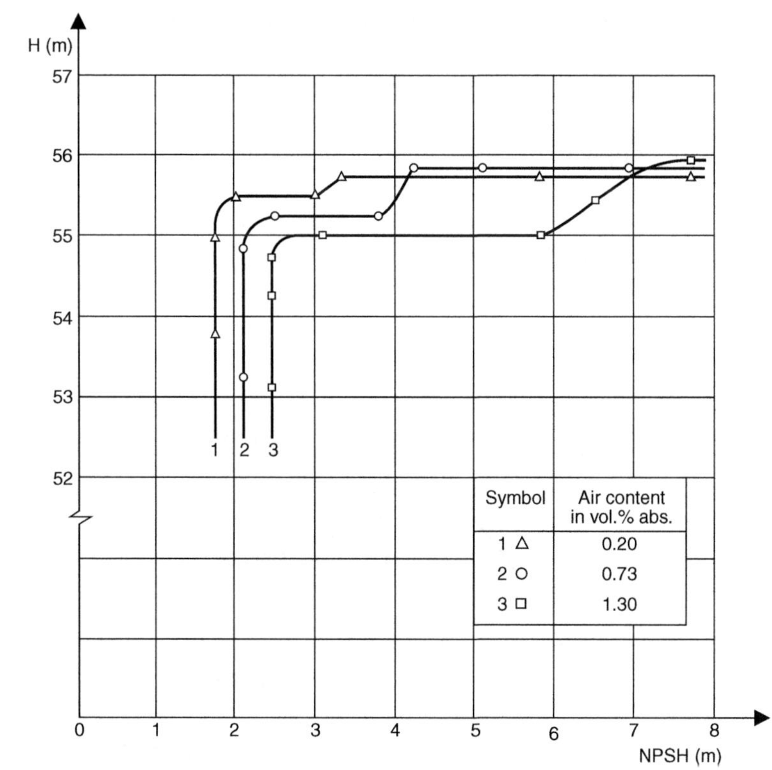

Figure 2.33 Influence of air content on NPSH (Q = constant, n = constant)

2.7.3 Pumping hydrocarbons

2.7.3.1 INFLUENCE ON SUCTION CAPACITY (NPSH)

The required NPSH for centrifugal pumps is normally determined only for handling clean, cold water. Nevertheless, test and accumulated experience have shown that pumps can handle hydrocarbons with less NPSH than would be necessary for cold water.

The expected NPSH reduction is a function of vapor pressure and the density of the hydrocarbons to be pumped. Figure 2.34 shows schematically the loss of head for cold, degassed water and for a defined hydrocarbon. With cold water, a much greater volume of vapor forms at the impeller entry at a given NPSH. This obstruction reduces the head. As the NPSH is reduced further, the growing vapor volume causes an abrupt loss of head. With hydrocarbons, there is less volume of vapor at low NPSH, so that the loss of head is delayed and more gradual.

Figure 2.35 depicts maximum admissible NPSH reductions for hot degassed water and certain gas-free hydrocarbons (according to the Hydraulic Institute). The NPSH reductions shown are based on operating experience and experimental data.

The validity of the diagram is subject to the following limitations:

1. The reduction may not exceed 50% of the NPSH necessary for cold water, or a maximum of 3 m, whichever is smaller.
2. It applies only to low-viscosity liquids and hydrocarbons free of air and gas.
3. It does not apply to transient operating conditions.
4. It applies when pumping petroleum products (mixtures of different hydrocarbons), provided the vapor pressure of the mixture is determined for the light fractions.

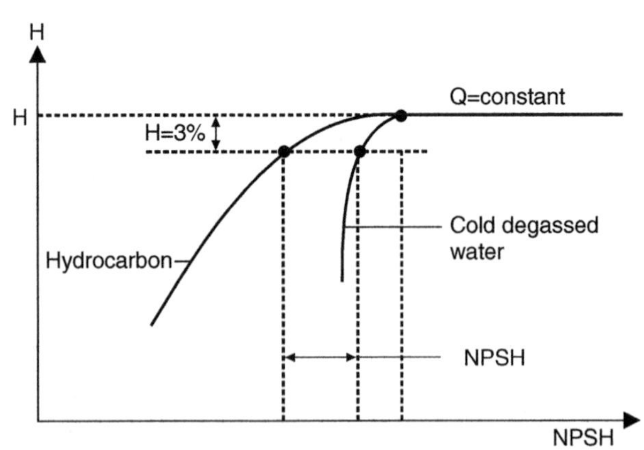

Figure 2.34 Loss of head with two different liquids

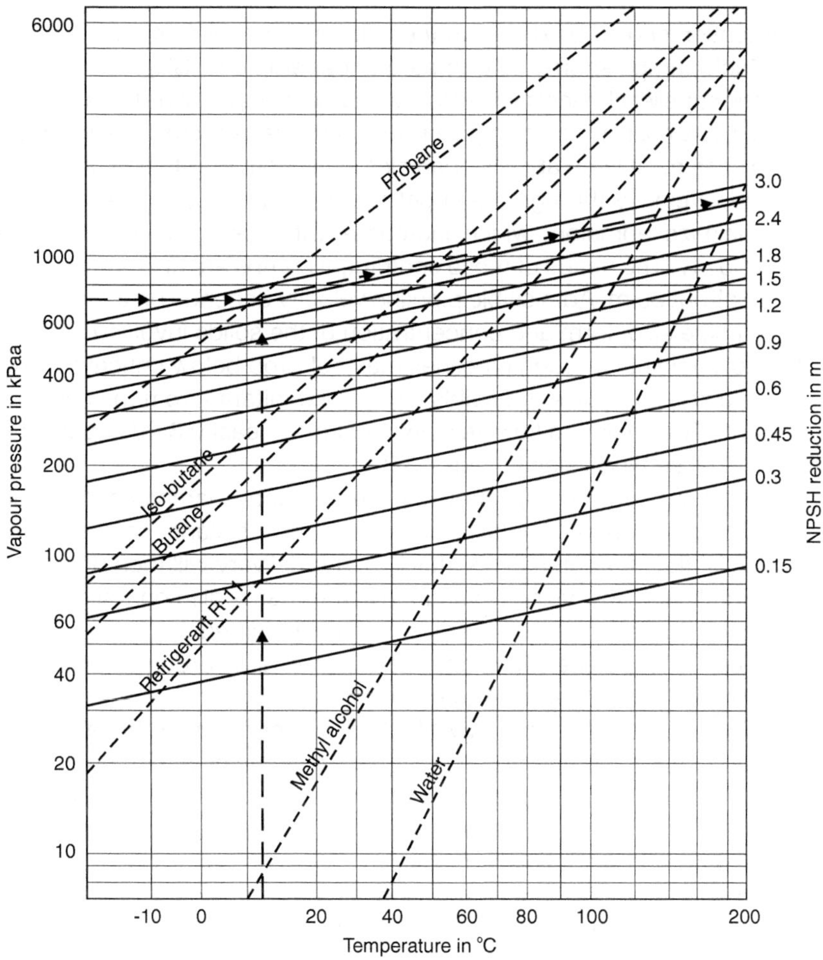

Figure 2.35 Maximum admissible NPSH reduction for hot, degassed water and certain gas-free hydrocarbons (Hydraulic Institute, 14th edition, 1983)

2.7.4 Handling solid–liquid mixtures with centrifugal pumps

By virtue of their simple design, uniform delivery, good ratio between delivery and size and low susceptibility to trouble, centrifugal pumps are also employed for handling mixtures of liquids and granular solids (termed "slurries"). For higher flow rates centrifugal pumps are also less expensive than positive displacement pumps for comparable pumping data.

A large variety of goods, including coal, iron ore, fly ash, gravel, sewage, fish, sugar beet, potatoes, wood pulp, paper stock and suspensions of lime, can be handled by centrifugal pumps. More details are given in Chapter 9 under the headings of solid-conveyance pumps, sewage pumps, pumps for the paper industry and pumps for flue gas desulphurization plants.

Attention should be drawn to the following special features:

1. Such pumps should be operated under ample $NPSH_A$.
2. The impeller channels and volute cross-sections must be large enough to allow passage of the biggest solid particles expected. Back shroud blades must keep abrasive particles away from the stuffing box.
3. Flow velocity through the impeller and volute must be kept low to minimize abrasion, which increases with the third power of the velocity. The heads of these centrifugal pumps are presently limited to about 80 m. Admissible pipeline velocity depends on the density, particle size, form and concentration of the solids involved.
4. Flow velocity through pipelines may not drop below critical settling velocity, otherwise the solids will be precipitated from the suspension. The minimum flow velocity must be at least 0.3 m/s above the critical settling velocity (Fig. 2.37).
5. The centrifugal pump should have a steep characteristic. If the concentration of solids is changed, the resistance curve alters too. If the characteristic is steep, pump delivery will then drop only slightly and flow velocity variation in the lines is also only slight (Fig. 2.36). Segregation of the solids is thus prevented.
6. Slurry pumps are often operated in series in order to meet the demand for a steep characteristic.
7. When the pipeline is partly blocked by solids which have settled upon shutdown, the pump must generate sufficient flow to get the solids moving. The higher resistances encountered at startup must be overcome, so that adequate flow velocity may be achieved to resume transportation process.
8. When pumping long fibers, open impellers should be used on account of the clogging risk. The inlet edges should be profiled and thickened so that fibers can slide off them more easily.
9. Slurry pumps are subject to heavy wear. This is reflected in losses of head, flow rate and efficiency, and must be taken into account when sizing the system, drivers and pumps.

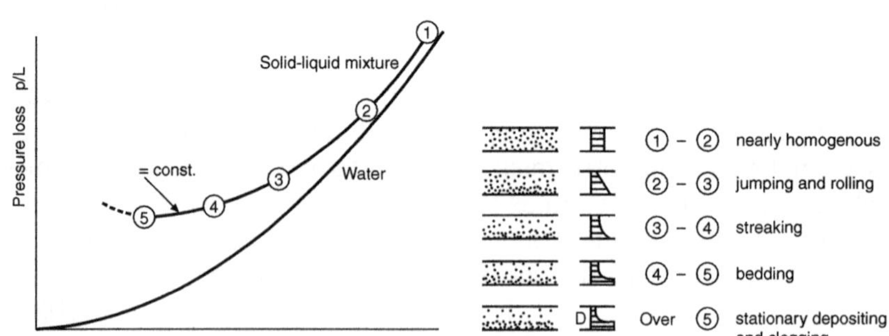

Figure 2.36 Segregation of solids as a function of flow velocity at constant mixture ratio μ

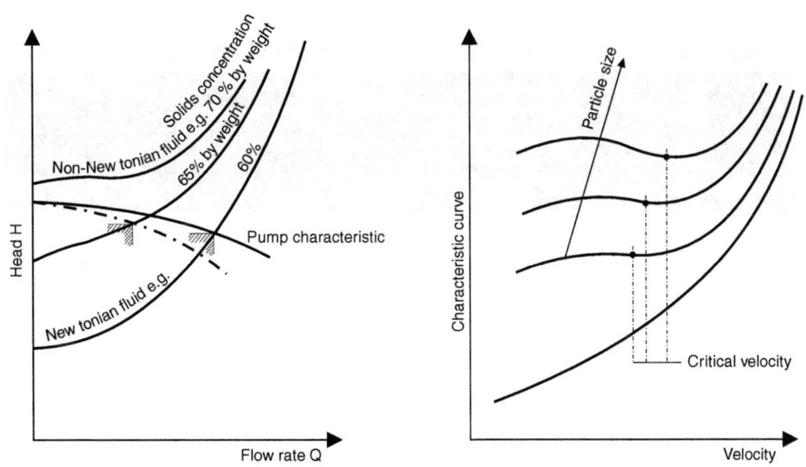

Figure 2.37 Pipeline characteristic as a function of solids concentration and particle size

10. The pump materials must be selected judiciously for the liquid and solids involved. Depending on its abrasiveness and/or corrosiveness, rubber, plastic or enamel linings can be provided, or alternatively particularly hard materials must be used.

Table 2.1 compares the Tyler screen sizes and mesh sizes in mm. Table 2.2 shows the settling velocities for various solids as a function of solid particle size.

Table 2.1 Comparison of Screen Sizes: Tyler and Mesh (mm)

	Tyler standard sieve series			Grade
	Mesh width		Mesh	
	inch	mm		
	3			
	2			
	1.5			
	1.050	26.67		Screen shingle gravel
Theoretical values	0.883	22.43		
	0.742	18.85		
	0.624	15.85		
	0.525	13.33		
	0.441	11.20		
	0.371	9.423		

(Continued)

Table 2.1 Comparison of Screen Sizes: Tyler and Mesh (mm) *(continued)*

	Tyler standard sieve series			Grade
	Mesh width		Mesh	
	inch	mm		
Theoretical values	0321	7.925	2.5	
	0263	6.68	3	
	0.221	5.613	3.5	
	0.185	4.699	4	
	0.156	3.962	5	
	0.131	3.327	6	
	0.110	2.794	7	
	0.093	2.362	8	
	0.078	1.981	9	Very coarse sand
	0.065	1.651	10	
	0.055	1.397	12	
	0.046	1.168	14	
	0.039	0.991	16	
	0.0328	0.833	20	Coarse sand
	0.0276	0.701	24	
	0.0232	0.589	28	
	0.0195	0.495	32	
	0.0164	0.417	35	Medium sand
	0.0138	0.351	42	
	0.0116	0.295	48	
	0.0097	0.248	60	
	0.0082	0.204	65	
	0.0069	0.175	80	
	0.0058	0.147	100	Fine sand
	0.0049	0.124	115	
	0.0041	0.104	150	
	0.0035	0.089	170	
	0.0029	0.074	200	Silt
	0.0024	0.061	250	
	0.0021	0.053	270	

(Continued)

Table 2.1 Comparison of Screen Sizes: Tyler and Mesh (mm) *(continued)*

	Tyler standard sieve series			Grade
	Mesh width		Mesh	
	inch	mm		
	0.0017	0.043	325	
Theoretical values	0.0015	0.038	400	pulverized
		0.025	500	
		0.020	625	
		0.010	1250	
		0.005	2500	
		0.001	12500	

Table 2.2 Sinking Velocities for Various Abrasive Materials

Diameter in mm	Mesh size US fine	Sinking velocity in m/s	ASTM	Soil grain size identification US Bureau of Soils USDA	MIT	International
0.0002		0.00000003	Clay	Clay	Medium-coarse clay	Fine clay
0.0006		0.00000028			Coarse clay	Coarse clay
0.001		0.0000070				
0.002		0.0000092			Fine silt	Fine silt
0.005		0.0017	Silt	Silt	Medium-coarse silt	Coarse silt
0.006		0.000025				
0.02		0.00028			Coarse silt	
0.05	270	0.0017	Fine sand	Very fine sand		
0.06	230	0.0025				
0.10	150	0.070		Fine sand	Fine sand	Fine sand
0.20	70	0.021				
0.25	60	0.026	Coarse sand	Sand	Medium-coarse sand	Medium-coarse sand
0.30		0.032				
0.50	35	0.053		Coarse sand		Coarse sand
0.60	30	0.063				
1.00	18	0.10		Fine gravel	Coarse sand	Very coarse sand
2.00	10	0.17				

2.8 MINIMUM FLOW RATE

Minimum flow rate is the lowest pump delivery that can be maintained for extended periods of operation without excessive wear or even damage. Minimum flow should be distinguished from minimum *continuous* flow which represents the lower limit of the preferable operation range, where the pump is allowed to operate for an indeterminate time.

Minimum flow rate of centrifugal pumps is determined by the following criteria:

- temperature rise due to the internal energy loss;
- increased vibration due to excessive flow separation and recirculation;
- increased pressure fluctuation at part load;
- increased axial thrust at low flow rates;
- increased radial thrust (especially with single-volute pumps).

2.8.1 Determining the minimum flow rate

2.8.1.1 ADMISSIBLE TEMPERATURE RISE

Centrifugal pumps convert only part of their power input into hydraulic energy. Some of it is converted into heat energy by friction. This internal energy loss depends very much on the flow rate through the pump, as is shown in Fig. 2.38. The external energy losses due to bearing and stuffing box friction are small and are ignored here. The lower the flow rate, the lower the efficiency and hence the greater the internal energy loss

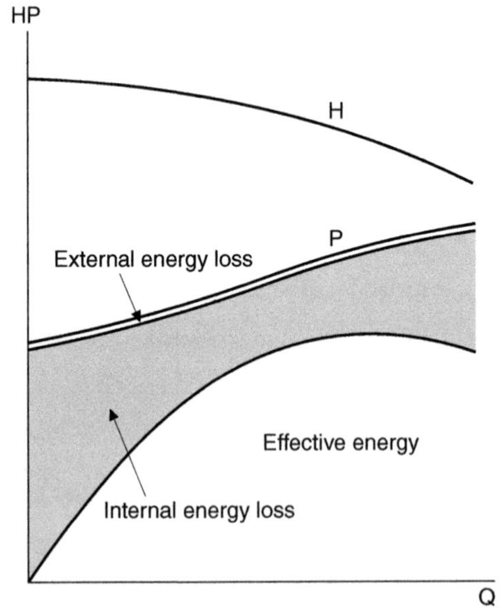

Figure 2.38 Effective energy and energy losses of a centrifugal pump

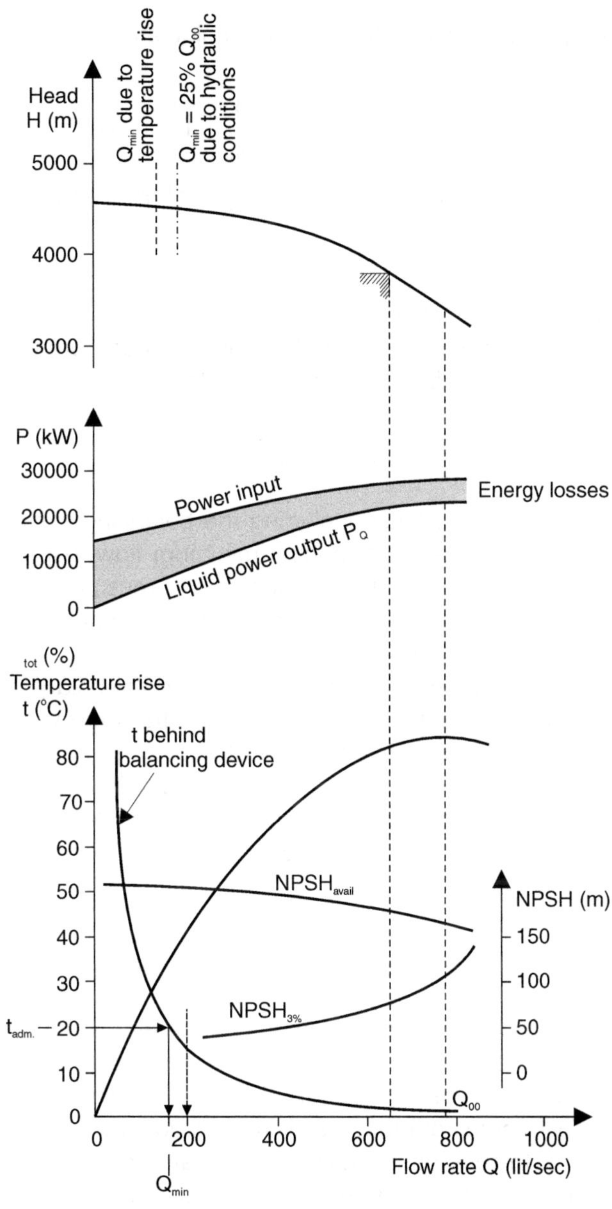

Figure 2.39 Temperature rise of the liquid in a pump and admissible minimum flow rate

converted into heat. The temperature of the water in the pumps rises asymptotically with diminishing flow (Fig. 2.39). To avoid overheating and possible evaporation in the pump, a certain amount of liquid (the so-called thermal minimum flow) passes through the pump.

The temperature rise in the pump is calculated for incompressible flow as:

$$\Delta t_D = \frac{0.00981}{c} \cdot H \cdot \left(\frac{1}{\eta_i} - 1\right) \text{ in °C}$$

H = total head in m $\quad \eta_i = \eta/\eta_m$ = internal efficiency

c in kJ/kg · K η_m = mechanical efficiency

for water: $c = 4.18$ kJ/kg · K

A temperature rise results from throttling an incompressible liquid in the clearance of the axial thrust balancing device:

$$\Delta t_D = \frac{0.00981}{c} \cdot \frac{\Delta H}{\eta_i}$$

The total temperature rise including the balancing system must be taken into account when determining minimum flow, especially where liquids are pumped close to their vaporization pressure.

More accurate calculations are advisable for compressible flows at very high pressures (>2000 m).

With multistage pumps, the pressure behind the balancing piston or disc must always be sufficiently far removed from the vapor pressure corresponding to the calculated temperature.

For smaller pumps running at a temperature sufficiently far from that corresponding to the vapor pressure, it will suffice to determine minimum flow by means of the following formula:

$$Q_{min} = \frac{P(\text{kW}) \cdot 3600}{\rho\left(\frac{\text{kg}}{\text{m}^3}\right) \cdot c\left(\frac{\text{kJ}}{\text{kg} \cdot \text{K}}\right) \cdot (t_E - t_s)} \left[\frac{\text{m}^3}{\text{h}}\right]$$

The admissible temperature rise $t_E - t_S$ is given by the existing temperature t_S at the suction nozzle and the admissible temperature t_E behind the balancing device. About 20°C may be taken as temperature difference. With small pumps, the balancing flow may be adequate as minimum flow. In such cases the Q_{min} calculated from the above equation must be equal to or less than the balancing flow. If the calculated Q_{min} exceeds the balancing flow, at least the difference between the two must be passed through a minimum flow bypass.

This minimum flow on small pumps may take the form of a continuous flow with a throttling device (with the disadvantage of reduced overall pump efficiency), or of flow controlled by means of an automatic recirculation control valve or a flow meter.

The minimum flow rate is either returned directly to the suction nozzle (but only if the available NPSH of the pump is well above required NPSH) or into the inlet tank. The latter arrangement is recommended, because the warmed water then mixes with the colder water in the tank. This is particularly advisable for hot water pumps, where the temperature rise at the impeller entry may cause a substantial loss of $NPSH_{avail}$.

2.8.1.2 POOR HYDRAULIC BEHAVIOR IN THE PART LOAD RANGE

The temperature rise discussed above is not the only criterion for determining minimum flow. For small pumps (up to about 100 kW), the minimum flow calculated by means of the above formula is sufficient. For pumps with a power input above about 1000 kW and with high specific speed, however, the forces due to flow recirculation at the impeller entry may, even at 25 to 35% of the best efficiency point flow rate, be so great that excessive vibration is excited in the pump and pipework. Higher minimum flow rates are necessary for such pump sizes (Fig. 2.40).

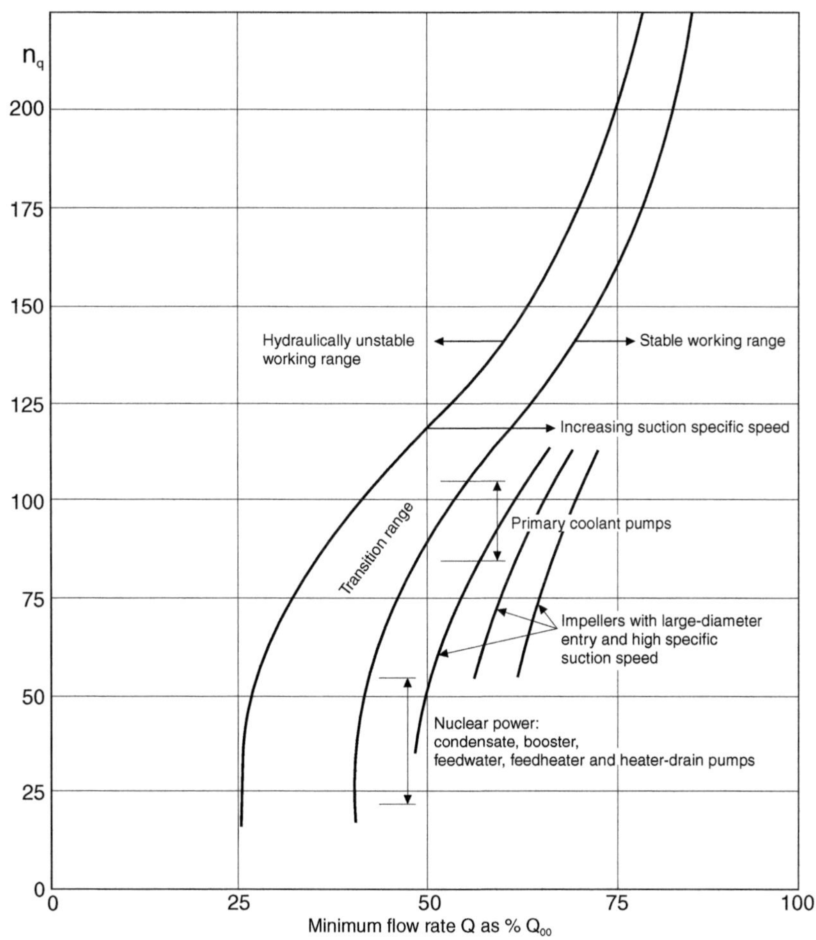

Figure 2.40 Minimum flow rate guidelines as a function of specific speed and pump type for power inputs exceeding 1000 kW

Special attention must be given to impellers with large eye diameters. Though they have a better $NPSH_{full\ cavitation}$, the $NPSH_{0\%}$ increases rapidly at part load. With such impellers, reverse flow at the entry already begins at around 40 to 60% of best point flow (Fig. 2.40).

2.8.1.3 INFLUENCE OF AXIAL AND RADIAL THRUST

The increasing axial thrust that occurs as flow diminishes in pumps operating at minimum flow for extended periods of time must be accommodated by suitable bearings. Single-volute pumps set up high radial thrusts under part loads, calling for appropriate precautions such as reinforcing the shaft and bearings.

Chapter | three

Acceptance Tests with Centrifugal Pumps

3.1 PRELIMINARY REMARKS

The purpose of an acceptance test is to demonstrate the fulfillment of the technical, hydraulic and mechanical guarantees agreed between the purchaser and the pump maker.

Every proof of guarantee entails costs and should therefore be confined to the data essential to satisfactory service. For standard pumps, only type tests are usually performed.

3.2 PUMP TESTS

These tests are classified as follows:

1. *The works or factory acceptance test* is performed on the manufacturer's test stand and is very accurate thanks to the technique of measurement employed and the reproducible conditions.
2. *Measurements on a pump* carried out in the field (field tests) are subject to the influence of the prevailing service conditions. Accuracy and reliability of the measured results depend largely on the instrumentation and measuring positions. Where contractually agreed, however, acceptance testing can be carried out in the field.
3. *Periodic field tests* serve to detect changes or wear on the pump and should be carried out to a fixed schedule using instruments of unvarying standard.
4. *Model tests* call for high measuring accuracy. The model test must be defined so that it serves as a substitute for the acceptance test on the full-scale pump. One hundred percent hydraulic model fidelity is a prerequisite.

Centrifugal Pump Handbook. DOI: 10.1016/B978-0-7506-8612-9.00003-6

3.2.1 Acceptance rules

The rules are laid down in standards in order to simplify understanding between pump maker and purchaser. They contain in general:

1. Definitions of all variables needed to define the functions of a centrifugal pump and to fix the guarantees for its hydraulic performance, flow capacity and total dynamic head and pump efficiency or power.
2. Definitions of the technical guarantees and their implementation.
3. Recommendations for preparing and performing acceptance tests to verify the guaranteed data.
4. Rules for comparing measured results with guaranteed data and for the conclusions to be drawn from the comparison.
5. Requirements for curves and test reports.
6. Description of the principal measuring techniques used for guarantee demonstrations.

In this connection, the terms "guarantee" and "acceptance" in the standards are to be understood in the technical and not in the legal sense.

Most standards applied nowadays contain two aspects:

1. The standards are intended for determining and analyzing measured results on a statistical basis and must ensure that the true value of the measured variable is ascertained with a statistical reliability of at least 95%.
2. The quality classes are defined in similar fashion in most standards, e.g. ISO 9906 accuracy grades *1 and 2*.

Note: Although standards such as ISO, API, HI and DIN are becoming increasingly harmonized we refer to them individually here as the industry continues to commonly use this terminology.

Top class (ISO 5198), only the smallest deviations are tolerated:

- Model pumps
  The ISO accuracy grade is employed chiefly for research, development and scientific work in laboratories, where very high measuring accuracy is demanded.
  In exceptional cases for very large pumps (e.g. >10 MW*) for energy generation, water transport and feed pumps, where the engineering grade 1 is too inaccurate.

Higher class (ISO 9906 accuracy grade 1), medium deviations are allowed:

- Pumps for conveying liquids, for injection and industrial power generation at medium outputs (0.5–10 MW*).

Lower class (ISO 9906 accuracy grade 2), larger deviations are possible:

- Standard pumps manufactured in series for industrial duties (e.g. with type testing).

In ISO 9906 the above accuracy standards define the required accuracy of the instruments used in the acceptance test.

No construction tolerances are laid down in the ISO acceptance rules. On the other hand, the overall tolerances are defined, taking into account the admissible measuring uncertainties and the construction tolerance.

3.2.2 Comparison of ISO, API 610 and Hydraulic Institute Standards with regard to guarantee points and measuring uncertainties

ISO standard accuracy grade	5198 precision grade top	9906 grade 1 middle	9906 grade 2 low

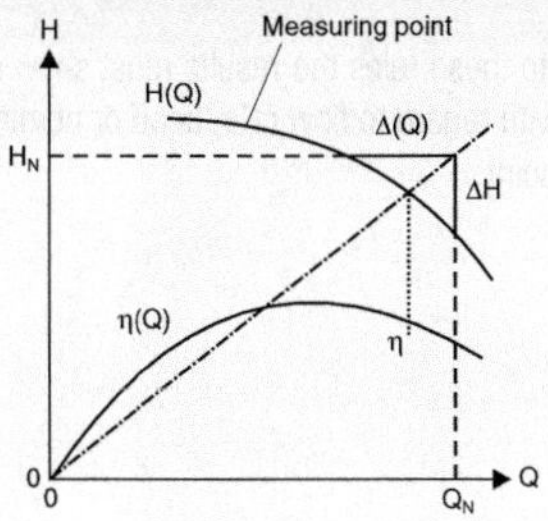

The construction tolerance is included in the increased overall measuring uncertainty and shown in the adjacent method for determining the service data and efficiency. To verify guarantee fulfillment a straight line is drawn through the points Q_N, H_N and $Q = 0$, $H = 0$, giving the intervals ΔQ and ΔH as horizontal and vertical distances from the measured head curve $H(Q)$.
A perpendicular through the intersection of the straight lines $0 - Q_N\ H_N$ with the head curve determines the efficiency η'.

The service guarantee is thus fulfilled

$$\left[\frac{HN \cdot XH}{\Delta H}\right]^2 + \left[\frac{QN \cdot XQ}{\Delta Q}\right]^2 \geq 1$$

	5198 precision grade top	9906 grade 1 middle	9906 grade 2 low
XH	(0.015)	± 0.03	0.05
XQ	(0.025)	± 0.045	0.08
$\eta' > x\eta \cdot \eta N$	(0.978)	0.97	0.95
XNPSH ~ 3% or max. NPSH deviation	(0.03 < 0.15 m)	0.03 < 0.15 m	0.06 < 0.3 m

Max. admissible measured value scatter based on 95% statistical accuracy:

Number of observations per measuring point	3	5	7	9	3	5	7	9	3	5	7	9
$\pm n$ in % (speed)	0.25	0.5	0.7	0.9	0.3	0.5	0.7	0.8	0.6	1.0	1.4	1.6
$\pm Q, H, P$ in % (service data)	0.8	1.6	2.2	2.8	0.8	1.6	2.2	2.8	1.8	3.5	4.5	5.8

Admissible total uncertainty for measuring instruments and measurement:
Values for measuring instruments (...)

	5198 precision grade top	9906 grade 1 middle	9906 grade 2 low
Flow rate	($\pm$1.0) $\pm$ 1.5%	($\pm$1.5) $\pm$ 2.0%	($\pm$2.5) $\pm$ 3.5%
Head	($\pm$0.5) $\pm$ 1.0%	($\pm$1.0) $\pm$ 1.5%	($\pm$1.5) $\pm$ 3.5%
Power input	($\pm$0.6) $\pm$ 1.3%	($\pm$1.0) $\pm$ 2.0%	($\pm$2.0)[1)] $\pm$ 4.0%
Efficiency	$\pm$ 2.25%	$\pm$ 3.2%	$\pm$ 6.4%
Speed	($\pm$0.1) $\pm$ 0.2%	($\pm$0.35) $\pm$ 0.5%	($\pm$1.4) $\pm$ 2.0%

[1)] *For determining pump efficiency*

Further to the ISO acceptance conditions, pumps used in the petrochemical industry are normally ordered to API 610 (ISO 13709) and other pumps may be ordered to the more generalized standard Hydraulic Institute.

API 610 (ISO 13709)			Hydraulic Institute Standards test code 1.6 and 2.6
at n_N and Q_N:			Pumps must lie within the following tolerances for acceptance to level A: At rated head: +10% of nominal flow rate or at rated flow rate/rpm +5% of head <152.4 m +3% of head >152.4 m Conformity with one of the above tolerances is required. Test tolerances: In tests according to these rules the results must show no minus tolerances with regard to flow rate, head or nominal efficiency at duty point.
Nominal differential head	Guarantee point	Shut off	
0–152.4 m	+5% −2%	±10%[1)]	
152.4–304.8 m	+3% −2%	±8%	
>304.8 m	±2%	±5%	
η_N	Not a rating value		
P_N	+4%		
$NPSH_{req}$	+0		
Admissible measuring uncertainty (instr.) fluctuation:			
The measuring uncertainties are included in the values quoted above.			Flow rate (±1.5) ±2% Head (±1.0) ±2% Suction head (±0.5) ±2% Speed (±0.3) ±0.3% Power input (±1.5) ±2%

1) If a rising Q/H curve is specified, the minus tolerance given here is allowed only if the test curve displays a characteristic rising.

3.2.3 Test beds

The pump must be set up on the test bed with proper suction and discharge piping, so that it corresponds to the actual installation layout. Most importantly, care shall be taken to assure proper inlet conditions.

Especially for vertical mixed and axial-flow pumps (possibly with inlet bends) operating with high flow rates and low heads the inlet configuration on the test bed must be maintained to assure proper inlet conditions and when possible matched to actual service conditions as closely as possible (Fig. 3.1).

Ideally the flow rate through the pressure measurement cross-section should satisfy the following requirements:

1. regular velocity distribution in the axial direction;
2. static pressure equalized in the measuring plane;
3. the inflow must be free of vortices.

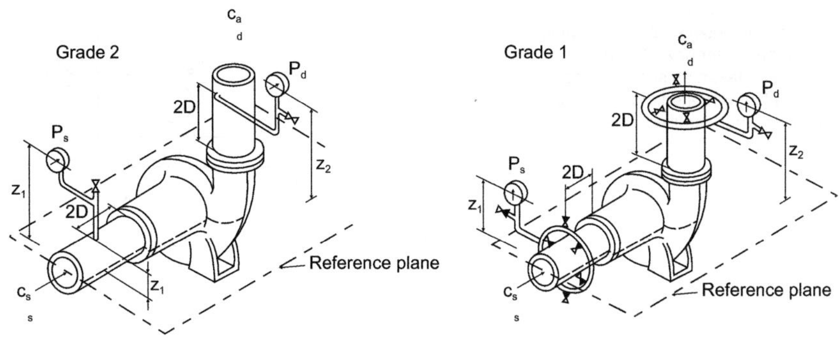

Figure 3.1 Measuring set-up for determining the pressure head

Often these conditions can hardly be met, but the important thing is to obtain a good velocity distribution at the pressure measuring point by suitable choice of pipe layout and if necessary by means of flow conditioners and straighteners.

Generally at least seven diameters of straight suction pipe are needed after a 90° bend or other pipe fitting to obtain an acceptable velocity distribution at the pump's inlet. Greater lengths should be used if a throttling valve is used in the suction pipe to calm the disturbance. To assure uniform velocity distribution at the pressure measuring point, flow conditioners or straighteners are fitted at sufficient distance from the measuring point.

The test stand configuration at the suction end depends on the nature of the pump inlet (see Fig. 3.2).

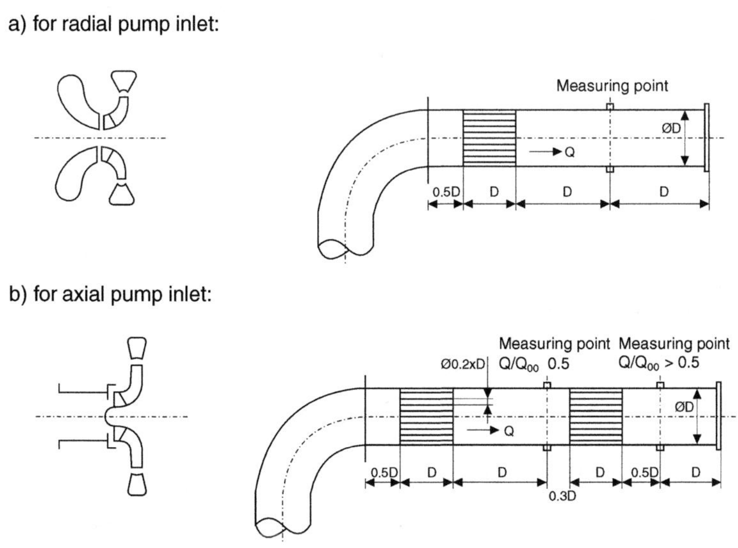

Figure 3.2 Test stand configurations

For example depending on the connection between pressure gauge and pipe (water or air filled), z_1 should be taken as shown in the illustration or related to the pipe tap location when air filled.

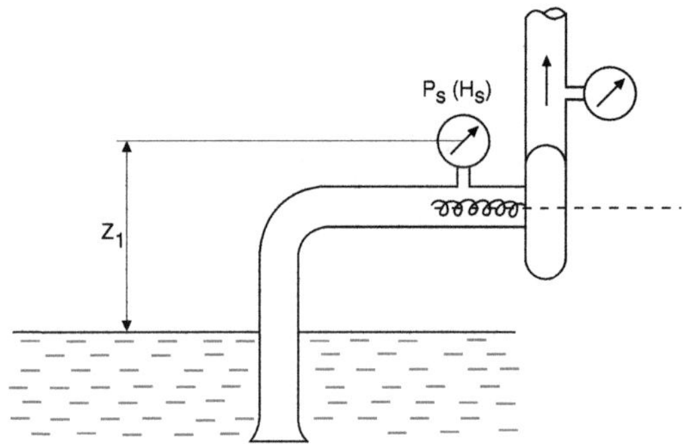

Figure 3.3

On pumps with an axial inlet it is necessary to suppress the pre-rotation set-up by the impeller by providing a second straightener for discharge rates $Q/Q_{00} < 0.5$, otherwise the static pressure measurement will be falsified.

For example, depending on the connection between pressure gauge and pipe (water or air filled), z_1 should be taken as shown in Fig. 3.3 or related to the pipe tap location when air filled.

On the other hand, the suction heads may be plotted against the square of the discharge rate. The tangent to the curve obtained in this way, which when discharge rate $Q = 0$ passes through the geodetic suction head, determines the suction head increase under part load flow rates (see Fig. 3.4).

The installations for measuring the NPSH of a pump are shown in Table 3.1.

For NPSH tests, depending on the inflow conditions, a booster pump may be required along with a throttling valve. Other test rigs are more

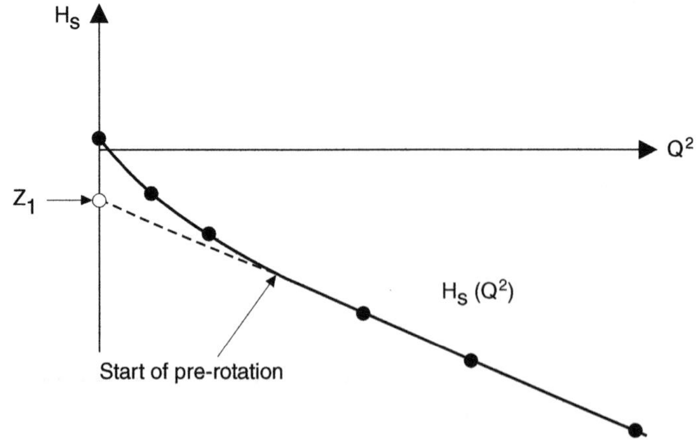

Figure 3.4 Start of pre-rotation

Table 3.1

System	Open systems		Closed systems
NPSH range	>2.5 m with suction throttle valve	2.5–8 m with controlled water level	Practically unlimited for closed loops
Installation principle	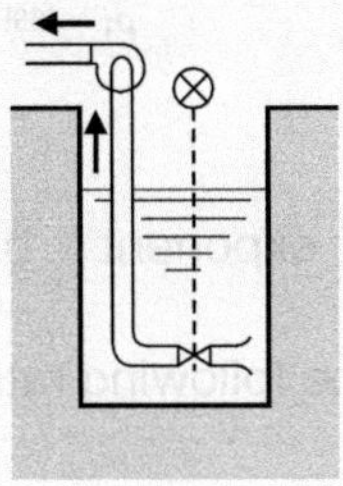	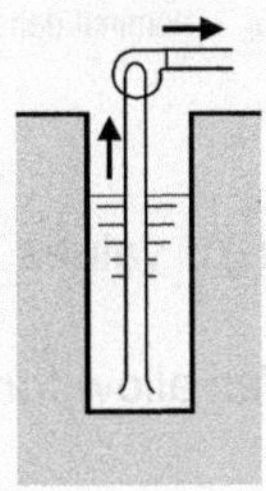	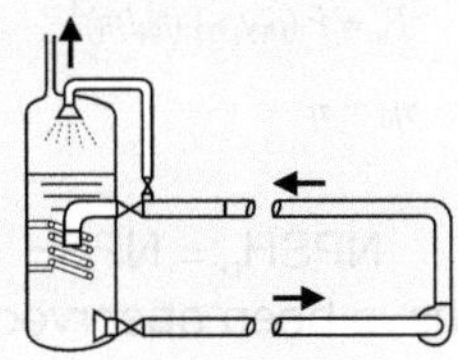
Flow rates	All flow rated	Low flow rates	Low, medium flow rates and heads
Control mode	Suction pressure throttling	Water level variation	System pressure variation
Evaluation method at constant discharge rates. Q (Q_G), H', n are the measured values	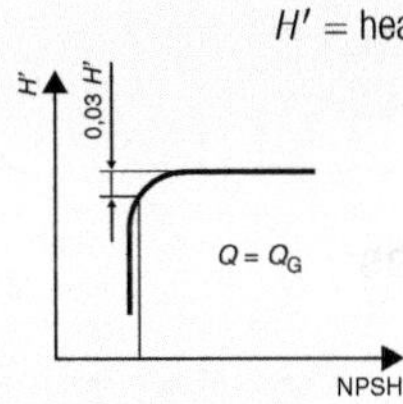	Q_G = constant flow rate Q = total flow rate H' = head first stage	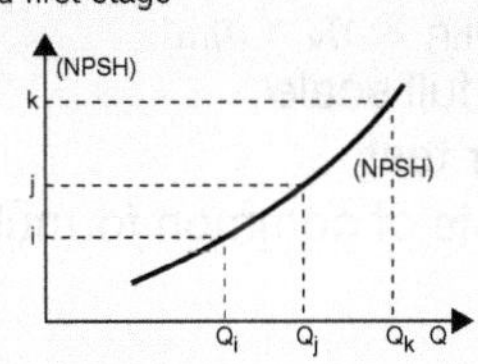

costly to operate. Furthermore, for higher pressure pumps (stage heads >600 m) detailed NPSH values and also bubble patterns for the suction stage may be determined in model tests.

3.2.4 Conversion of test results

All measured results determined at a speed n and with test fluid deviating from the nominal speed n_N and fluid density ρ_N must be converted to the latter. *The admissible speed deviations* according to HI and ISO are given in the table below.

Speed range				
Test values	HI	HI >225 kW	ISO 9906	According to experience
Service data efficiency	80 to 120%	60 to 140%	50 to 120% +20%	±20% ÷ −50% ±20%
NPSH −3%	80 to 120%	60 to 140%	80 to 120%	±20%

Converting the measured values (index u denotes converted values):

$Q_u = Q\,(n_N/n_T)$	n_N = nominal speed	n_T = test speed
$H_u = H\,(n_N/n_T)^2$		
$P_u = P\,(\rho_N/\rho_T)\,(n_N/n_T)^3$	ρ_N = nominal density	ρ_T = test density
$\eta_u = \eta$		

$NPSH_u = NPSH\ (n_N/n_T)^x$; values of exponent x between 1.3 and 2 have been observed.

The correction formulas allow for the following ratios:

- test and actual speed;
- test and actual fluid density;
- viscosities of the pumped liquids;
- model and full-scale diameters.

To effect efficiency correction formulas exist which relate to pump efficiency η, hydraulic efficiency η_h or η_{hR} as defined in section 1.2.

All correction methods are applied solely to the efficiency at best point:

$\eta_{OO} = \eta_{hR} \times \eta_v \times \eta_m$

η_{OO} for full scale

η_{TOO} for test

Table of common formulas:

Origin	Formula	Notes
Sulzer	$\frac{1-\eta_1}{1-\eta_2} = \left(\frac{D_2}{D_1}\right)^{0.224} \times \left(\frac{N_2}{N_1}\right)^{0.07} \times \left(\frac{\nu_1}{\nu_2}\right)^{0.07}$	Stage efficiency (only for efficiencies greater than 25%)
Karassik	$\frac{1-\eta_1}{1-\eta_2} = \frac{\eta_1}{\eta_2} \times \left(\frac{N_2}{N_1}\right)^{0.17} \times \left(\frac{\nu_1}{\nu_2}\right)^{0.07}$	Pump overall efficiency

3.2.5 Measuring instrument uncertainty

The abstracted guidelines below for measuring uncertainties of particular values apply to acceptance tests class 1 (ISO 5198 Annex A).

Measuring the flow rate Q	f_q
1. with tank and clocked measuring time	
$t \geq 50$ s	$\pm\frac{0.3}{\Delta z}\%$
(measuring the level difference Δz in m by siphon gauge when the filling jet is swung in and out)	

2. with differential pressure instruments	$\pm 1.0 \div \pm 1.5\%$
(according to ISO 5167)	
3. magnetic, ultrasonic and mass flow meters	$\pm 1.5\%$
(according to ISO 9104)	
Measuring the pressure head $p/(\rho \cdot g)$	f_p
($p/\rho \cdot g$) is to be inserted below)	
1. with liquid columns between 0.1 and 1.5 m	$\pm \frac{0.1}{p/(\rho . g)}\%$
and Δh fluctuations in m $\leq \pm\ 10^{-3}$ m, with wider fluctuations of Δh in m	$\pm 10^2 \cdot \frac{\Delta h}{p/(\rho \cdot g)}$
2. deadweight manometer	$\pm 0.1\%$
3. with calibrated spring manometer of	$\pm 0.6 \frac{\text{final value}}{\text{read-out value}}\%$
accuracy class 0.6	
4. pressure transducer	$\pm 0.1\%$
Measuring the velocity head $c^2/2gf_e$	
with circular cross-sections (exactly measurable)	$\pm 1.5 f_q\%$
Measuring the speed n	f_n
1. with hand revolution counters	$\pm 0.5\%$
2. with electronic counters	$\pm 0.1\%$
Measuring the power input P	f_p
1. power input to 3-phase motor:	
$P_N \leq 25$ kW	$\pm 1.5\%$
at nominal power:	
25 kW $< P_N \leq 250$ kW	$\pm 1.0\%$
$P_N > 250$ kW	$\pm 0.8\%$
2. from torque and speed:	
a. swivel motor	$\pm\sqrt{1.0^2 + fn^2}$
b. torque meter	
$\alpha \geq 0.75\alpha_{max}$	$\pm\sqrt{1.2^2 + fn^2}$
$0.5\alpha_{max} < \alpha < 0.75\alpha_{max}$	$\pm\sqrt{1.5^2 + fn_n^2}$
Determining the density ρ	f_ρ
of water at temperatures of up to 100 °C	$\pm 0.1\%$

The measuring uncertainties for the service data and efficiency are obtained from the individually measured values:

1. for the flow rate Q: $f_Q = f_q$

2. for the head H:

$$f_H = \sqrt{\left(\frac{Z_d - Z_s}{H}\right)^2 \cdot f_z^2 + \left(\frac{P_d}{\rho \cdot g \cdot H}\right)^2 \cdot f_{pd}^2 + \left(\frac{P_s}{\rho \cdot g \cdot H}\right)^2 \cdot f_{ps}^2 + \frac{(C_d - C_s)^4}{(g \cdot H)} \cdot f_e^2}$$

For pumps with high heads and low ps/pd values it may be assumed that:

$$f_H = \approx f_p$$

3. for the efficiency: $f_\eta = \pm\sqrt{f_q^2 + f_H^2 + f_p^2} \cong \pm\sqrt{f_q^2 + f_H^2 + f_p^2}$

If, for example, the head curve H (Q) of measured service data is converted to nominal speed, the measuring uncertainty is

- for the flow rate Q_u: $f_{Q_u} = \sqrt{f_Q^2 + f_n^2}$
- for the head H_u: $f_{H_u} = \sqrt{f_H^2 + 4f_n^2}$

3.2.6 Fulfillment of guarantee

3.2.6.1 ALLOWING FOR THE MEASURING UNCERTAINTIES IN THE TEST EVALUATION

In a diagram each measuring point appears as an ellipse if the uncertainties for both coordinate variables are taken into account. The major axes of these ellipses emerge from the product of uncertainty and measured value for the variable in question, i.e.:

• for the flow rate	$f_{Qx} \cdot Q_x$
• for the head	$f_{Hx} \cdot H_x$
• for the power input	$f_{Px} \cdot P_x$
• for the efficiency	$f_{\eta x} \cdot \eta_x$

However, it is sufficiently accurate to replace the area of each measuring uncertainty ellipse by the four corners of its two axes. The characteristics H (Q) and η (Q) thus emerge as bands bounded by continuous upper and lower curves, which must be plotted, so that each of the two limiting curves intersects or touches at least one of the two axes at each measuring point (see Figs 3.5a and 3.5b).

3.2.6.2 FULFILLING GUARANTEES WITH RADIAL, MIXED AND AXIAL-FLOW PUMPS ACCORDING TO ISO

The performance guarantee is fulfilled if the head curve passes through the tolerance cross representing the guaranteed requirements in accordance with section 3.2.2 (see Figs 3.6a and 3.6b). The head, efficiency and power curves are to be developed using a polynomial curve fitted to the third or fourth order or a spline curve fitted for comparison to the guarantee limits.

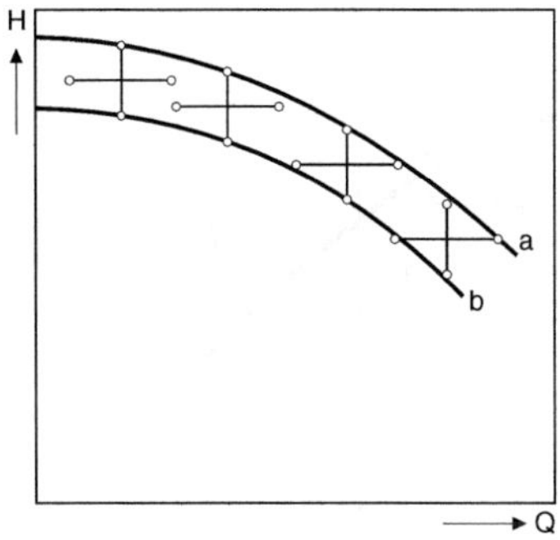

Figure 3.5a Allowing for the measurement uncertainties in the *H* (*Q*) curve

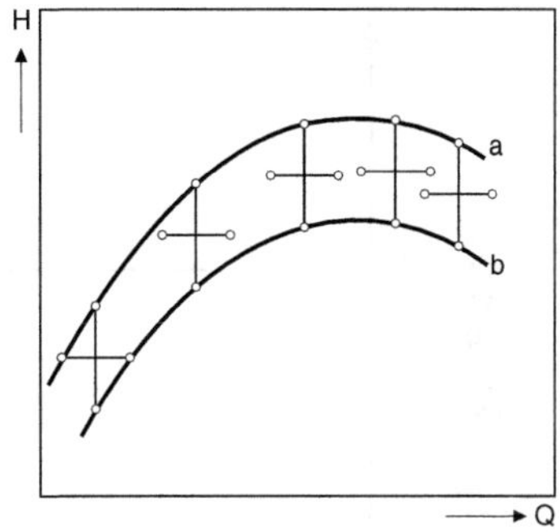

Figure 3.5b Allowing for the measurement uncertainties in the *η* (*Q*) curve

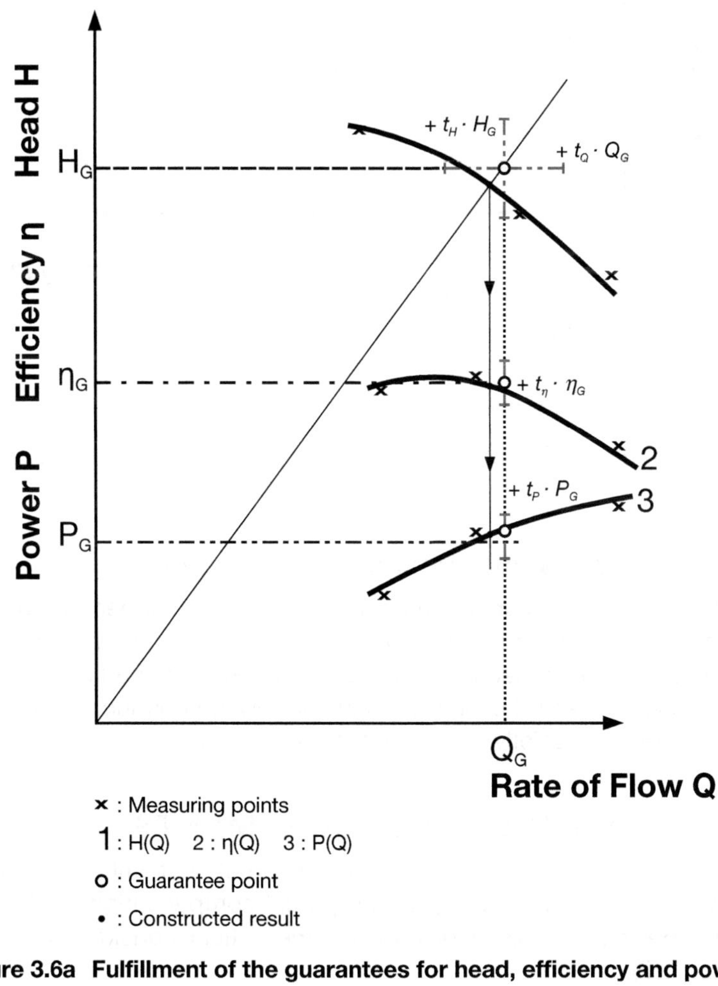

Figure 3.6a Fulfillment of the guarantees for head, efficiency and power

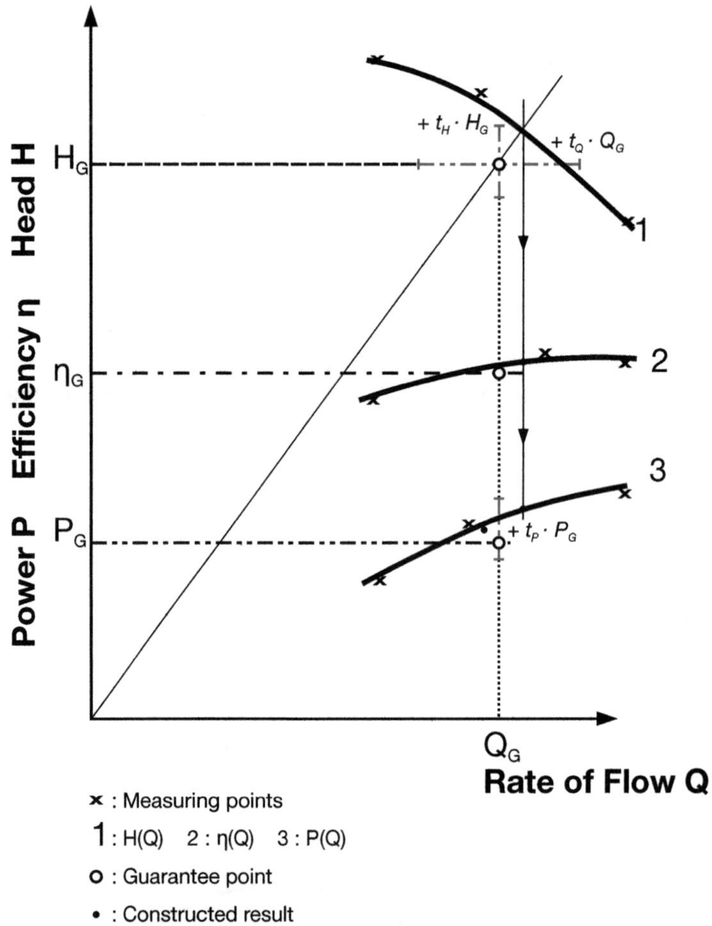

Figure 3.6b Fulfillment of the guarantees for head and power only

3.2.7 Test beds and measuring instruments

In all Sulzer manufacturing plants there are pump test beds for acceptance tests and in-house development tests (Fig. 3.7). Their size and equipment are governed by the product ranges of the particular facility. Basic research and the development of new pumps including model development are chiefly carried out in Winterthur.

3.3 AXIAL THRUST MEASUREMENTS

Here axial thrust denotes the residual thrust that still acts on the thrust bearing after allowing for the partial compensation by the balancing device (usually a piston) (for theoretical considerations see section 1.7).

Figure 3.7 One of four test beds at Sulzer Pumps' UK facility

Hydraulic axial thrust is the resultant of all forces acting on the rotor. These are mainly:

- the forces induced by the pressures on the suction and delivery-side shrouds of the impeller;
- the impulse force (deflection of the flow);
- the forces resulting from the pressures in front of and behind the shaft seal;
- the forces acting on the balancing device.

3.3.1 Reasons for measuring the axial thrust

- For larger pumps (special versions) a guarantee is often given for the direction and maximum amount of the axial thrust. The measurement must demonstrate fulfillment of this guarantee. It is usually performed on the test stand at the same time as the acceptance test for the *Q/H* characteristic. If, however, the test stand conditions deviate too much from the actual installation with regard to speed and temperature, field measurement is possible.
- Using the measured values, the calculated values can be checked and the existing computer programs confirmed and expanded.
- If the pump is equipped with new hydraulic components not previously measured in full scale, or with new impeller configurations, the axial thrust is measured on the first-off pump.

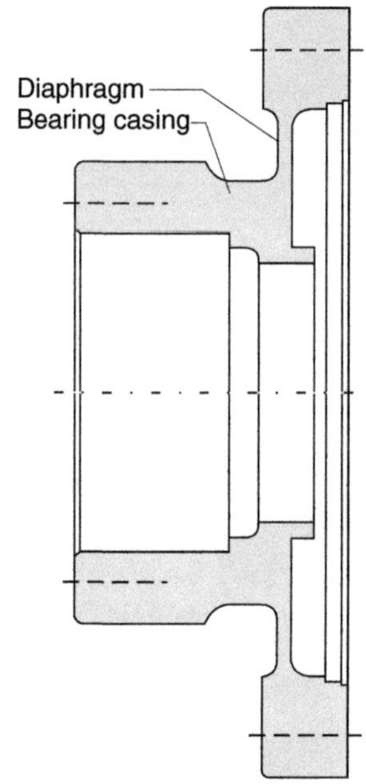

Figure 3.8 Diaphragm as axial load transducer

- On all model pumps, the axial thrust is as a firm principle measured on the final variant intended for standardization or full scale-up.

3.3.2 Axial force transducers employed

When selecting axial force transducers, a fundamental distinction is made between models and full-scale pumps. With a model pump the design can be adapted largely to the demands of the measuring technique used. On the other hand, on a full-scale pump more concessions must be made to the design, i.e. it must be possible to fit the transducer in the bearing housing with minimal alterations.

In either case the element fitted must be sufficiently elastic to provide an adequately strong electrical signal via strain gauges.

Figure 3.8 shows a *diaphragm* with integrated bearing housing for measuring the axial thrust on model pumps. With this device the thrust can be measured in both load directions. By means of suitable strain gauge circuitry the influence of temperature can be eliminated almost entirely, giving a nearly constant sensitivity throughout the measuring range.

Full-scale pumps are usually fitted with annular load transducers as in Fig. 3.9. This type measures only compressive forces. The rings are fitted behind the supporting rings of the axial thrust bearing in the normal bearing housing (see Fig. 3.10).

If the thrust acts in one direction only, it will suffice to fit one annular load transducer.

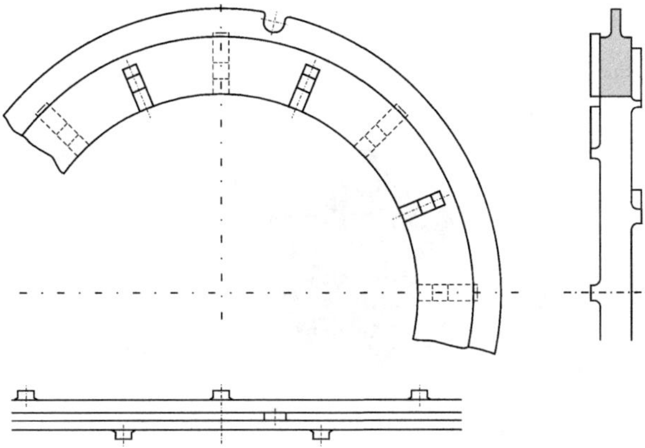

Figure 3.9 Annular load transducer for fitting to a full-scale pump

The strain gauges are attached to the rings so that the following requirements are met optimally:

- total load is indicated even if it acts extremely eccentrically;
- low temperature sensitivity;
- sensitivity variation within ±1% throughout the measuring range.

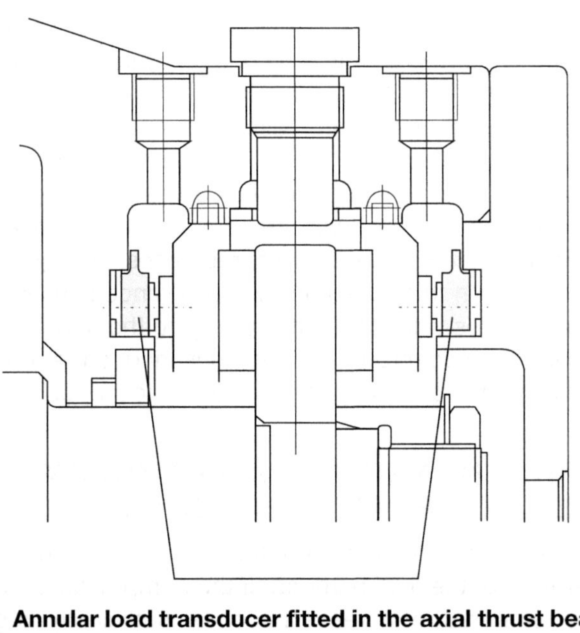

Figure 3.10 Annular load transducer fitted in the axial thrust bearing

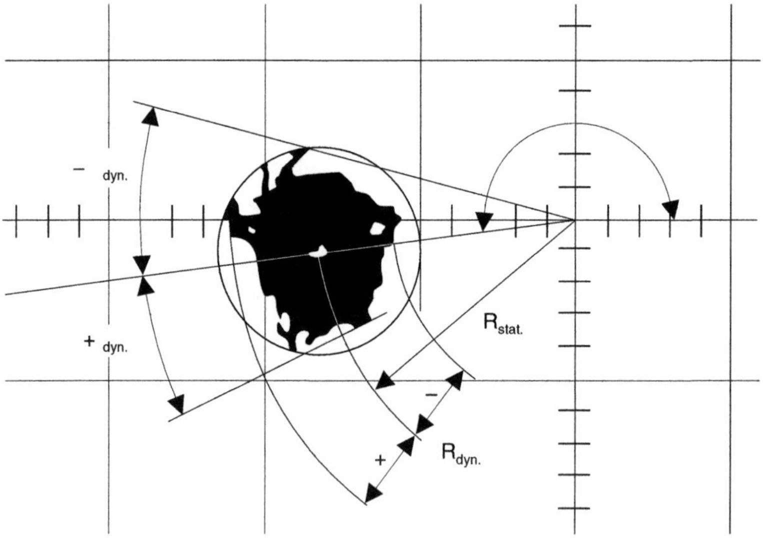

Figure 3.11 Oscilloscope display of radial thrust measurement

3.4 RADIAL THRUST MEASUREMENTS

(For theoretical considerations see section 1.8.)

The radial thrust is measured by means of stay rings fitted between the bearing and its housing. Stay rings are fitted on the bearing near to the impeller for single-entry pumps and on both bearings for double-entry pumps. The stays are sized to facilitate determination of deformation by means of strain gauges. The strain gauges on two opposite stays are connected to form a Wheatstone bridge, giving a high output signal and good temperature stability. The signals from the two stay pairs at right angles to each other are displayed on an X/Y oscilloscope. Figure 3.11 shows an example of the force sustained by a stay ring.

The stay ring may be calibrated statically or dynamically. For *static calibration* the ring is loaded via the shaft in different directions. This method demands great expenditure of resources. The *dynamic method* is less expensive. In this case a known unbalance is used to generate a force that revolves with the shaft. By varying the unbalance or speed the force can be varied throughout the measuring range. The measuring rig is satisfactory if a circle appears in the oscilloscope.

3.5 EFFICIENCY DETERMINATION BY THE THERMOMETRIC MEASUREMENTS

Losses and a slight compression of the fluid cause a small increase in the temperature of the fluid pumped. If the pump head exceeds 100 to 200 m this temperature increase becomes sufficiently large to allow

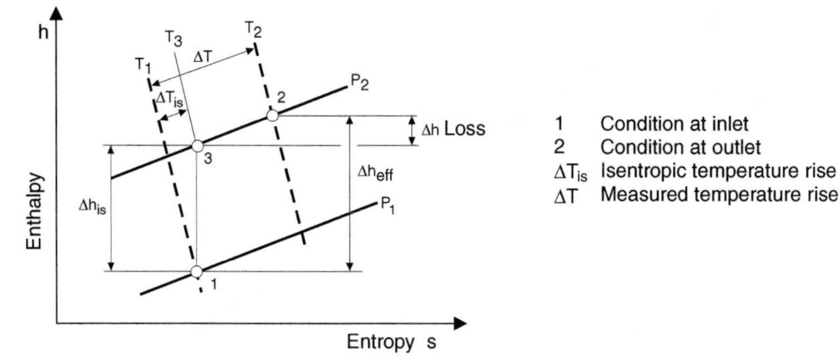

Figure 3.12 Enthalpy diagram

determination of the inner efficiency η_i by very accurately measuring the temperature difference between inlet and outlet of the pump (DIN 1944).

Figure 3.12 shows the pumping process in an enthalpy/entropy diagram. Point 1 represents suction, point 2 discharge condition. The enthalpy rise $\Delta h_{is} = gH$ from 1 to 3 is due to the compression of the fluid (isentropic process, no losses) while the increase from 3 to 2 is due to the inner losses of the pump (these include all except the mechanical losses, see section 1.2). The inner efficiency is given by:

$$\eta_i = \frac{\Delta h_{is}}{\Delta h_{eff}} = \frac{\Delta h_{is}}{\Delta h_{is} + \Delta h_{loss}} \tag{1}$$

If p_1, p_2, T_1 and T_2 are measured, enthalpy difference in equation (1) can be determined from appropriate tables (water/steam table in the case of water).

For cold water the physical properties can be taken from Fig. 3.13 and inserted into equation (2):

$$\eta_i = \frac{1}{B + C\frac{T_2 - T_1}{P_2 - P_1}} \tag{2}$$

The relation between pump efficiency and internal efficiency is given by the following equation (see section 1.2 for definitions of efficiencies):

$$\eta = \eta_i \cdot \eta_m$$

This presupposes that the balancing water from the balancing device, disc or piston (if provided) is fed back to the suction branch.

3.5.1 Methods of measuring temperature

When measuring the water temperature in the inlet and delivery pipes, it is necessary to make sure that the measured values represent the actual mean temperature in the measuring cross-section.

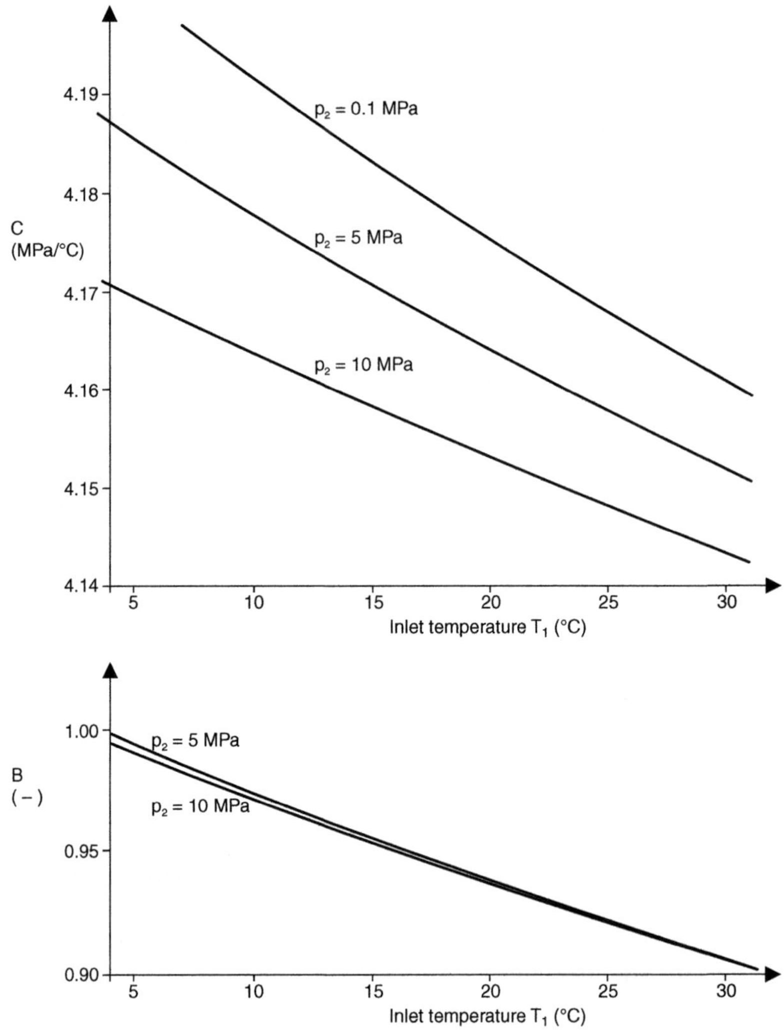

Figure 3.13 Physical characteristics *B* and *C* for water

3.5.1.1 MEASURING WITH SAMPLING PROBES AND QUARTZ THERMOMETER

Probes are inserted at both measuring cross-sections and a small amount of liquid is bled off through them. This flow through the probe ensures that the sensor of the quartz thermometer is exposed to the actual temperature at the sampling point. By throttling the sampling rate to nil the total pressure can be determined at the same time. The best immersion depth for the probe is about 30% of the pipe radius. Figure 3.14 shows a typical probe.

The temperature difference can be read off directly from the quartz thermometer with a resolution of 0.0001°C.

On hot water pumps the two sensors of the quartz thermometer are inserted in welded-in temperature sockets.

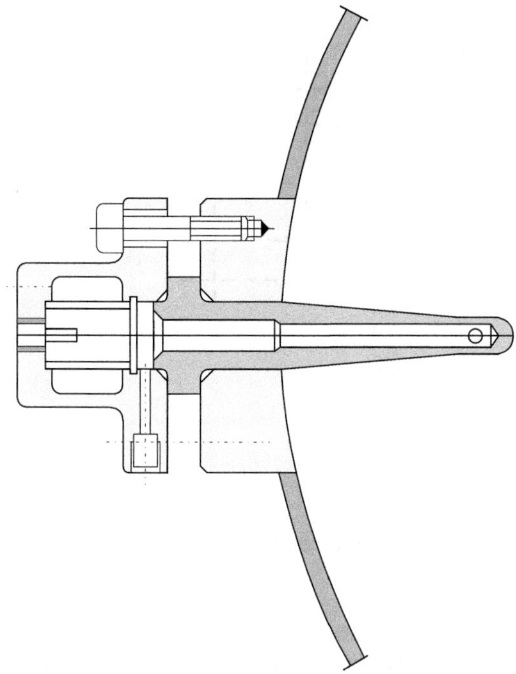

Figure 3.14 Probe for measuring with quartz thermometer

The probe method is also suitable for fast measurements in the plant provided the pipe cross-section is not too large and the temperature distribution in the measuring cross-sections is sufficiently uniform.

3.5.1.2 MEASURING THE OUTSIDE PIPE TEMPERATURE

A simpler but less accurate method is to measure the temperature on the outer wall of the suction and discharge pipe. The great advantage of this method is that no parts have to be welded into the pipe and that the measurement can be performed in a very short time. The procedure is also well suited to monitor a decrease in pump efficiency during service (e.g. due to wear).

By means of thermocouples wired in opposition, the thermoelectric voltage between delivery and suction is measured directly. Adequate insulation must be provided to ensure that the outside of the pipe wall assumes the temperature of the water.

Figure 3.15 shows thermocouples in a differential circuit (thermopile) with three elements.

The approximate temperature rise (and thermoelectric voltages) for different efficiencies are plotted as a function of head in Fig. 3.16.

- Power data are based on empirical values.
- It may, however, be agreed in the supply contract that the service data Q and H may be converted according to the above formulas, and the efficiency according to a known correction formula for hydrodynamic

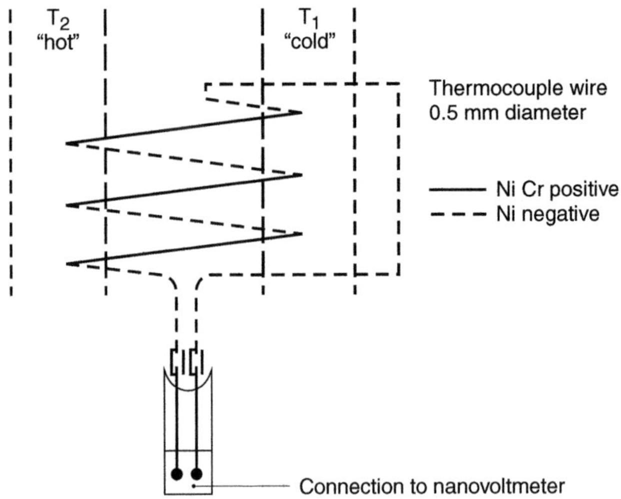

Figure 3.15 Thermopile with three elements

machines, even if the speed deviations are greater than the standards allow. A few common uplift formulas are shown in the table on page 76. However, they hold good only if the following conditions are satisfied by both the model and the full-scale machine:

- $Re_D > 3 \times 10^6$ (related to impeller outside diameter)
- Impeller diameter of model $D_2 \geq 350$ mm.

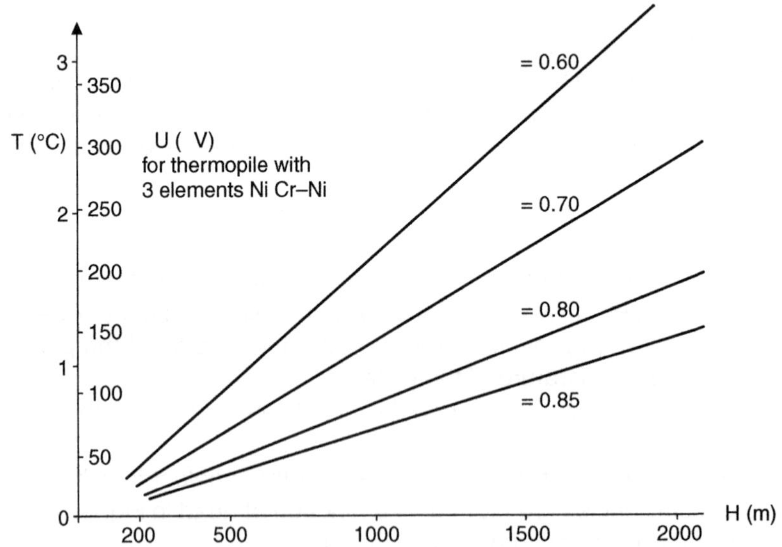

Figure 3.16 Approximate temperature rise in a pump (water temperature below 40°C)

Chapter | four

Special Data for Planning Centrifugal Pump Installations

4.1 INTAKE DESIGN FOR VERTICAL PUMPS

4.1.1 Flow conditions

The purpose of the intake design (approach flow and pump bay) is to ensure undisturbed approach flow of the medium to the pump impeller. If this requirement is not satisfied, capacity and efficiency losses and damage due to vibration and cavitation may be sustained.

The effects of disturbed approach flow on operational behavior increase with the high specific speed and the size of the pump. The basic conditions for undisturbed approach flow are set out below.

4.1.1.1 VELOCITY DISTRIBUTION

The *velocity distribution* at the impeller entry must be as uniform as possible. This is achieved by suitable design of the approach flow and pump bay along the entire flow path (absence of vortices), followed by acceleration in the suction bell preceding the impeller or in the inlet bend. If the local meridional velocity deviates from the mean nominal velocity at the impeller inlet by more than 10%, uneven blade loading may impair the efficiency, and increase the radial forces, vibrations and wear of components.

4.1.1.2 PLANT NPSH

The *plant NPSH* is determined from the required $NPSH_R$ for the pump or the necessary minimum submergence of the entrance to the suction bell or inlet bend. The minimum submergence given in the design rules below must be observed.

4.1.1.3 FORMATION OF VORTICES

Formation of vortices in the approach flow upstream of the pump must be prevented. This applies to all kinds of vortices, i.e. air-entraining or hollow vortices proceeding from the water surface, bottom or wall vortices and vortices resulting from flow separation or skewed velocity profiles in the approach flow.

Centrifugal Pump Handbook. DOI: 10.1016/B978-0-7506-8612-9.00004-8

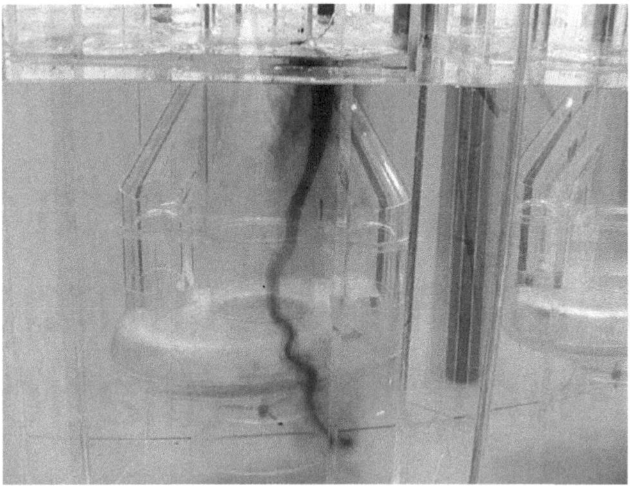

Figure 4.1 Vortex entering the suction bell of a vertical pump model

Air-entraining vortices are to be avoided by suitable configuration of the pump bay immediately upstream of the entry (suction bell or bend – see Fig. 4.1) and adequate immersion depth (minimum submergence).

Vortices due to flow separation in the approach flow (mostly in the transition zone from approach flow to the pump bay) can be prevented by proper execution of the intake design.

The disturbing influence of air-entraining vortices on the operational behavior of the pump may produce the following effects:

- alteration of *Q–H* characteristics and impairment of efficiency;
- excitation of vibration when the impeller blades cut through the vortex;
- increased radial forces on the impeller;
- aggravated risk of cavitation erosion on the impeller blades, as the chopping of the vortex may cause local pressure drops down to vapor pressure, leading possibly to the formation of bubbles.

4.1.1.4 PREVENTING PRE-SWIRL IN THE PUMP INLET

Two sources of pre-swirl need particular attention:

- Asymmetric approach flow can induce a pre-swirl and vortices. This source of pre-swirl must be prevented by proper design of the intake structure.
- Every impeller is subject to flow recirculation at the impeller inlet during part load operation. The fluid recirculating from the impeller induces a strong pre-swirl which can lead to the formation of cavitating or non-cavitating vortices. In order to avoid detrimental effects on performance, cavitation and vibrations, this type of pre-swirl must be suppressed to a large extent by swirl-breaking elements upstream of the impeller such as ribs or floor splitters. The details of these design features are discussed in the following.

4.1.2 Design guidelines

The design of the intake for vertical pumps usually is a compromise between the structural constraints and the hydraulic requirements. Only close collaboration between project engineer and pump maker can lead to a reliable and low-cost installation.

The requirement for undisturbed approach flow to the impeller can be met either with a pump bay and suction bell or with an inlet bend. With careful design, both solutions may be regarded as hydraulically equivalent. For larger capacities (big stations involving heavy construction costs) the decision for one or the other must be reached individually on the strength of an economic assessment taking the entire installation costs into account. In case of doubt, model tests should be carried out for both alternatives, observing the Froude number for all operating conditions, in order to arrive at the highest possible plant reliability.

An inlet bend is recommended for:

- simultaneous installation of several identical pumps, using the bend formwork repeatedly;
- pumps with volute or annular casing executed in concrete;
- if construction constraints or costs for civil engineering favor this option.

Within the present scope it is not possible to enlarge upon all possible alternatives. Only basic configurations can be dealt with.

4.1.2.1 INTAKE DESIGN (FIG. 4.2)

The intake structure consists of the approach flow section followed by the pump bay for wet pit installation or an inlet bend for the dry pit arrange-

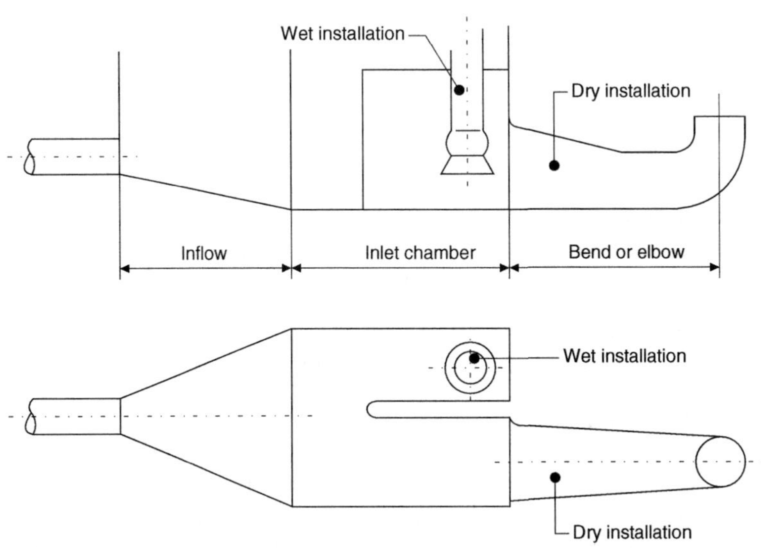

Figure 4.2 Intake general arrangement

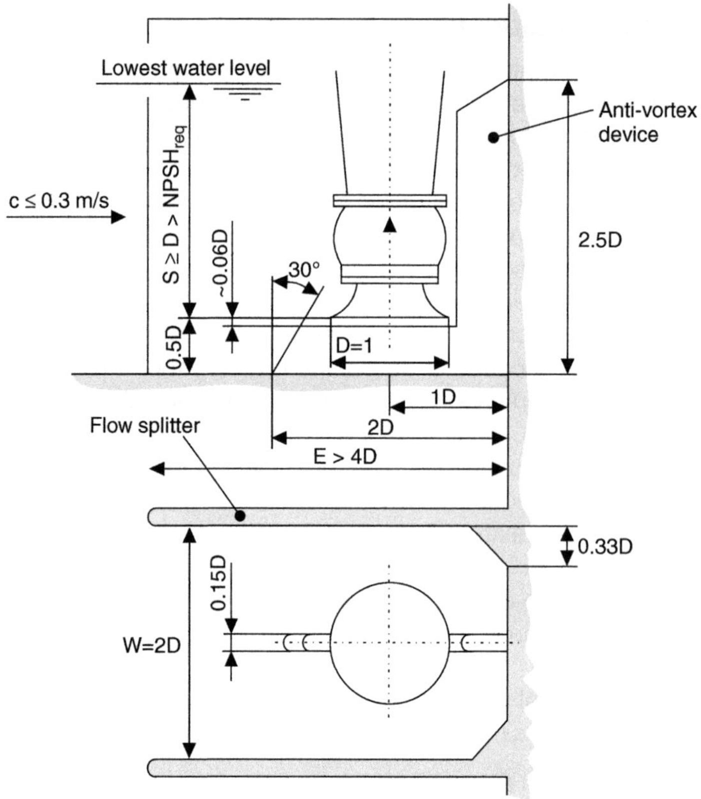

Figure 4.3 Open pump bay, normal installation for flow rates $\leq 20\ m^3/s$. Minimum submergence $S \geq 2D \geq NPSH_A$

ment. In Figs 4.3–4.5 $NPSH_A$ stands for that water level which is required in order to provide sufficient NPSH (i.e. $NPSH_3$ times the margin according to section 1.5.3). The approach flow section distributes the water uniformly under a variety of service conditions to the actual pump bay.

4.1.2.2 WET PIT INSTALLATIONS

Individual pump bay dimensions

The installation according to Fig. 4.5 is employed only in large intake reservoirs, with pumps arranged in series (see Figs 4.6–4.8).

Parallel installation

Refer to Fig. 4.6 for parallel operation pump bay dimensions.

Series installation

The pump spacings *X* and *Y* are a function of the flow velocity. The last pump is installed according to the normal pump bay configuration (Figs 4.3 and 4.4). Pump arrangements as in Figs 4.7 and 4.8 should be avoided, if at all possible, because undesirable flow influences due to the series installation can be circumvented only with difficulty.

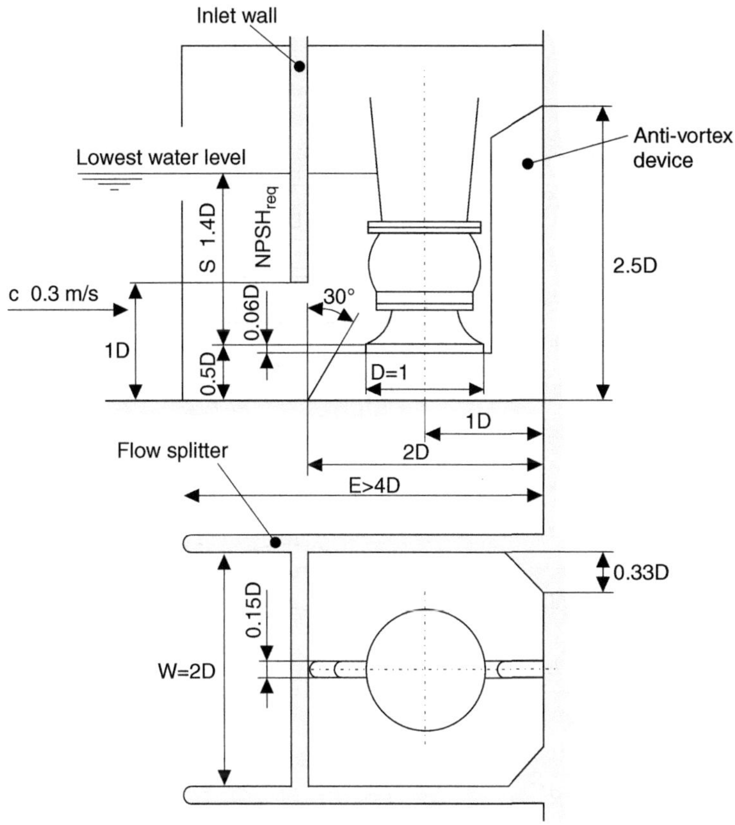

Figure 4.4 Open pump bay with inlet wall for flow rates $\leq$20 m^3/s. Minimum submergence $S \geq 1.4D \geq \text{NPSH}_A$

If series installation is unavoidable, the stated duct width, maximum flow velocity 0.3 m/s at the first pump and pump spacing must be observed without fail. The only difference between the baffle cylinder and anti-vortex device lies in the duct width required ($W = 3.5D$ instead of $3D$).

4.1.2.3 DRY PIT INSTALLATION

Of the numerous inlet bend configurations developed in model tests and confirmed in service, the following bend geometry is recommended:

c = velocity in the pump bay
c_B = velocity into inlet bend $\leq$1.6 m/s
$\Delta c_B = c_B - c$
D_B = bend inlet diameter
With rectangular entry cross-section:

$$D_B = \sqrt{\frac{4 \cdot B \cdot H}{\pi}}$$

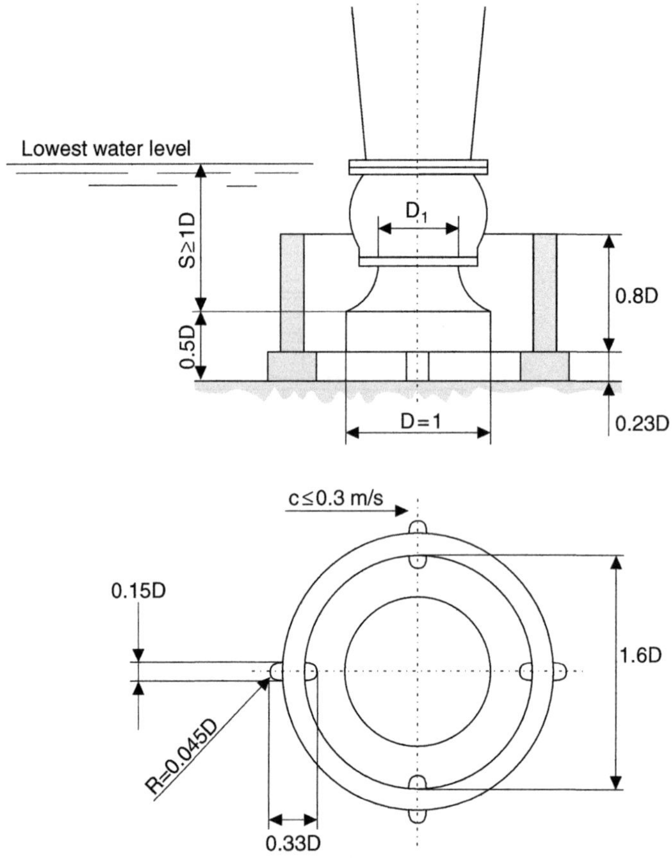

Figure 4.5 Installation with baffle cylinder for flow rates $\leq 20\ \mathrm{m^3/s}$. Minimum submergence $S \geq 1D \geq \mathrm{NPSH_A}$

Dimensions B, H or D_B may be obtained from Figs 4.10 and 4.11. The bend is executed is in Fig. 4.9.

D_1 = impeller entry diameter

S = minimum submergence (observing the necessary $\mathrm{NPSH_A}$)

g = gravitational acceleration (9.81 m/s^2)

Calculation of submergence S:

Steps:

1. Determination of impeller entry diameter D_1.
2. Determination of pump spacing in station and bend inlet with B. B = pump spacing minus partition thickness.
3. The inlet height H_c (inlet bend/connection to pump bay) is calculated from the approach flow velocity c:

$$H_c = \frac{Q_{\text{pump}}}{B \cdot c}$$

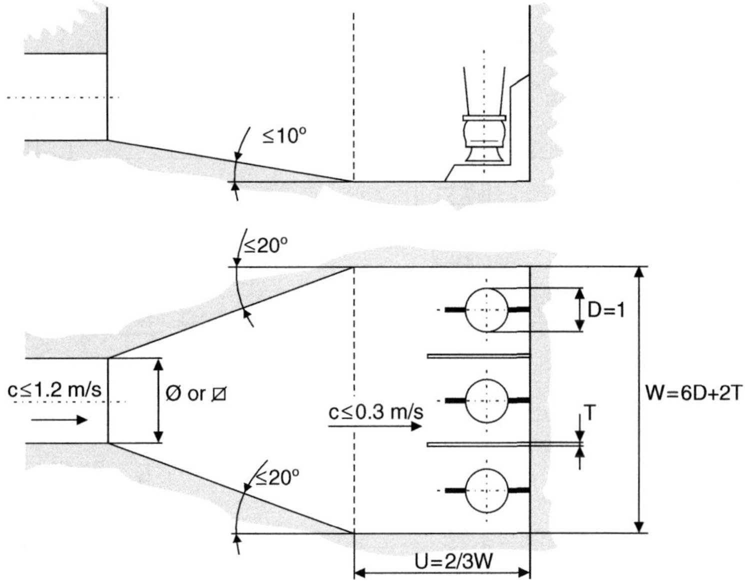

Figure 4.6 The dimensions of the pump bay and the arrangement of the pump correspond to Figs 4.3 and 4.4. The additional data concern the approach flow and chamber structure. If the maximum admissible diffuser angle (≤20°) is exceeded, grids must be provided at a distance "*U*"

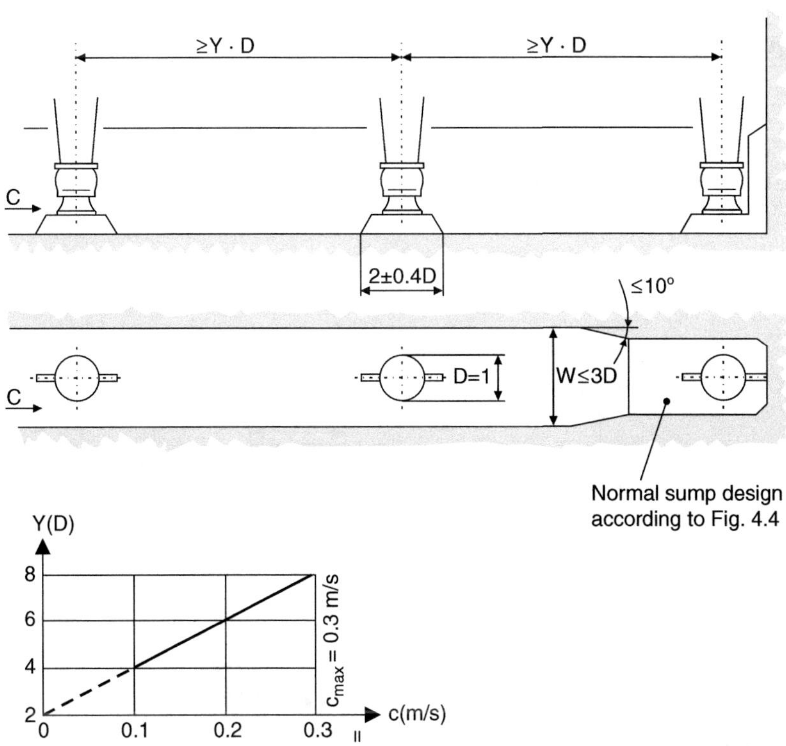

Figure 4.7 Normal series installation

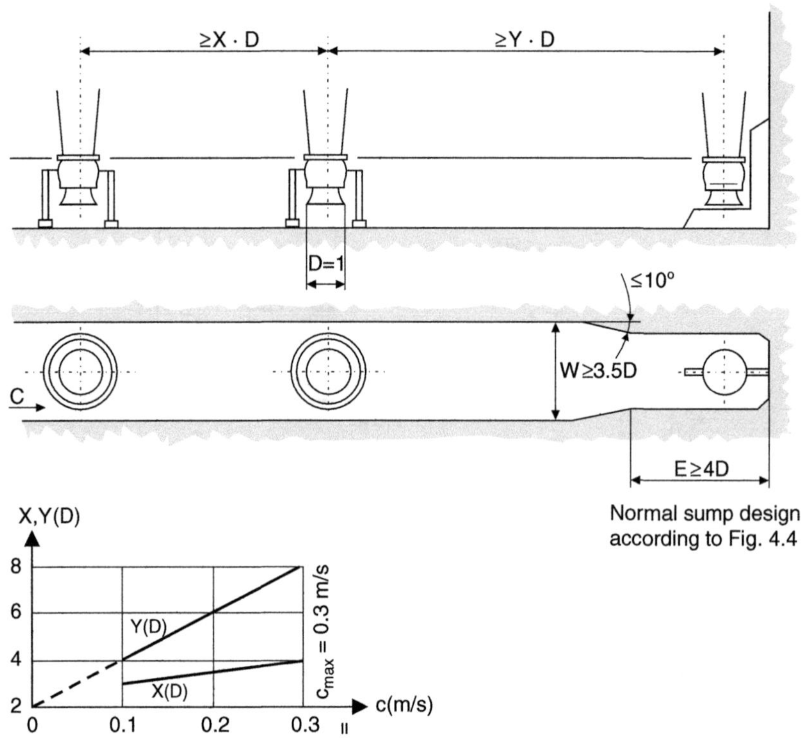

Figure 4.8 Series installation with baffle

$c \leq 0.7$ m/s with screen before the bend
$c \leq 1.6$ m/s without screen before the bend.

4. Draft bend inlet between ① and ② with maximum reduction angle $\leq 30°$.

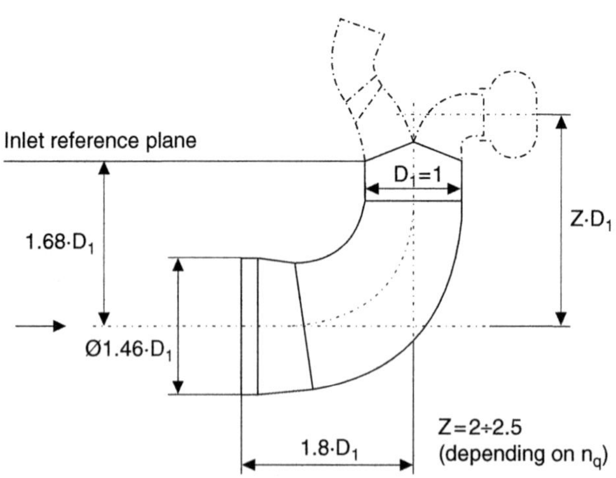

Figure 4.9 Bend geometry

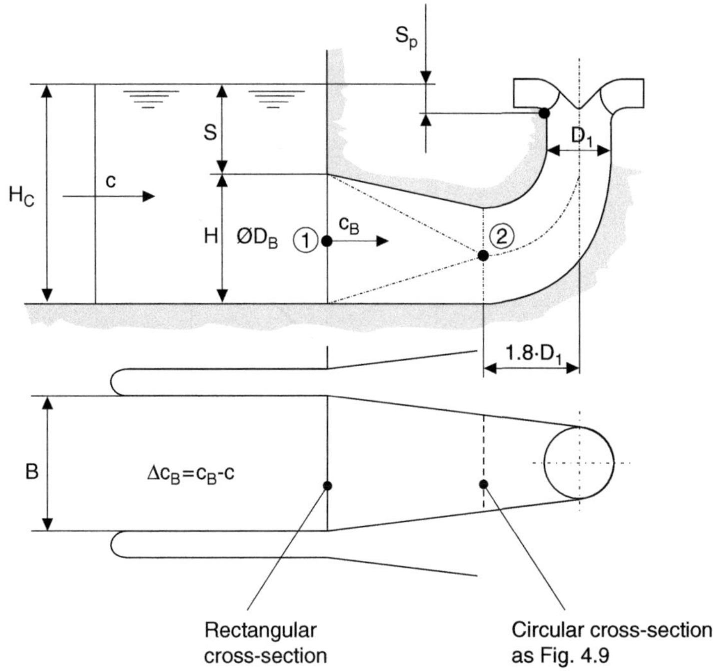

Figure 4.10 Transition from inlet bend to pump bay

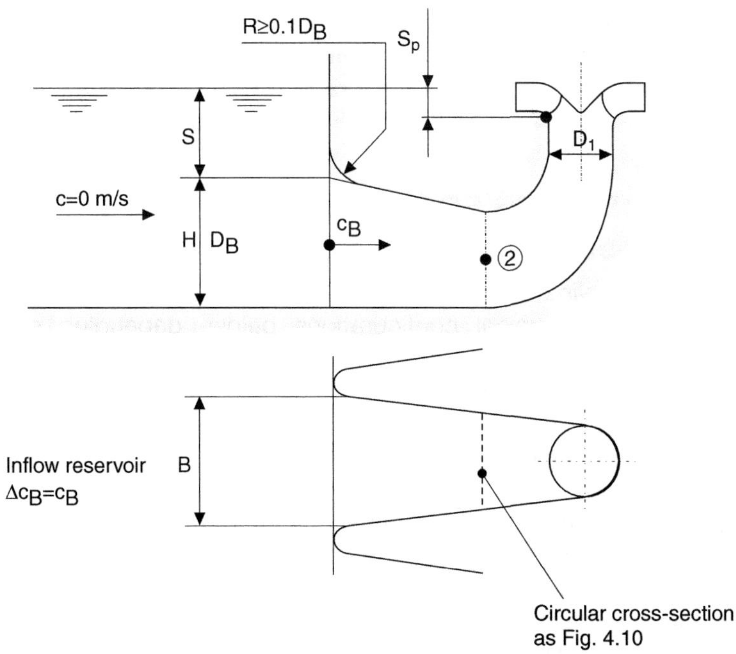

Figure 4.11 Transition from inlet bend to approach flow reservoir

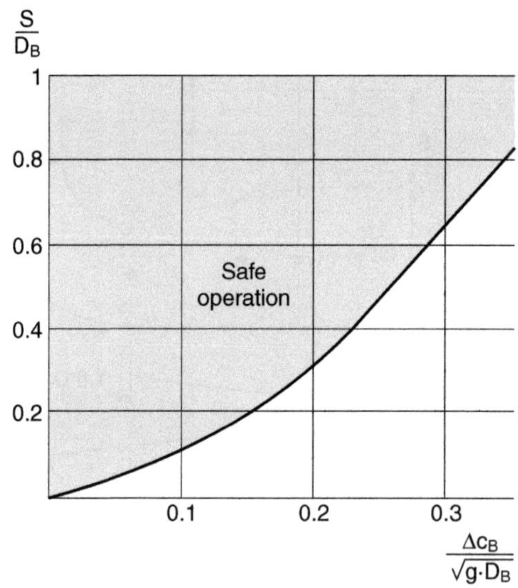

Figure 4.12 Minimum submergence *S*

5. Verify minimum submergence *S* in conjunction with pump $NPSH_A$.
6. Verify minimum submergence *S* for the inlet bend according to the curve in Fig. 4.12 (to avoid vortex formation).
7. If the submergence *S* is inadequate (below the curve, see Fig. 4.12), the intake geometry must be modified.

Recommendations for sound hydraulic design or improvements to existing installations may be found in Figs 4.13 and 4.14.

4.2 GUIDELINES FOR INLET PIPELINES

4.2.1 Connection of pipe to suction reservoir

A horizontal pipeline may be connected to the suction or inlet reservoir according to the typical configurations below, depending on the capacity:

- for $Q < 0.5\ m^3/s$ (see Fig. 4.15);
- for $Q < 1.0\ m^3/s$ (see Fig. 4.16);
- for $Q < 5.0\ m^3/s$ (see Fig. 4.17).

Recommended data:

- Inlet velocity in pump bay ≤ 0.5 m/s
- Inlet velocity in suction bell (range 1.2 to 2.1 m/s) 1.7 m/s
- Velocity in pipeline D_L to pump ≤ 4.0 m/s
- Dimensions: $D \approx 1.75 D_L$

  $S \geq 1.0D \geq NPSH_A$

  $W = 2.0D$.

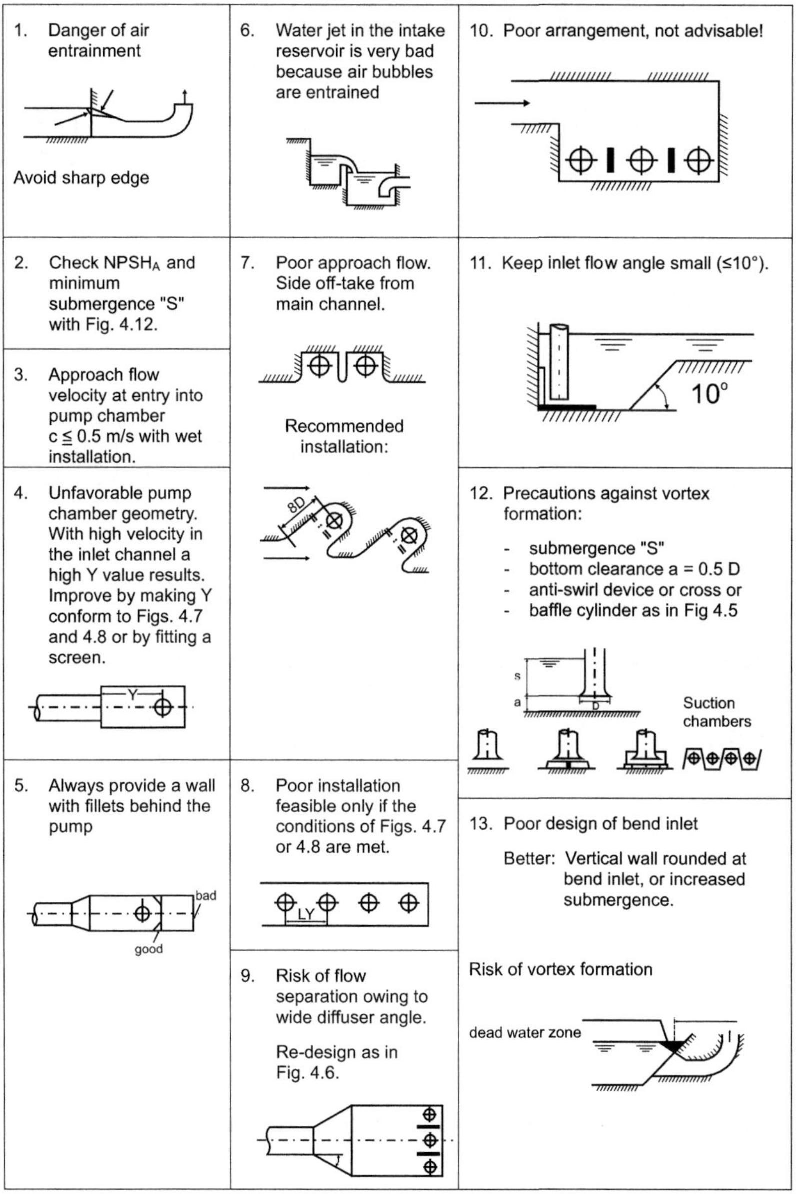

Figure 4.13 Hints for designing pump bays

Recommended data:

- Inlet velocity in chamber ≤ 0.5 m/s
- Inlet velocity in suction bell (range 1.2 to 2.1 m/s) 1.7 m/s
- Velocity in pipeline D_L to pump ≤ 4.0 m/s
- Dimensions: $D \approx 1.75 D_L$

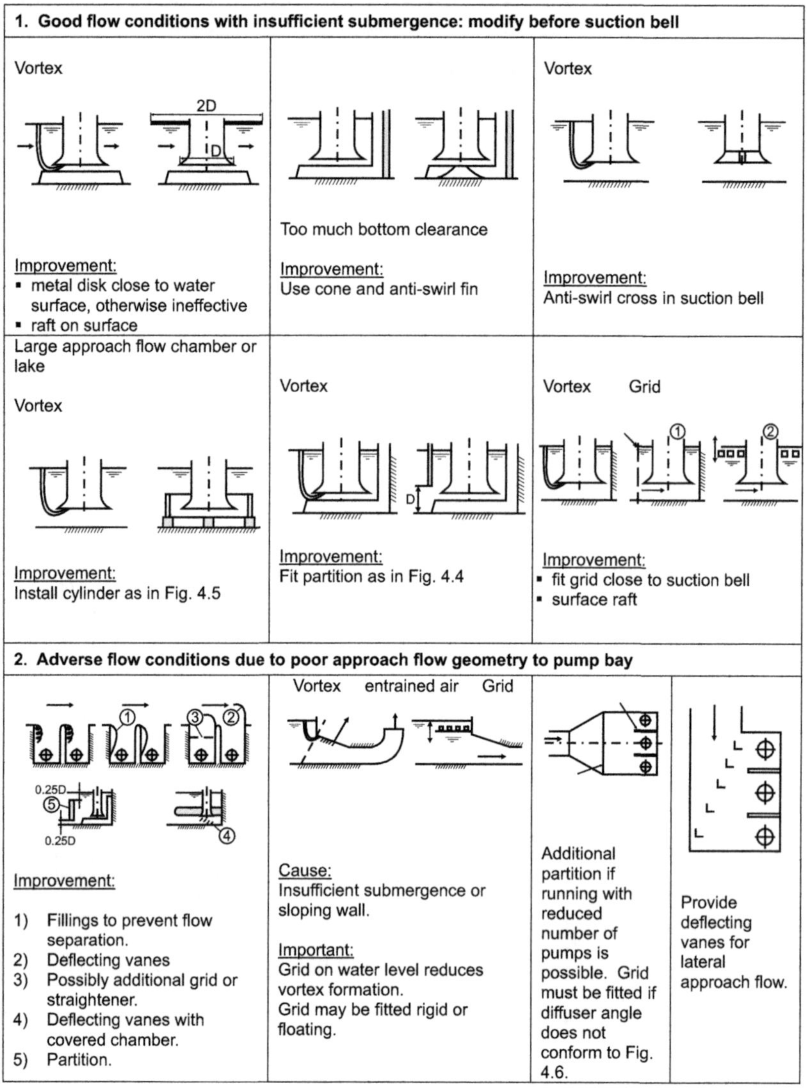

Figure 4.14 Hints for improving the flow in existing installations

$R \geq 1.5D$
$S \geq 1.5D \geq \text{NPSH}_A$
$a = 0.5D$
$b = 1.0D$
$W = 2.0D$
$E = 4.0D$
$e = 0.33D.$

Recommended data:

- Inlet velocity in chamber ≤ 0.5 m/s

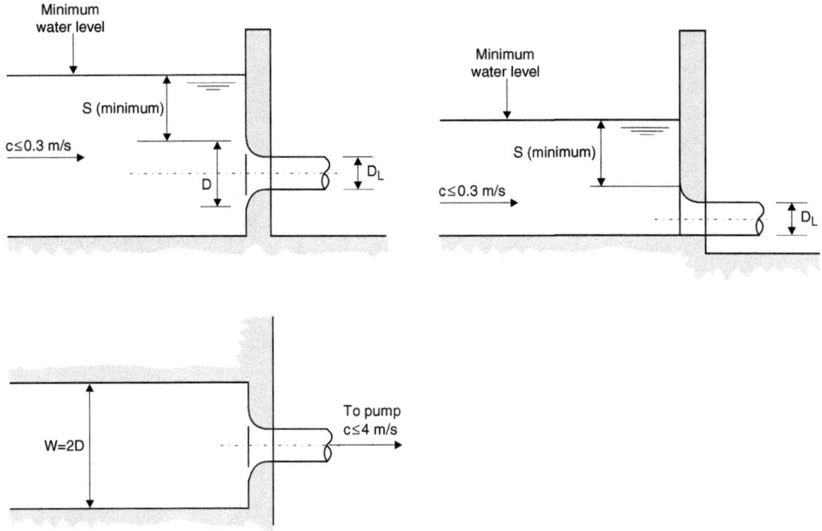

Figure 4.15 $Q < 0.5\,m^3/s$

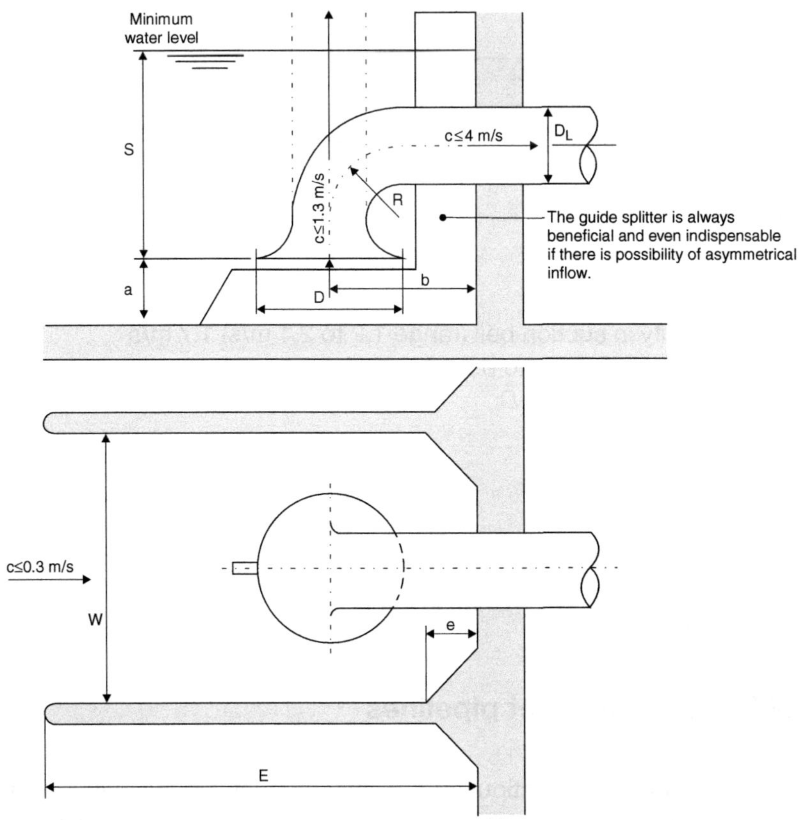

Figure 4.16 $Q < 1.0\,m^3/s$

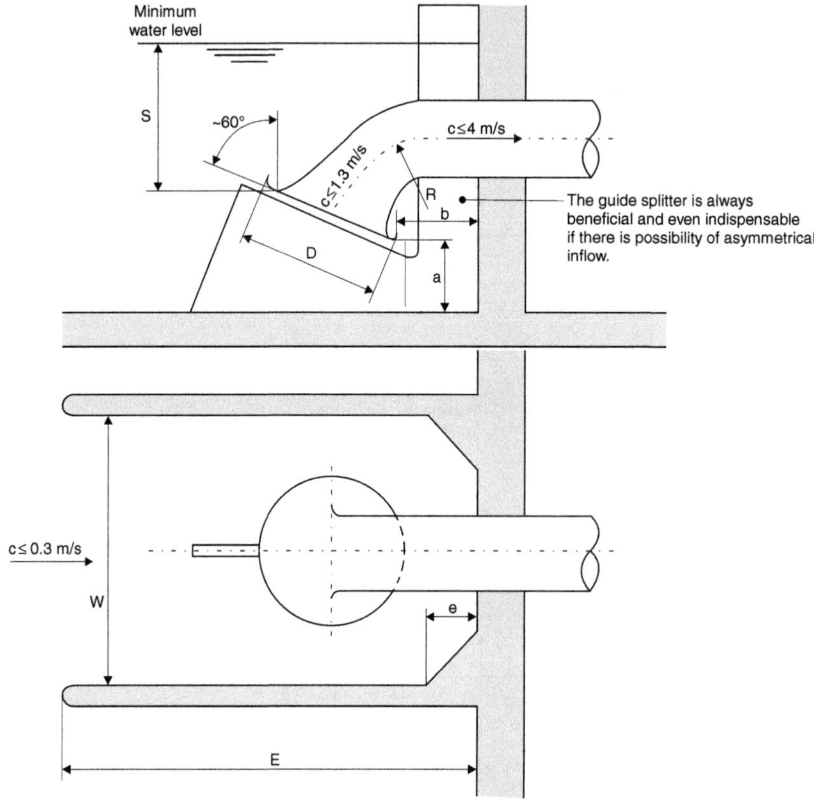

Figure 4.17 $Q < 5.0\ \mathrm{m^3/s}$

- Inlet velocity in suction bell (range 1.2 to 2.1 m/s) 1.7 m/s
- Velocity in pipeline D_L to pump $\leq$4.0 m/s
- Dimensions: $D = \sim 1.75 D_L$

  $R \geq 1.5D$

  $S \geq 1.5D \geq \mathrm{NPSH_A}$

  $a = 0.5D$

  $b = 1.0D$

  $W = 2.0D$

  $E = 4.0D$

  $e = 0.33D$.

4.2.2 Designing inlet pipelines

4.2.2.1 PRELIMINARY REMARKS

The same requirements stipulated for intake structures in section 4.1 to assure undisturbed approach flow to the pump must be regarded as fundamental when designing suction and inlet pipes.

As already mentioned, flow disturbances in the form of unequal velocity distribution and vortex at the pump inlet are detrimental to pump performance, cavitation behavior and smooth pump running.

Attention should therefore be given to the reasoning and guidelines below for avoiding such troubles.

The approach flow pipe should be engineered so that sources of disturbance like bends, branches or valves are as few as possible. Such components may cause uneven velocity distribution and vortex formation.

If bends are arranged in short succession, the velocity pattern will be worsened further. Consequently, all influences leading to asymmetrical approach flow must be avoided. Any bends upstream of the pump inlet nozzle should, if possible, be placed only in the plane of symmetry of the pump. If the approach flow symmetry is disturbed, appropriate precautions must be taken to minimize such disturbances.

Recommendation:

- Straight pipe sections of adequate length ($L \geq 7D$) before the suction nozzle
- Flow acceleration
- In critical cases a flow straightener may be required
- Bends with ribs

The data given below are minimum requirements which should be adhered to.

4.2.2.2 DESIGN GUIDELINES

End suction pumps

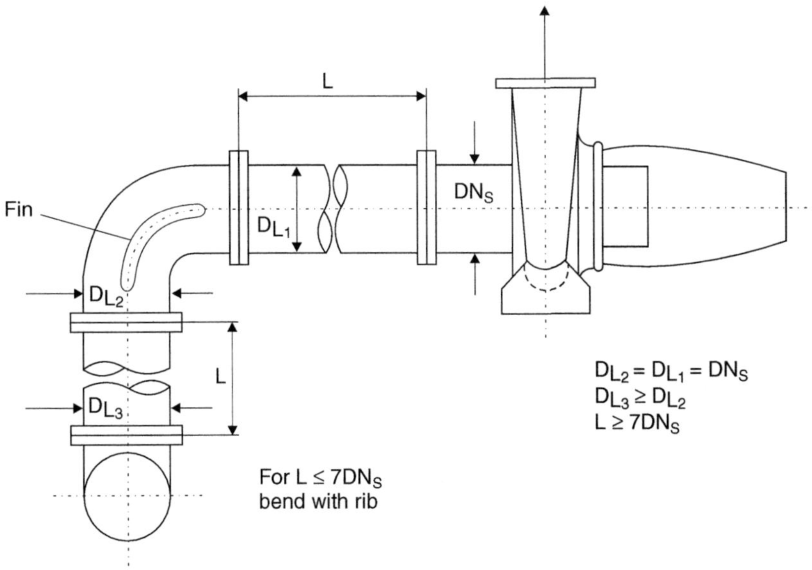

Figure 4.18 Pumps of moderate size and specific speed

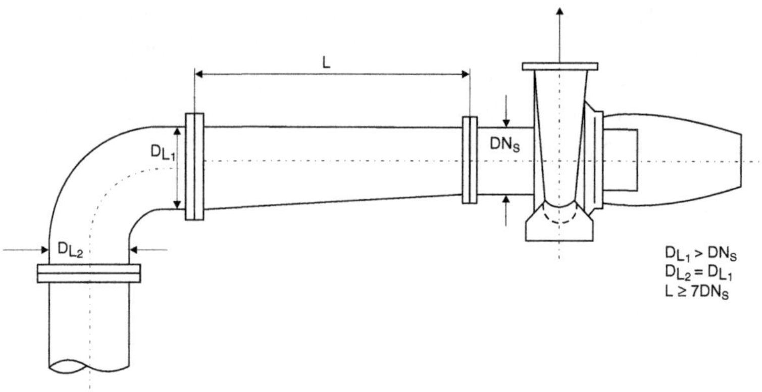

Figure 4.19 Pumps with $n_q \geq 55$ or risk of gas separation

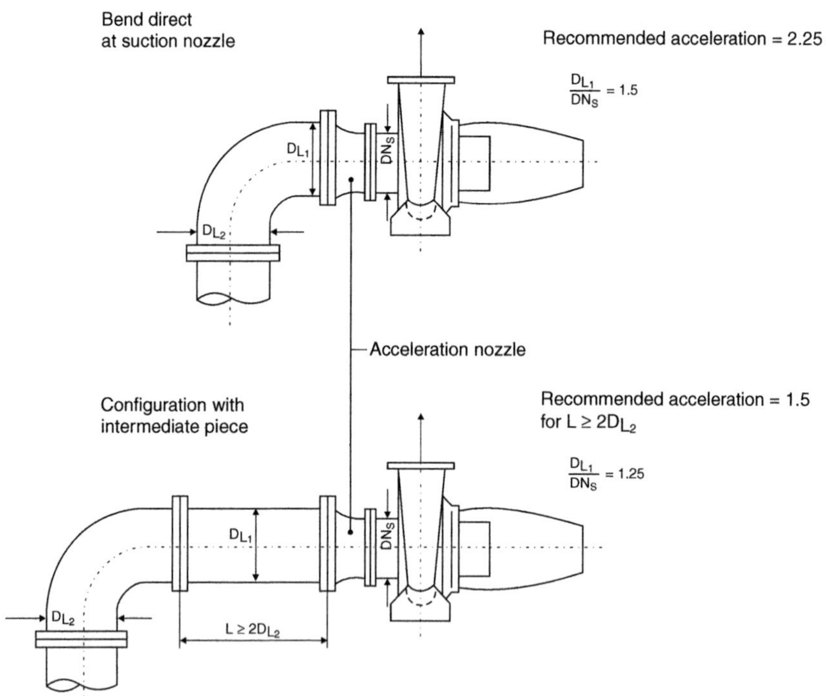

Figure 4.20 Flow acceleration upstream of suction nozzle

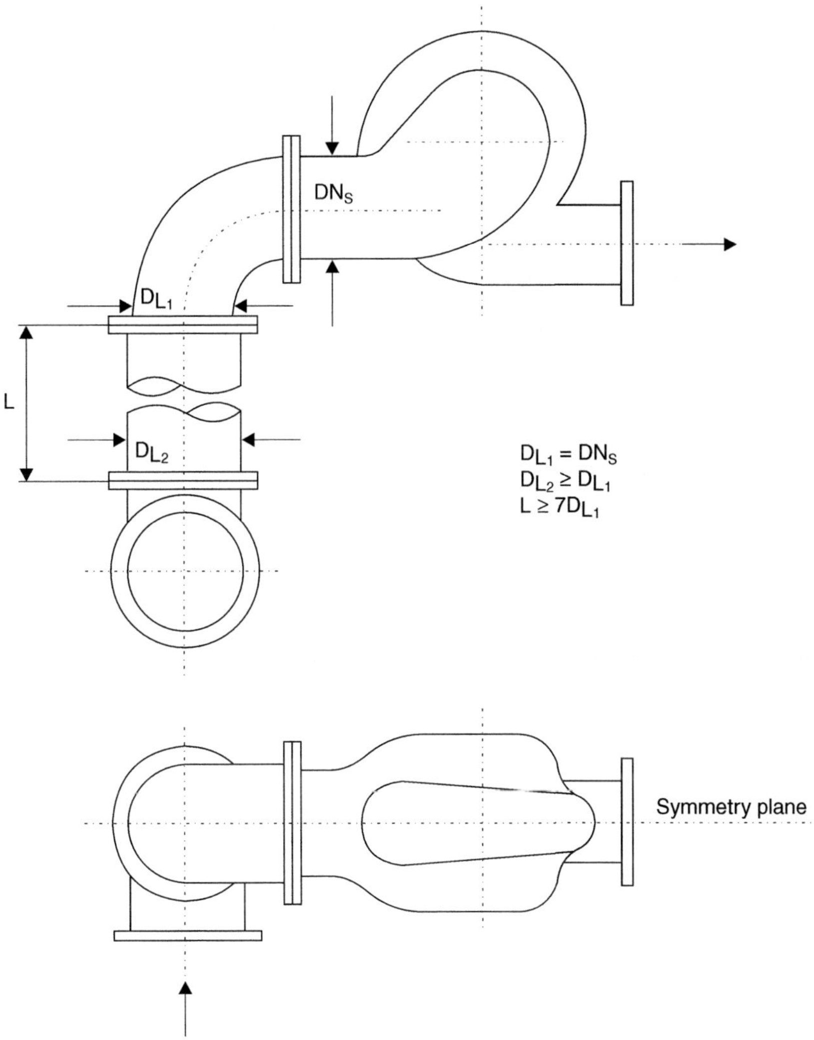

Figure 4.21 Bend in plane of symmetry

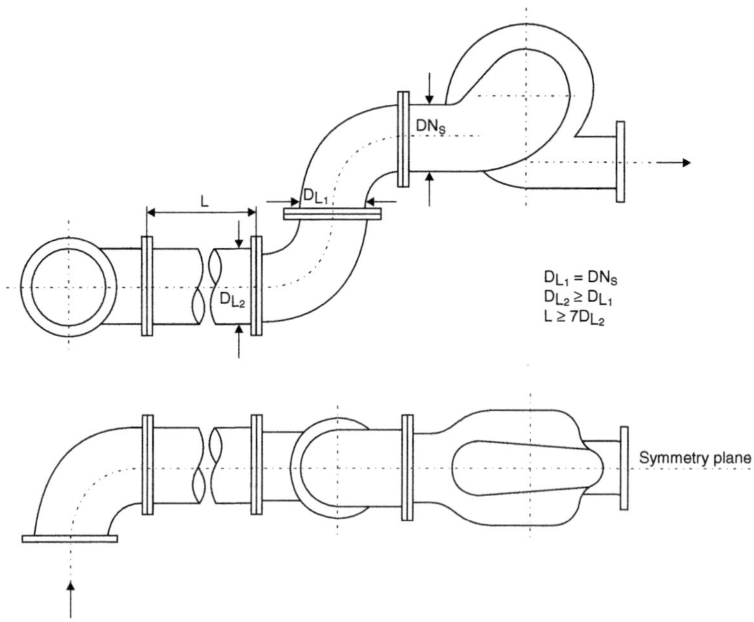

Figure 4.22 Double bend in plane of symmetry

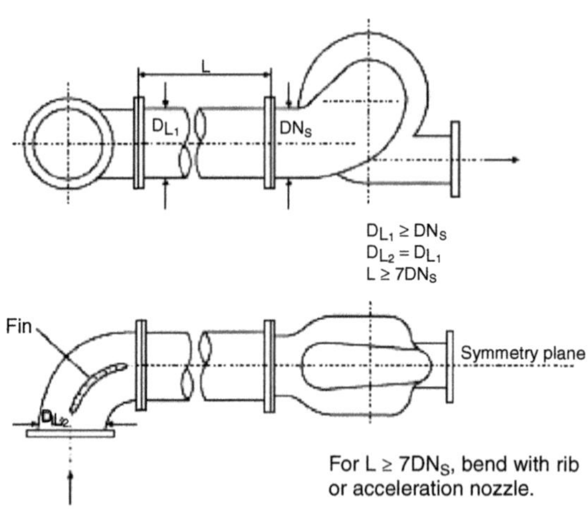

Figure 4.23 Bend at right angles to plane of symmetry

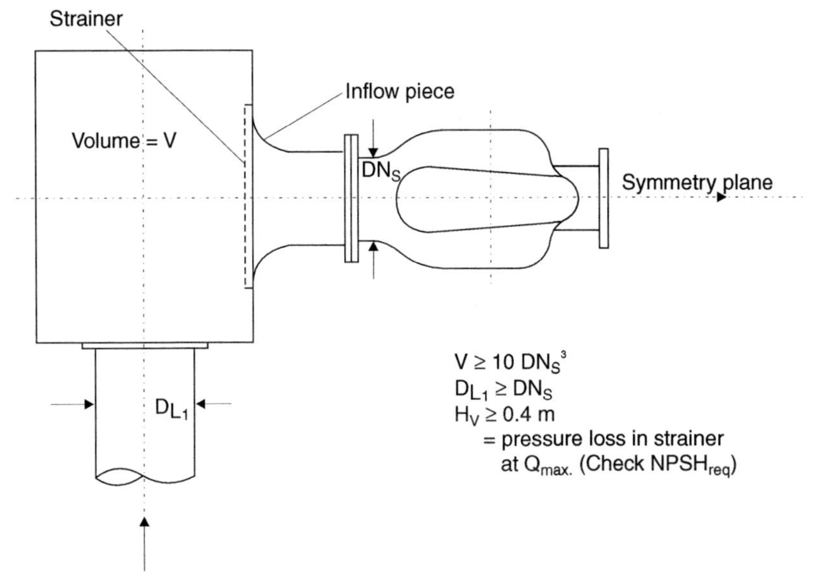

Figure 4.24 Inlet box with approach flow at right angles to the in plane of symmetry

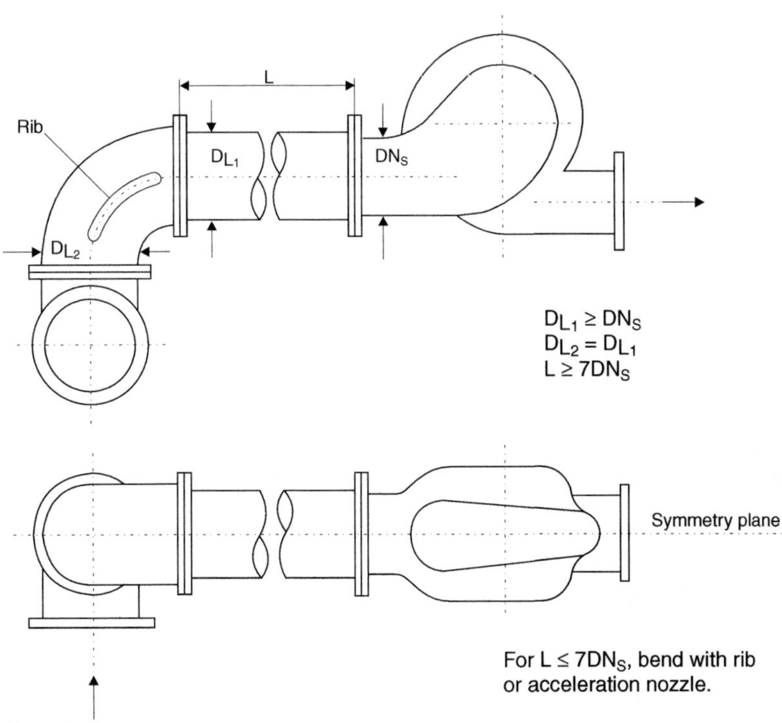

Figure 4.25 Double bend in different planes

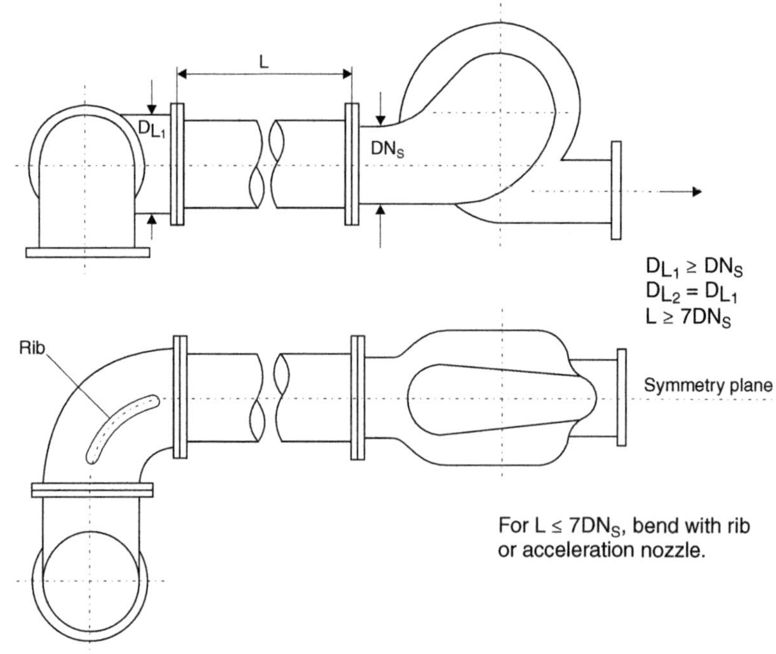

Figure 4.26 Double bend in different planes

Pumps with double-entry impellers

Figure 4.27 shows the negative influence on the approach flow to a double-entry centrifugal pump using a wedge gate valve that does not give absolutely free passage in the open position (such valves are to be strictly avoided) and without the necessary straight pipe length of $\geq 7D$ or more upstream of the pump inlet.

The distribution of the flow rate between the two impeller halves is disturbed. Performance is impaired, pressure pulsations may increase and the hydraulic axial forces are no longer balanced.

Improvement can be obtained by turning the valve by 90°, giving a more equal distribution of the flow rate between the two sides of the impeller.

In general, attention should always by given to the hydraulically proper fitting of valves, especially in suction or inlet pipes.

Inlet pipe connection to the suction reservoir

Hydraulic disturbances in the operational behavior of pumps (see also section 4.2.2.1) are often caused by:

- inadequate design of the connection at the inlet of feed water tank;
- insufficient submergence "S" (liquid level too low);
- excessive inlet velocity.

This may lead to vortices in the approach flow, and hence disturb the smooth operation of the pump.

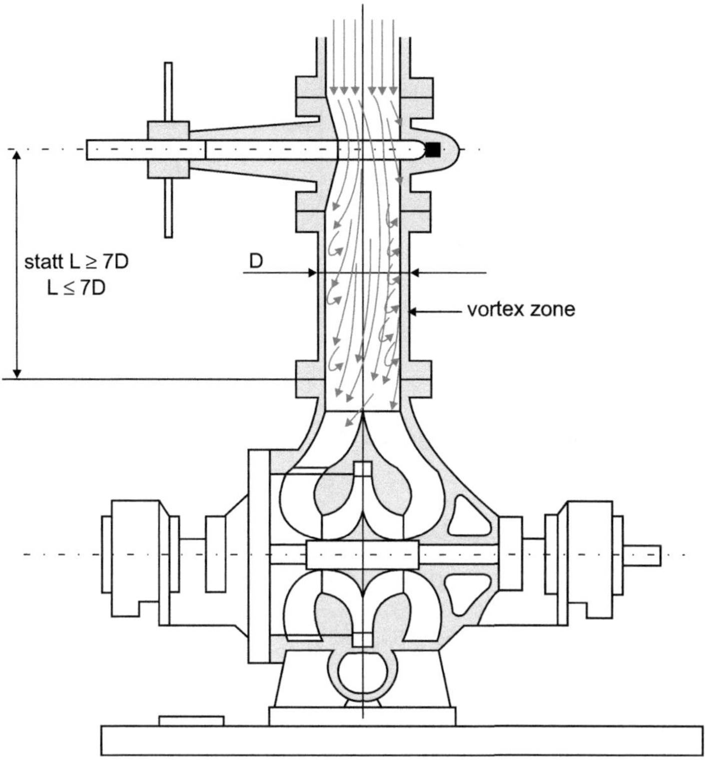

Figure 4.27 Effect of a badly fitted valve on the approach flow to the pump

Figure 4.28a and b offers guidelines for connecting pump approach flow pipes to inlet tanks such as are often used, for example, in the chemical and petrochemical industries.

The velocity at the branch connection must not exceed about 2.5 m/s, to keep the submergence "*S*" within acceptable limits.

The inlet cross in Fig. 4.28b is fabricated with four or eight axial guide vanes. It supports the guide plate at the same time. This inlet cross ought to be provided whenever possible, to prevent vortices and swirl.

Figure 4.29 gives guidelines for connecting inlet pipelines to feedwater tanks.

The inlet pipeline is connected to the feedwater tank via a cone, to keep the inlet velocity as low as possible. The inlet has four axial vanes to prevent vortex formation. The rounding on the inside of the tank is intended to avoid sludge from being drawn into the pump.

Above the inlet cross is a perforated plate, enabling vapor bubbles to be separated in the tank in the event of sudden pressure drop (load change). Starting from a minimum of $L = 2.8D$, the length L of the inlet too must be greater than the velocity head $c_D^2/2g$ in order to avoid evaporation at the inlet. The actual submergence S is not taken into account here but is regarded as an additional safety margin.

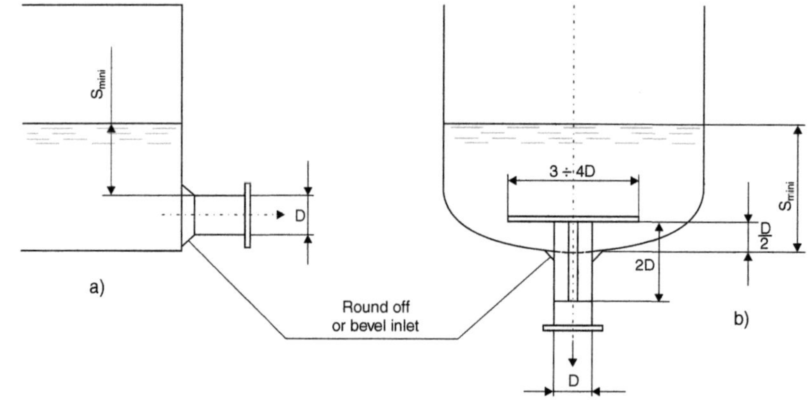

Figure 4.28 Minimum submergence "S"

	Fig. 4.28a	Fig. 4.28b
For liquids at vapor pressure	$1.75D \geqq S_{mini} \leqq 2\frac{C_D^2}{2g}$	$2D \geqq S_{mini} \leqq 2\frac{C_D^2}{2g}$
For liquids at atmospheric pressure with $p_v < 0.5p$	$1.75D \geqq S_{mini} \leqq 2\frac{C_D^2}{2g}$	$2D \geqq S_{mini} \leqq 2\frac{C_D^2}{2g}$

Where several pumps are installed, the connections to the feedwater tank must be spaced as far apart as possible ($L > 10D$ at least).

4.3 PRESSURE SURGES IN PIPELINE SYSTEMS

4.3.1 Causes of a pressure change

In pipeline systems any change in the operating state leads to dynamic pressure changes, which must be taken into account in the planning and operation of installations. Pressure surges (events of "water hammer") are capable of causing serious damage.

Water transport and supply systems are often extremely varied in nature, so that it is impossible to find by approximation procedures a solution that will reliably preclude overstressing of components in any system.

In steady operation, the flow velocity is constant in time and space. By contrast, in transient flow, the velocity varies in time and location. Transient flows occur at every change from an existing steady operating state to a new one. Pressure surges result from transient flows.

Consequently the following operation modes in pumping stations are associated with surging problems:

- starting a pump
- stopping a pump

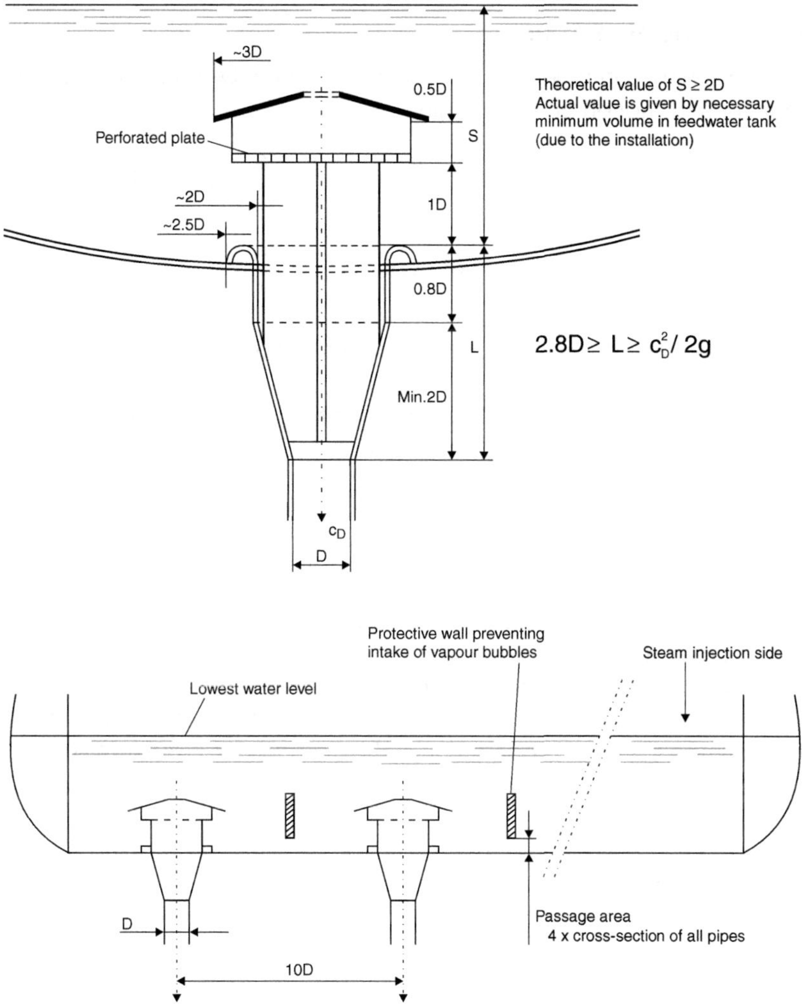

Figure 4.29

- switching over pumps
- altering the valve setting
- altering the speed
- power failure on one or more pumps
- inexpert operation.

With the exceptions of power failure and inexpert operation, all of them are deliberate actions which can be performed properly. On the other hand, power failures are unintentional and generally constitute the most severe case. Additional safety devices must be provided against power failure in particular.

4.3.2 Fundamentals

Pressure surges in pipelines are described by the two partial differential equations of motion and continuity.

Differential equation of motion : $\frac{\delta c}{\delta t}+g\frac{\delta H}{\delta x}=0$

Differential equation of continuity : $\frac{\delta c}{\delta x}+\frac{g}{a^2}\cdot\frac{\delta H}{\delta t}=0$

Besides the mass inertia of the liquid, its elasticity and that of the pipe wall are taken into account.

For a sudden change in the flow velocity, Joukowsky found that the maximum pressure change amounts to:

$$\Delta H_{\text{nax}}=\frac{a}{g}\cdot\Delta c\ (\text{m})$$

where:
a = speed of propagation of the disturbance or speed of sound (m/s)
g = gravitational acceleration (m/s^2)
Δc = change in flow velocity (m/s).

For cast iron, steel, concrete or asbestos cement pipes and water as medium, the speed of propagation of the disturbance is 800 to 1200 m/s. For plastic pipes it may be much lower on account of the lower Young's modulus of the material.

From its point of origin the disturbance travels through the pipeline system at the propagation speed. In the example shown in Fig. 4.30 the reservoir is the reflection point. Here the disturbance is reflected.

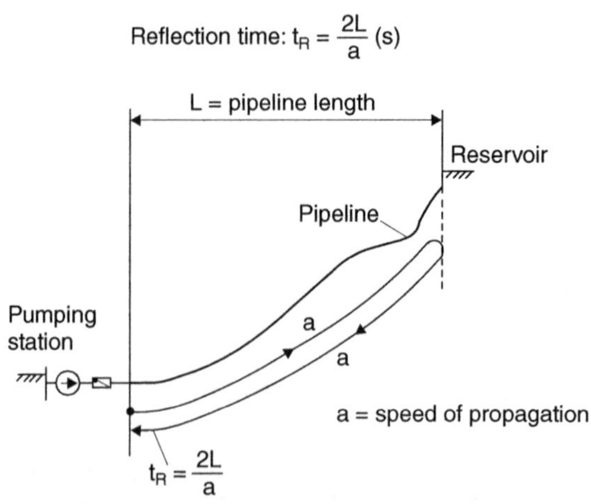

Figure 4.30 Reflection time t_R

The reflection time is the time taken by the disturbance to return to its point of origin:

$$\text{Reflection time}: \; t_R = \frac{2L}{a} \; (\text{s})$$

where L = the total pipeline length.

The pressure at the time t and the point x in the system is always the sum of all pressure waves that have passed the point x at the time t.

Using the example in Fig. 4.31 the behavior of the pressure wave after pump trip will be examined at selected points in time. It is assumed that

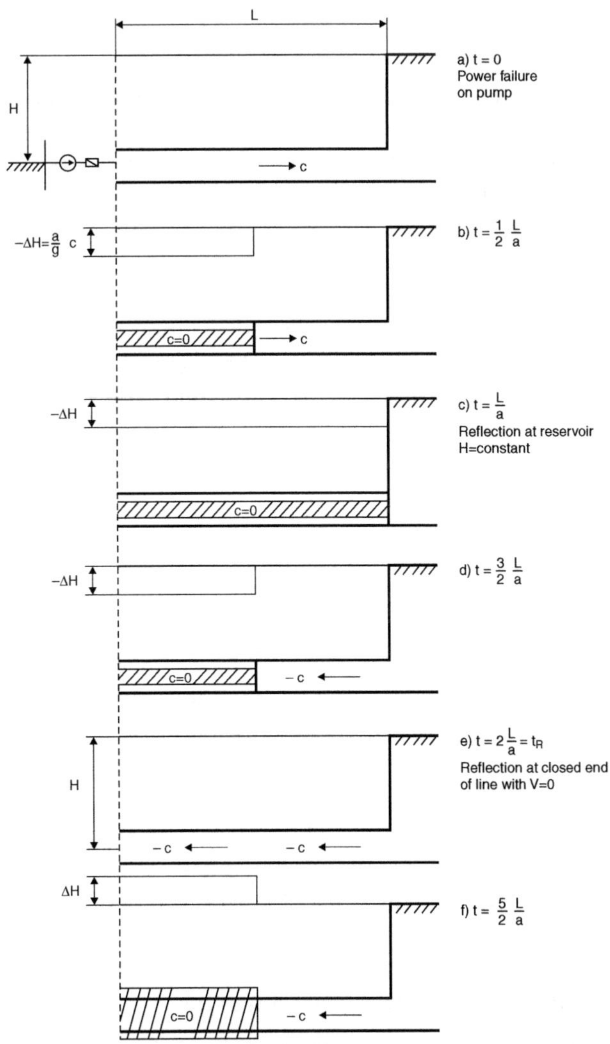

Figure 4.31 Propagation of the pressure wave along the pipe

the mass moment of inertia of the revolving parts (motor, coupling and pump) is zero. Consequently for this theoretical consideration there is a sudden speed change from the steady-state value c to 0. The pipe friction is taken as zero as well.

In the steady state (Fig. 4.31a) there is a constant pressure H and constant velocity c along the entire pipe length L:

$$t = 0$$

At the time $t = 0$ the power supply for driving the pump fails, causing the speed to fall to zero. According to the Joukowsky equation there is a pressure drop of ΔH:

$$t = \frac{1}{2} \cdot \frac{L}{a}$$

At the time $t = \{1/2\} \times L/a$ the pressure wave reaches half the pipe length (Fig. 4.31b). Ahead of the pressure wave the original velocity still prevails, whereas the velocity behind the pressure wave is zero. Behind the pressure wave the pressure has dropped by ΔH:

$$t = \frac{L}{a}$$

At the time $t = L/a$ the pressure wave reaches the reservoir. In the entire pipe there is now zero velocity and a pressure reduced by ΔH. At the reservoir there is a reflection at $H =$ constant. The pressure drop ΔH cannot maintain itself, consequently there is a pressure rise and a flow velocity c emerges:

$$t = 2\frac{L}{a} = t_R$$

At the time $t = 2L/a = t_R$ the pressure wave reaches the closed check valve at the pump. The entire pipe is now under the original pressure H but the velocity is directed towards the pump. Reflection at the closed end of the pipe now causes a pressure rise. The cycle described above is repeated, though this time with a pressure increase:

$$t = 4\frac{L}{a} = 2t_R$$

After the time $t = 4L/a = 2t_R$ a full cycle has been completed. Figure 4.32 plots the pressure against time at the beginning of the pipe for the model described.

4.3.3 Calculating the pressure surge

For calculating this transient phenomenon there are two methods available.

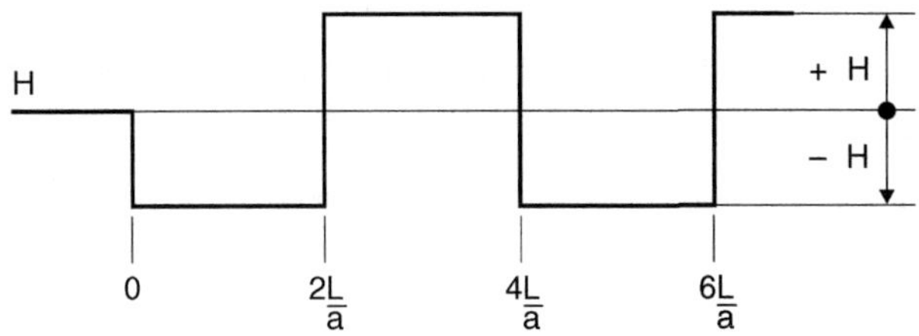

Figure 4.32 Pressure against time at beginning of the pipeline

Approximation

Approximation may be used for particularly simple systems to size a flywheel or a surge tank (as surge-reducing devices). The application limits must be borne in mind, otherwise false conclusions may be drawn.

Computer-aided calculation

This method yields accurate results for any system in a short time. It is often employed with simple tools, as a complement to approximation.

If the calculation is to supply meaningful results, the system data used as inputs must also be defined precisely. Table 4.1 sets out all data required.

If calculations are carried out in the project planning stage, the following variables must be known as a minimum:

- pump characteristic
- maximum flow rate
- profiles of all pipelines
- pipe diameters (material and wall thickness if possible).

4.3.4 Necessity for pressure surge calculation

Basically any system may be at risk, though low-pressure installations are more endangered than high-pressure systems, because the maximum pressure change given by the Joukowsky equation is independent of the system pressure.

The purpose of every pressure surge investigation is to determine the size of a protective device so that no limits are exceeded anywhere in the plant, even in the most extreme transient case (this is usually power failure). The limiting factors are as follows:

- *Minimum pipeline pressure*. The pressure in the pipe must not be allowed to drop to an extent that the vapor pressure is reached, otherwise water column separation must be expected, with a subsequent extreme pressure rise. In many cases it is not permissible for the atmospheric pressure to be undershot by more than a few meters if the pipe is not to be crushed by the external pressure. With drinking water pipelines the atmospheric pressure must not, as a rule, be undershot. Otherwise there is a risk of impurities getting into the system. With plastic pipes, special attention must be given to the admissible external pressure.

Table 4.1 Checklist of Data Needed for Calculating Pressure Surges

Pumping station

1. Type of pump, number, maximum number in parallel operation
2. Pump data, mass moment of inertia of pump and motor (J)
* 3. Pump and system characteristics
* 4. Maximum flow rate
5. Maximum flywheel admissible for motor (mass, J)
6. Normal starting and stopping (against closed valve or check valve only)

System

*11. Pipe profiles of approach flow and discharge piping; plans showing the arrangement of the piping and the components

*12. Pipe diameters, material and wall thicknesses

13. Boundary conditions:

– where does the water come from (reservoir, ring main. . .)

– where does the discharge pipe lead to (reservoir, ring main, mains. . .)

14. Maximum pipe pressure admissible

15. Minimum pipe pressure admissible

16. Type and closing law of the valves

17. Throttling characteristic of stop valve (pressure loss as a function of opening)

**Minimum data required at project planning stage.*

- *Maximum pipeline pressure*. The nominal pressure rating of the pipes and the design pressure of the components must not be exceeded.
- *Maximum reverse speed*. This is important in pumping installations with controlled valves and also in turbine installations, on account of mechanical stressing of the pump, motor or turbine.
- *Pipeline profile*. In order to correctly assess the extreme pipe pressures this must be known exactly.

The pressure surge investigation often reveals that a better pipeline profile will allow much smaller protective devices. However, this advantage can be exploited only if the transients have already been examined at the project planning stage.

4.3.5 Protective measures

If a preliminary calculation shows a need for additional protective devices, an attempt is always made first to reduce the primary pressure change. This may be achieved by the following measures.

The speed change is reduced by:

- *A flywheel*. This prolongs the pump rundown time. The limits for employing a flywheel are systems up to about 2 km pipeline length. For standard versions the maximum flywheel size is dictated by the accelerating capacity of the motor and/or the load capacity of the bearings.

  For special versions with separate intermediate flywheel bearings, only the acceleration is of importance.
- *A surge tank installation*. This takes over full pipeline flow surge with virtually no time lag. With a suitable surge tank, the minimum admissible pressure is maintained in most systems. Low-pressure systems are frequent exceptions to this.

Figures 4.33 and 4.34 schematically show a surge tank installation. The water level is monitored automatically. If the level in the surge tank exceeds a maximum (due to leakage or defective compressor), the installation must be taken out of service till the level can be kept normal again.

Reducing the active length of the system. This results in one or more additional boundary conditions (reflection points):

- *Surge tank or standpipe*. At a pronounced high point in the pipe system a boundary condition is created with H = constant. Both options have the great advantage of being effective even with the smallest pressure wave. Standpipes up to 82 m in height have been built.
- *One-way surge tank*. If the use of a standpipe is ruled out, a one-way surge tank is often provided (Fig. 4.35). However, this takes effect only when the pressure in the pipe drops to the level in the tank. All disturbances that have passed the junction up to that time cannot be influenced.

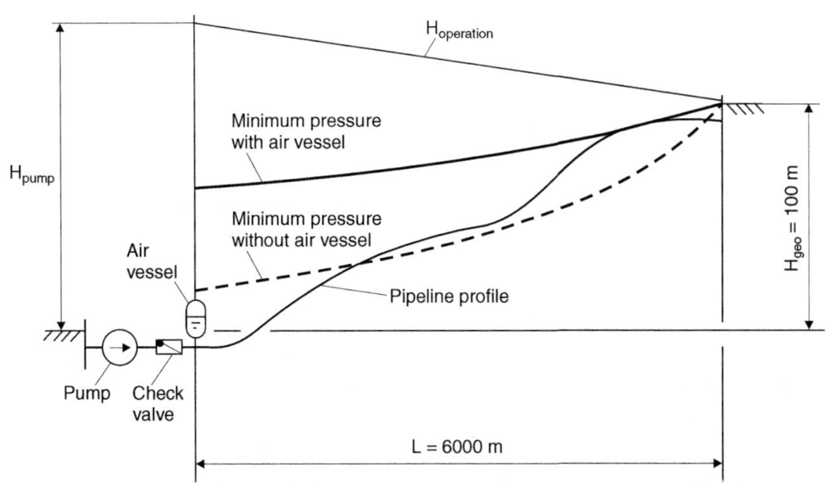

Figure 4.33 Effect on a surge tank on the minimum pipeline pressure

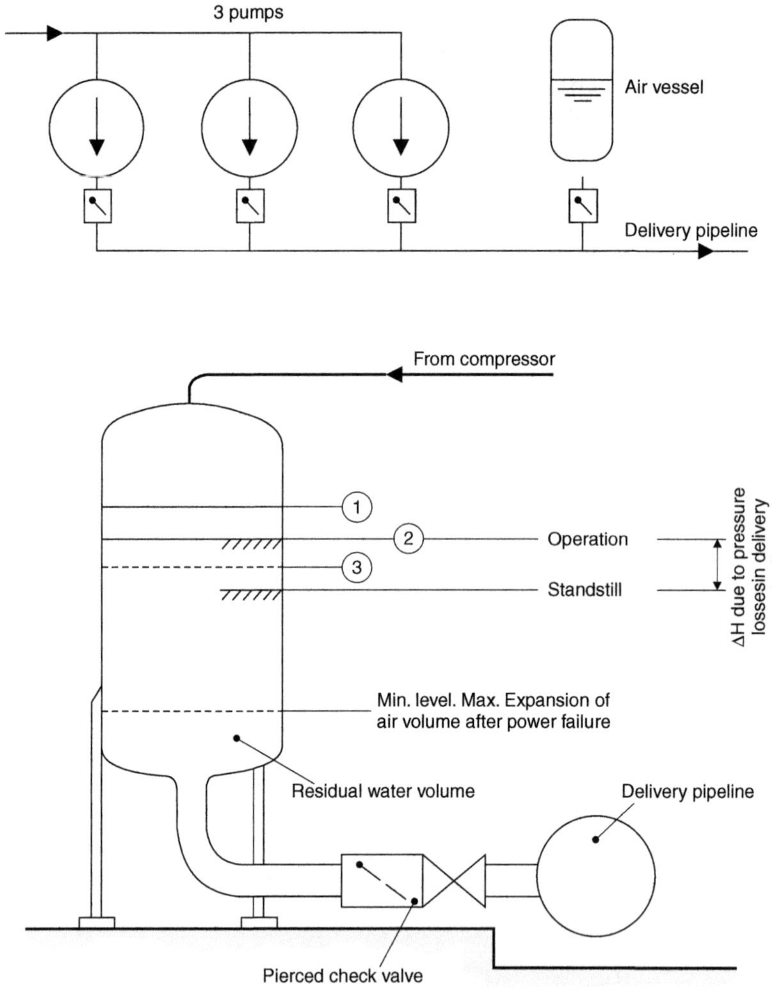

① Level too high: emergency, pumps are shut down normally.
② Normal operating level: compressor is switched on when it is exceeded.
③ Maximum air volume reached: compressor off (normally by timing relay).

Figure 4.34 Arrangement of a surge tank in pumping system and monitoring air volume

- *Air valves*. These too are effective only from the moment when atmospheric pressure is reached in the pipe at the point where the valve is fitted. Substantial volumes of air may be drawn into the system; consequently air valves are ruled out for water supply systems.

Problems may arise in operation if greater quantities of air remain trapped in the system (unstable operation, higher pressure losses). Good venting must therefore be assured.

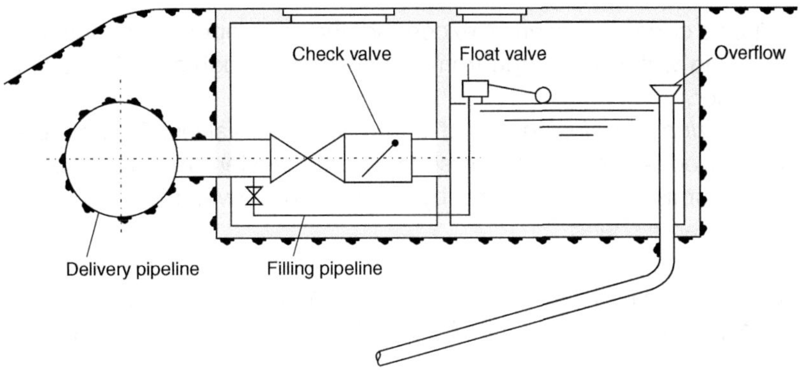

Figure 4.35 One-way surge tank

While a *bypass* between the approach flow and discharge sides creates no new boundary condition, in low-pressure systems it can significantly reduce the pressure loss through the pumping station and thus influence the velocity change, albeit to a small extent only.

If only the maximum pressure is important, or if it cannot be reduced sufficiently by primary protection measures, the following *secondary protective devices* may be considered:

- *Pressure relief valve* (Fig. 4.36). This must operate with minimum inertia and sufficient damping, otherwise pressure pulsations may be excited.

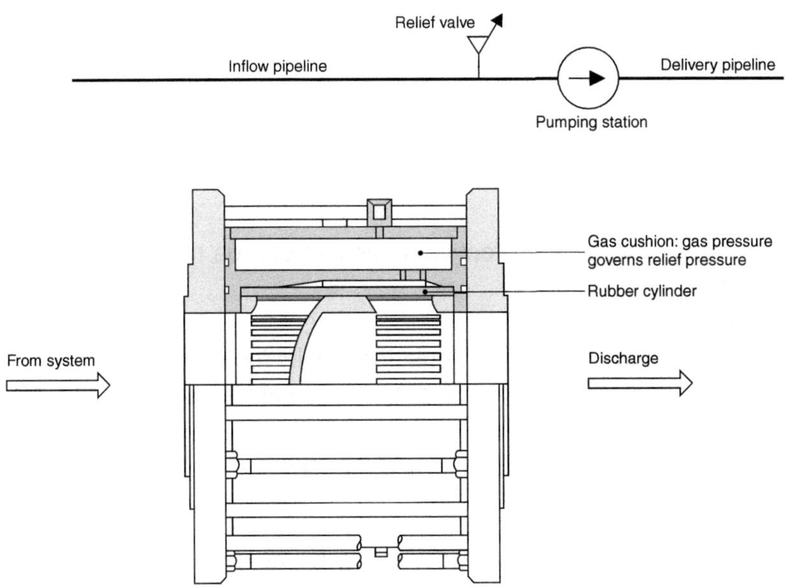

Figure 4.36 Pressure relief valve

- When a certain pressure is exceeded, the valve opens to atmosphere, allowing water to escape from the system through slots in the valve cylinder.
- *Optimizing the closing law* of controlled stop valves. This effects a controlled reduction of reverse flow in the system to zero. Usually a compromise must be found by calculation between pressure surge and reverse speed.
- *Check valve* in main pipe. In the event of reverse flow, this divides the pipe system into sections with different maximum pressures. Lower-level pipe sections can then be built for a lower nominal pressure.

4.3.6 Flap hammer

With a properly sized surge tank the pressure in the pipe is kept as high as is necessary for the system. The higher the pressure in the system, the shorter the pump rundown time t_A (i.e. the time taken to reach zero flow).

If the check valve of the pump is still partly open at the time t_A, reverse flow begins. This reverse flow causes rapid closing of the check valve. The result is a considerable pressure peak, accompanied by a very loud noise (in extreme cases a report). The condition for preventing flap hammer is: $t_S \leq t_A$.

t_S is the shortest possible closing time of the check valve, conditioned by the design.

Flap hammer can be prevented in the following ways:

- Selection of a check valve with a closing time conforming to the specified t_S.
- Increasing the mass moment of inertia of the revolving parts with a flywheel, so that the pump rundown time is prolonged.
- Installing the surge tank at some distance from the pump. Limits to this are set by the building, unless a separate building is provided for the surge tank installation, which must, however, remain within the control range of the pumping station.

4.3.7 Examples

The effects of various protective measures can be made clear by two examples.

Example 1

Figures 4.37 and 4.38 show calculation results for a system with 8000 m pipeline length and 90 m geodetic head. A total capacity of 1.11 m^3/s is provided by three pumps.

The minimum pressures are plotted against the pipeline profile in Fig. 4.37 for the following states:

- without additional protection;
- with the biggest flywheel possible (maximum size is governed by motor acceleration);
- with a surge tank.

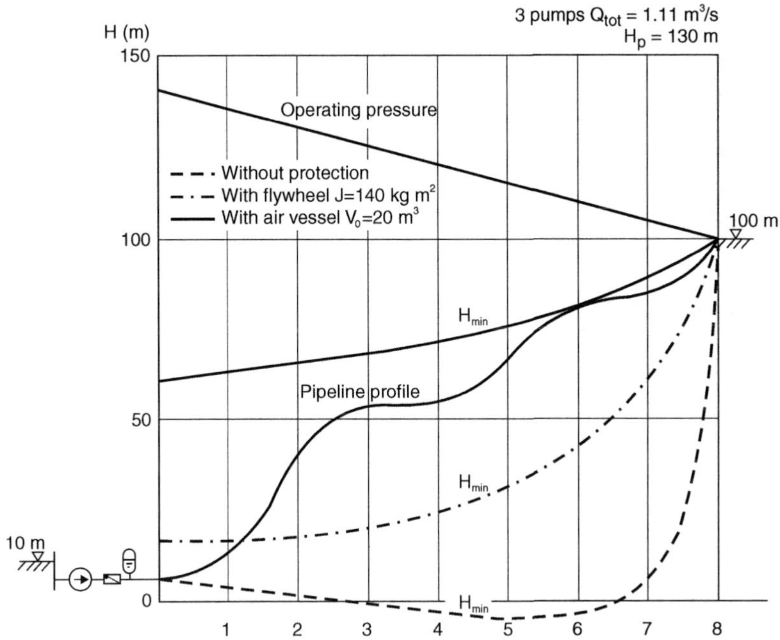

Figure 4.37 Minimum pressure along the pipeline after power failure

From the minimum pressure curves it is evident that only the surge tank prevents unacceptably low values being reached.

Figure 4.38 plots the pressure and flow rate at the beginning of the pipe against time.

Without protection and with a flywheel, the flow curve is very steep, causing an extremely steep pressure drop. Only with the surge tank is a slow, steady change of flow rate possible, resulting in a sufficiently flat pressure curve. The necessary surge tank size is 50 m^3, with an air volume of 20 m^3 under operating conditions.

Example 2

This example shows the results of measurements on a system with 2670 m pipe length, 515 m geodetic head and 0.080 m^3/s capacity operating with one pump (Figs 4.39–4.42). Figure 4.39 shows the pipeline profile and the extreme pressures. Without additional protection the admissible pipe pressure of 600 m is seriously exceeded (Fig. 4.40), owing to the water column separation between 1600 and 2550 m pipe length. An attempt was made to prevent column separation in the critical part of the pipe by means of air admission valves. As Fig. 4.41 shows, however, this did not cure the problem. Only the provision of a surge tank brought success (Fig. 4.42). At the beginning of the pipe the maximum pipe pressure is now only 585 m.

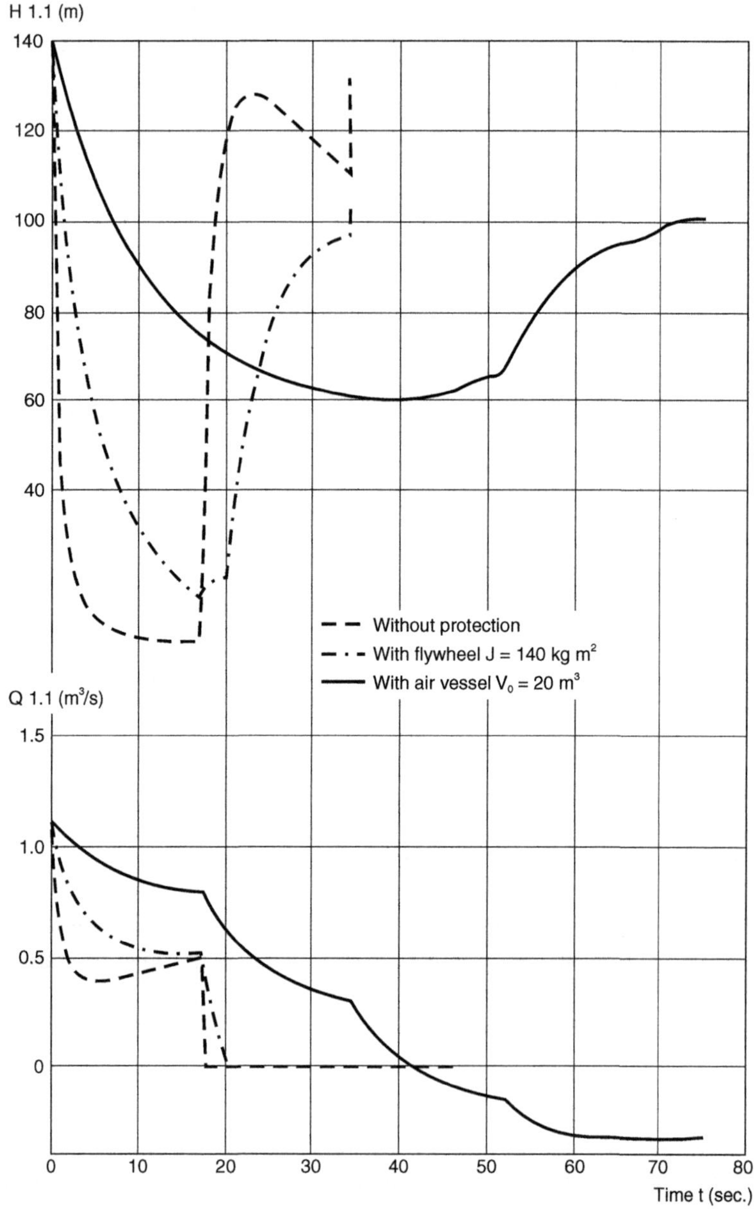

Figure 4.38 Transient curve for *H*1.1: pressure at beginning of pipe and *Q*1.1: flow at beginning of pipe

4.3.8 Guarantee and acceptance test

Based on the guarantees of the Swiss Codes for Hydraulic Machines (SEV 3055.1974) the accuracy of the calculated pressure variations is guaranteed as follows:

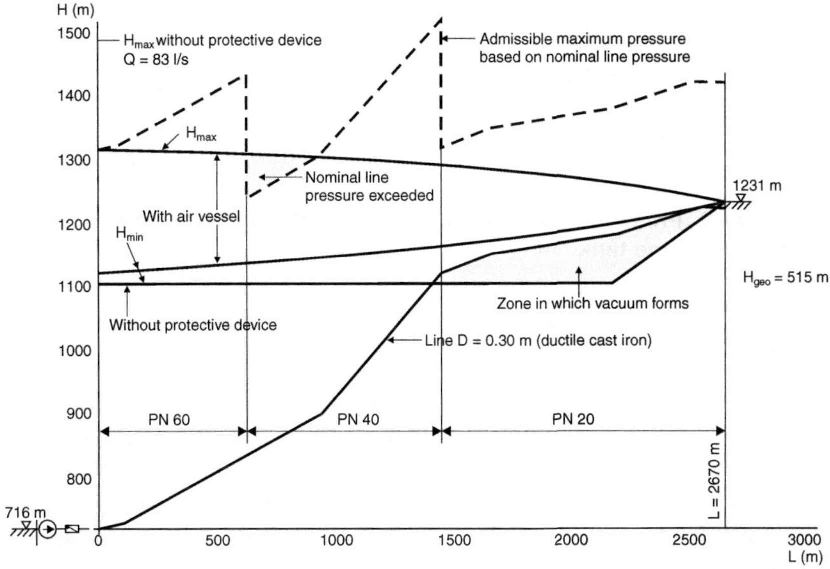

Figure 4.39 Pipeline profile and extreme pressures

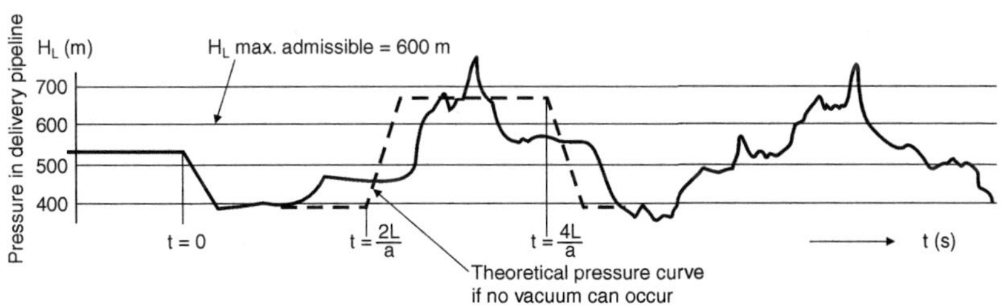

Figure 4.40 Pressure curve in the pumping station after power failure, without additional protection

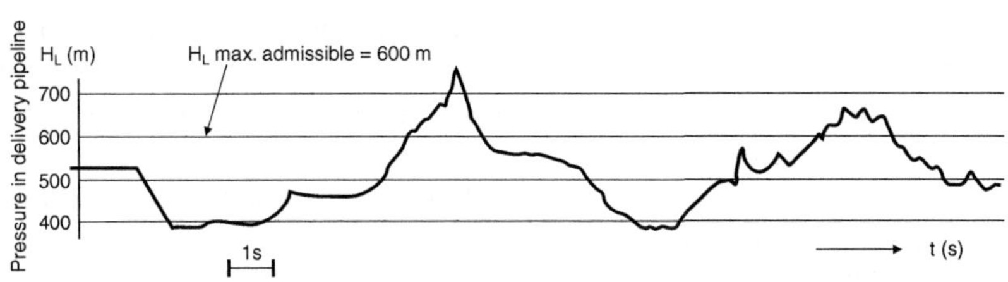

Figure 4.41 Pressure curve in the pumping station after power failure. Protection: air admission valves at pipeline lengths 1600 and 2200

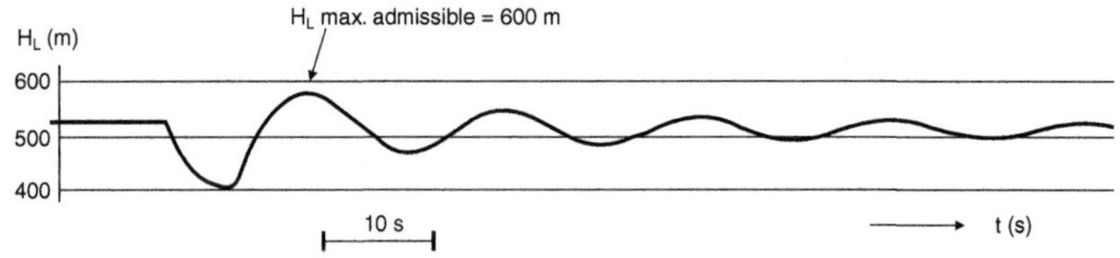

Figure 4.42 Pressure curve in the pumping station after power failure. Protection: surge tank

Pressure fluctuations at the beginning of the pipe (where the disturbance originates) may exceed the calculated values by not more than 20% (minimum 3 m).

To confirm fulfillment of the guarantee, shutdown tests are to be performed with the data on which the calculation was based, not later than at the conclusion of commissioning. All data of relevance to steady-state operation must be recorded. During the shutdown test the approach flow pressure, pressure on the discharge side, pump speed and stop valve opening must be recorded with electronic instruments.

4.4 MONITORING AND INSTRUMENTATION FOR CENTRIFUGAL PUMPS

The choice of instrumentation depends on the purpose, mode of operation and protection philosophy for the planned installation:

- *The basic rule is: "avoid everything superfluous"*. Additional devices by no means enhance reliability. On the contrary, they clutter the installation and increase the number of possible trouble sources.
- Partial or full automation should be justifiably economical. Minimum attendance and maintenance costs should be aimed at.
- Only a small selection of the possible measuring and monitoring instruments for protecting centrifugal pumps can be listed here (see examples in Fig. 4.43).
- Additional external conditions for assuring pump protection are:
  - Before starting: the pump must be primed and must stay primed during operation.
  - A minimum flow must be assured in operation to avoid excessive rise of the liquid temperature (with risk of evaporation) due to the internal energy losses in the pump. At zero flow, depending on the pump type and specific speed, 50% or more of the driver input at BEP is converted into heat (see section 2.8.1: Determining the minimum flow rate).
  - The protective devices employed must be absolutely dependable. Operational reliability is paramount.

  Remarks:
- When choosing instruments the ability of the suppliers to render *service on-site* must be verified.

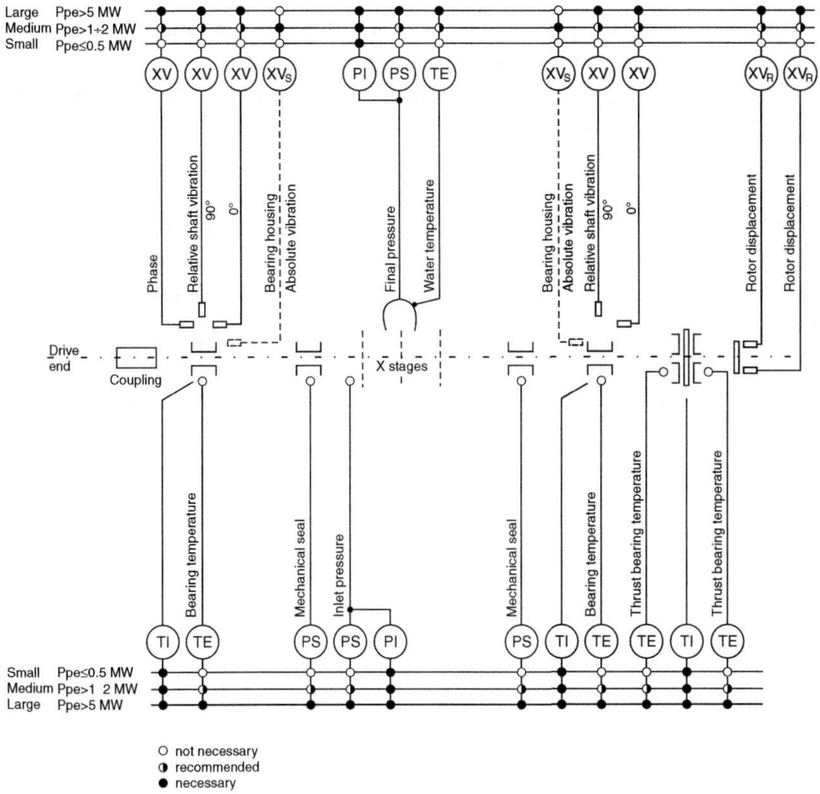

Figure 4.43 Examples for selecting measuring and monitoring instruments for horizontal water transporting pumps

- A clearly defined *functional description* of the entire pumping unit (pump, motor, control system, protective devices and power supply) must be available prior to commissioning.
- For instruments the constraints affecting three-phase motors mentioned in section 7.1.3 must be noted in particular.
- As far as the installation allows, whenever possible *the pumps should be arranged* so that they are always under inlet pressure, i.e. remain primed when idle. This makes for simple starting and higher reliability.

4.5 NOISE EMISSION FROM CENTRIFUGAL PUMPS

4.5.1 Basic acoustic terminology

4.5.1.1 ACOUSTIC QUANTITIES

Sound is defined as mechanical oscillations in an elastic medium. Depending on the medium, a distinction is drawn between airborne, structure-borne and liquid-borne sound:

- *Sound pressure* is the quantity perceived by the human ear.

- *Sound power* is the acoustic power emitted by a sound source.
- *Sound intensity* is the sound power per unit area.
- *Structure-borne sound* is the sound propagated in a solid medium or at its surface. Vibrations on the surface may radiate airborne sound.

4.5.1.2 LEVELS

In order to quantify conveniently the wide dynamic range of the sound intensities perceptible by human hearing, from the audibility threshold to the threshold of pain, a logarithmic scale is used. The various levels are defined as:

$$L_p[\text{dB}] = 10 \cdot 1g\frac{p^2}{p_0^2} = 20 \cdot \log\frac{p}{P_0} \quad \text{Sound pressure level (or noise level)} \tag{1}$$

$$P_0 = 2 \cdot 10^{-5}\frac{\text{N}}{\text{m}^2} \quad \text{Reference value}$$

$$L_w[\text{dB}] = 10 \cdot 1g\frac{P}{P_0} \quad \text{Sound power level} \tag{2}$$

$$P_0 = 10^{-12}\text{W Reference value}$$

$$L_1[\text{dB}] = 10 \cdot 1g\frac{I}{I_0} \quad \text{Sound intensity level} \tag{3}$$

$$I_0 = 10^{-12}\frac{\text{W}}{\text{m}^2} \quad \text{Reference value}$$

$$L_v[\text{dB}] = 10 \cdot 1g\frac{V}{V_0} \quad \text{Structure-bone sound level} \tag{4}$$

$$v_0 = 5 \cdot 10^{-8}\frac{\text{m}}{\text{s}} \quad \text{Reference value}$$

The reference value for sound pressure corresponds to the audibility threshold of the human ear at a frequency of 1000 Hz. Typical sound pressure levels are:

	L_p (dB)
Quiet speech	40
Conversational speech	65
Motor truck	90
Pneumatic hammer	115

4.5.1.3 FREQUENCY SPECTRUM

Sound levels are expressed as a function of frequency. Depending on the kind of frequency filter, distinctions are drawn as follows:

- octave spectrum
  interval between the center frequencies of the octaves, with ratio 2:1
- third-octave spectrum
  three third octaves per octave
- narrow-band spectrum

With the octave and third-octave spectra the bandwidth is proportional to frequency.

4.5.1.4 WEIGHTING CURVES

The perceptive capacity of the human ear is not the same at all frequencies. Its maximum sensitivity is around 4 kHz, falling off steeply towards low frequencies (below 500 Hz) especially. In order to adapt sound measurements to hearing sensitivity, various frequency weightings for acoustic signals have been introduced. Weighted sound pressure levels A, B or C are then referred to. Weighting A allows for aural sensitivity to moderate sound pressure levels, weighting B for medium and weighting C for high sound pressure levels. Weighting curves A, B and C are shown in Fig. 4.44.

Nowadays the A-weighted sound level L_{PA} is used almost exclusively for defining the magnitude of industrial noise. This is the total level of the A-weighted frequency spectrum.

4.5.1.5 SUMMATION OF SOUND LEVELS

To add sound levels the corresponding power quantities (p^2, p) must be summed up, i.e. the levels must be converted back from the logarithmic form:

$$L_3 = 10 \cdot 1g\left(10^{\frac{L_1}{10}} + 10^{\frac{L_2}{10}}\right) \text{ Total level of } L_1,\ L_2 \qquad (5)$$

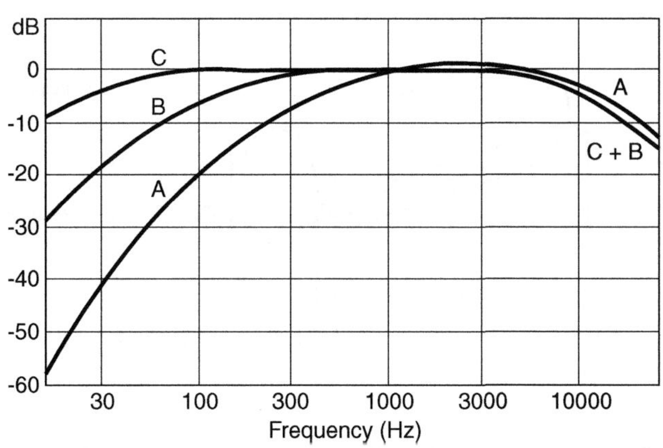

Figure 4.44 Weighting curves

Accordingly, doubling sound power, for instance, means raising the level by 3 dB.

A sound level increase of 3 dB is just perceptible subjectively. A 10 dB increase is perceived as a doubling of the loudness.

4.5.1.6 CORRELATION BETWEEN SOUND POWER, SOUND PRESSURE AND SOUND INTENSITY

The sound power radiated by a pump is determined by measuring the sound intensity or sound pressure on a measuring surface (Fig. 4.45). The following equation holds:

$$L_W = \bar{L}_1 + 10 \cdot 1g \cdot \frac{S}{S_0} \tag{6}$$

where:
L_W = sound power level (dB)
$\bar{L}_1$ = sound intensity integrated over the measuring surface (normal component) (dB)
S = measuring surface (m^2)
S_0 = reference surface (1 m^2)

With free sound propagation in the remote field of a sound source (no extraneous noise), the following applies also:

$$L_W = \bar{L}_p + 10 \cdot 1g \cdot \frac{S}{S_0} \tag{7}$$

where $\bar{L}_p$ = mean sound pressure level (dB) on measuring surface.

These simple equations hold true because the reference levels are selected appropriately.

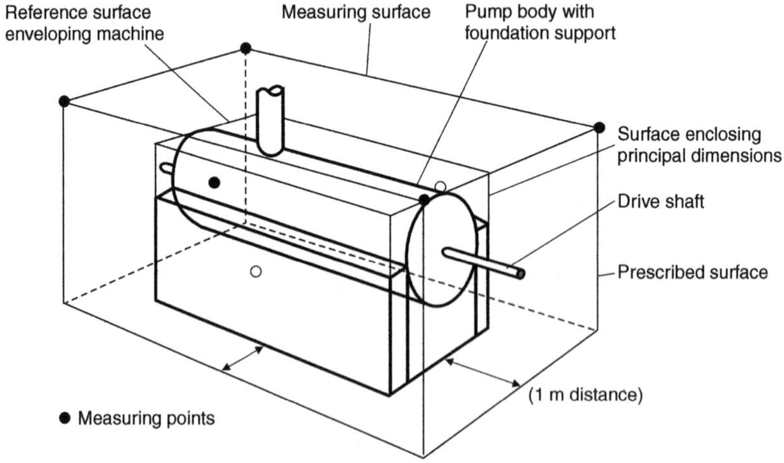

Figure 4.45 Schematic pump with measuring surface (reference surface) and arrangement of measuring points

4.5.2 Origins of noise from pumps

Mechanical

Mechanical noises are caused mainly by:

- bearings
- imbalance.

Hydraulic

Two different mechanisms are distinguished in hydraulic noises:

- Turbulent flow causes a broad-band noise.
- The velocity profile at the impeller exit is not uniform around the circumference. This results in periodic excitation through impingement on the diffuser vanes or volute cutwaters. Narrow-band tones are set up with the frequency of the blade passing at the diffuser. These are called blade passing frequencies:

  $f_D = n \cdot Z_2 \cdot k$ = blade passing frequency of rotor
  $f_D = n \cdot Z_3 \cdot k$ = blade passing frequency of stator
  $f_D = n \cdot Z_2 \cdot Z_3 \cdot k$ = blade passing frequency of rotor/stator interference
  f_D [Hz] = blade passing frequency
  n [s^{-1}] = speed
  z_2 = number of impeller blades
  z_3 = number of diffuser vanes or volute cutwaters
  k = 1st, 2nd, 3rd...harmonics.

The blade passing frequency due to the number of impeller blades predominates mostly. "k" denotes the harmonics. Most important are the 1st, 2nd and 3rd harmonics.

Further noises may result from operating the pump in unfavorable conditions, e.g. with cavitation at part load or overload. Generally, the noise is lowest near the best efficiency point.

4.5.3 Measuring noise

To start with, it is necessary to define the object to be measured as the source of sound emission. The pump maker considers only the pump itself, with pump body, suction and discharge nozzles and baseplate pads. The user, however, is interested in the overall noise level of the pump installation. This includes additionally the noise from the motor and any gearing plus emissions from pipes and surfaces excited by structure-borne noise. Because the latter depend on the particular installation, they cannot be included in the noise data for a given pump type.

The following methods may be used for measuring pump noise:

1. Sound pressure measurement, prescribed surface method;
2. Sound pressure measurement, reverberation room method;
3. Sound pressure measurement with reference sound source;

Figure 4.46 Pump with acoustic enclosure

4. Structure-borne noise measurement;
5. Sound intensity measurement.

Figure 4.45 shows schematically a pump with measuring points arranged for the prescribed surface method.

All these measuring procedures are standardized. Standards in force are tabulated in section 4.5.6.

For *sound pressure measurements*, limiting the extraneous noise often requires considerable expense. A sound-absorbing enclosure for the electric motor and the discharge pipe for sound measurement may be needed on the pump test stand.

The *structure-borne noise method* is insensitive to extraneous noise. However, problems arise from the inexactly known radiation efficiency and from radiating openings where no transducers (e.g. accelerometers) for structure-borne noise can be fixed.

Intensity measurement is the latest method and measures directly the required physical quantity, i.e. sound intensity, which is directly related to sound power. Extensive sound power measurements have been made on Sulzer pumps, so that sound emission values based on measurements (as described below) can be quoted for a number of pump types.

4.5.4 Sound data from measurements

Some results from sound pressure measurements by the prescribed surface method are given below.

Figure 4.47 shows the measured sound power in dB(A) for feed pumps as a function of pump output. For comparison, the curve from

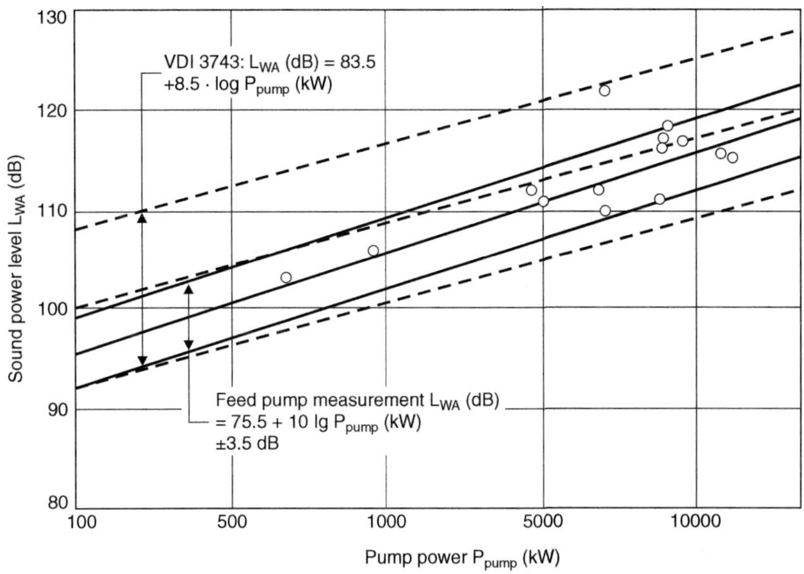

Figure 4.47 Sound power levels from feed pumps (compared with VDI 3743)

VDI guideline No. 3743 for volute and annular casing pumps is also shown. Note that compared with the sound power levels the sound pressure levels are reduced by the measuring surface (see equation (6)). A typical frequency pattern is shown by the octave spectrum in Fig. 4.48.

Figure 4.49 plots the rise of the sound level as a function of speed. Besides data measured in a Sulzer laboratory a curve from Tourret is plotted.

Finally, Fig. 4.50 shows measured sound levels as a function of pump load. It is clear from this that the noise generation is lowest in the vicinity of the best efficiency point (Q_{00}).

4.5.5 Methods of abating noise

Besides the primary possibilities taken into account by the pump designer, the conditions of installation in the plant can affect the noise level considerably. Generally speaking, hydraulically well-designed pumps tend to behave well with regard to noise also. Otherwise it must be remembered that the pump itself often generates relatively little sound power. Other major noise sources are:

1. drive motor, gears;
2. excitation of structure-borne noise in adjacent light structures (e.g. chequered plate);
3. radiation from the discharge pipe, and to a lesser extent from the suction pipe as well.

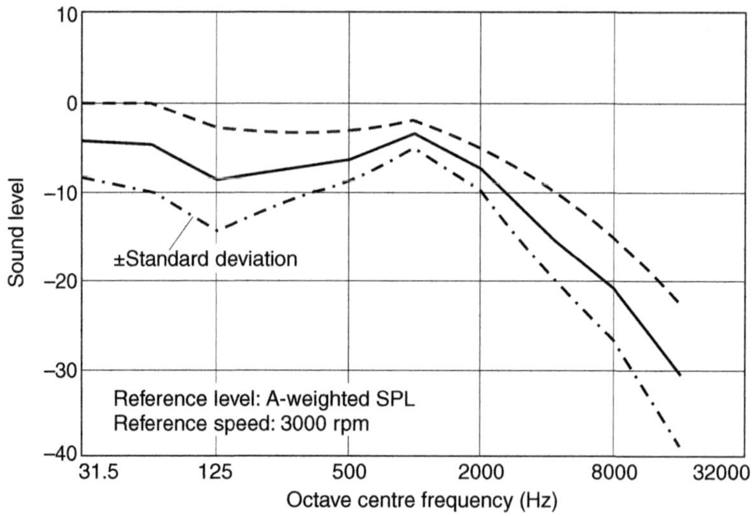

Figure 4.48 Relative octave spectrum

4.5.5.1 PUMP DRIVE

During the acoustic planning of the installation the gearing and especially the motor should be selected with an eye to the noise abatement requirements. Water-cooled motors are generally much quieter than air-cooled machines.

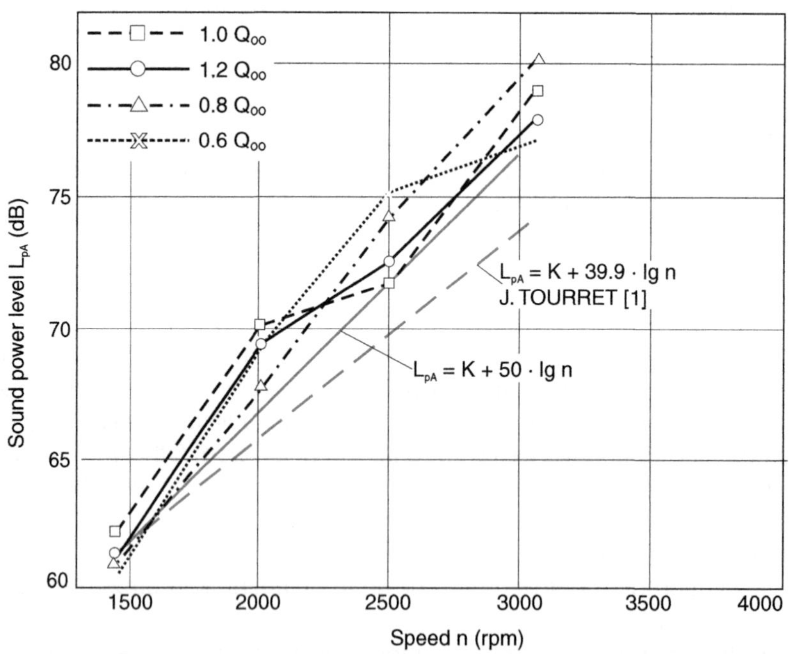

Figure 4.49 Sound pressure level as a function of speed

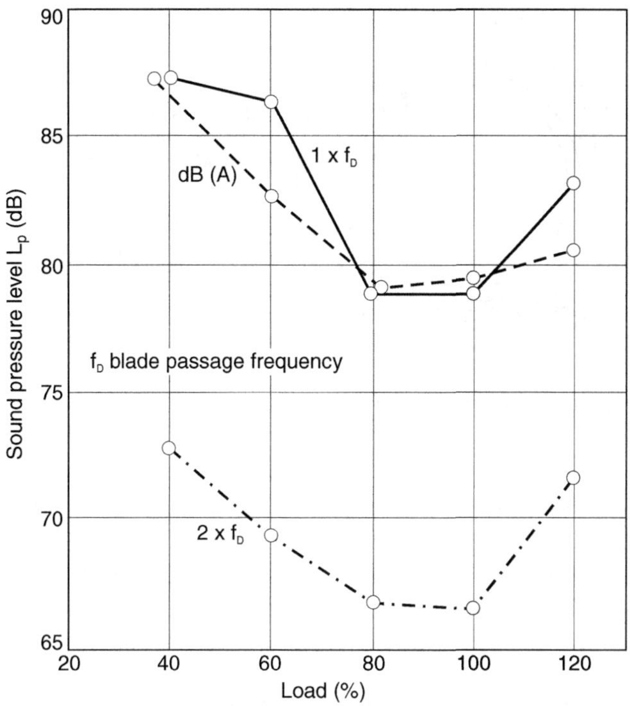

Figure 4.50 Sound pressure level as a function of load

4.5.5.2 STRUCTURE-BORNE NOISE

Care must be taken to ensure that the structure-borne noise generated by the machinery (pump gearing, motor) is radiated as little as possible by other adjacent components. Elastic mounting would be best, but this is usually not possible. In any case light structures (thin sheets) which are not part of the pump fixing should not be joined to foundation supports or via other elements to the pump or driving machinery.

4.5.5.3 PIPING

Ideally the pipes ought to be isolated acoustically by means of compensators and liquid silencers. As a secondary measure sound-absorbent insulation of the pipes may be considered. Such insulation can reduce noise from pump pipelines by 5 to 15 dB (Fig. 4.51).

4.5.5.4 REDUCED EMISSION FROM THE PUMP

If the admissible noise levels are exceeded by radiation from the pump itself there exist two differently effective approaches to reducing it:

1. *Sound-absorbent cladding* of the pump (possibly providing thermal insulation at the same time). Attainable level reduction up to about 5 dB.
2. *Encapsulation* (sound hoods). Sound hoods bring abatements up to 30 dB and more, depending on their design. Usually it is advisable to

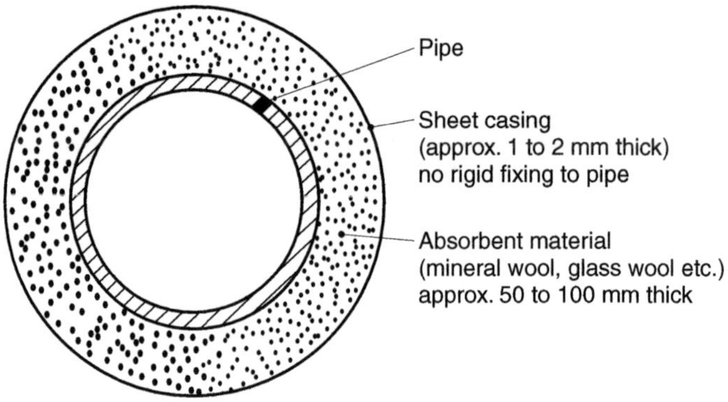

Figure 4.51 Acoustic insulation of a pipe

enclose the motor and any gearing as well. Figure 4.52 shows the basic design.

It must be borne in mind that the level reduction attainable in the machine room depends also on the radiation from the components mentioned previously (pipelines, structure-borne noise).

4.5.6 List of standards and guidelines

Sound level meters

IEC 651, 1990 Precision sound level meters.

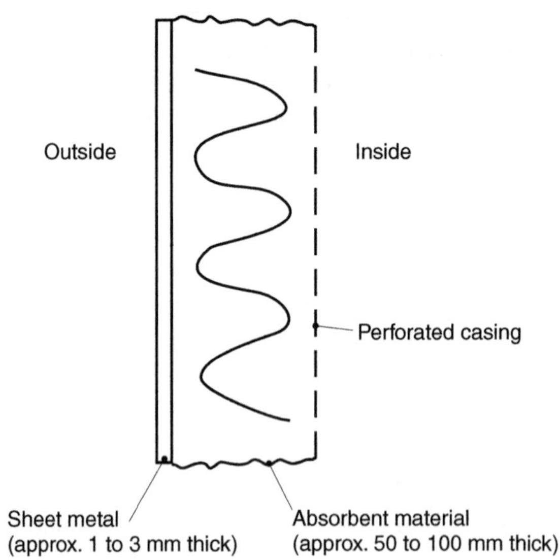

Figure 4.52 Schematic wall cross-section of sound hood

Noise measurements

DIN 45635	Measurement of airborne noise emitted by machines
Part 1, 1984	Enveloping surface method; basic requirements for 3 accuracy classes
Part 2, 1987	Reverberation room method; basic measurement method
Part 3, 1985	Engineering methods for special reverberation test room
Part 8, 1985	Measurement of structure-borne noise; basic requirements
Part 24, 1980	Enveloping method; liquid pumps
ISO 3741–3746	Acoustics – Determination of sound power levels of noise sources
3741, 3742	Reverberation room method
3744	Measuring outdoors in large rooms
3745	Laboratory, anechoic room
3746	No special test environment
ISO 4871	Acoustics – Noise labeling of machinery and equipment
ISO TR 7849	Acoustics – Estimation of airborne noise emitted by machinery using vibration measurement
ISO 9614	Acoustics – Determination of sound power levels of noise sources using sound intensity
Part 1, 1993	Measurement at discrete points
Part 2, 1996	Measurement by scanning
ISO 11203	Acoustics – Noise emitted by machinery and equipment; determination of emission sound pressure levels at the work station and other specified positions from the sound power level.

Sound emission

VDI 3743	Emission characteristics of technical noise sources
sheet 1, 1989	Centrifugal pumps

Sound insulation

VDI 2711, 1978	Noise reduction of enclosures

4.6 VIBRATION ON CENTRIFUGAL PUMPS

4.6.1 Introduction

In every pump, unsteady forces of mechanical and hydraulic origin are present and a certain level of vibration is therefore inevitable. To ensure the safety of the pump and associated plant components the vibration must be kept within specific limits. If the mechanical state of the pump and its drive are good, the approach flow conditions are reasonably uniform and the duty point is within the allowable range, these limits can be observed without difficulty. However, if these limits are exceeded or if vibrations markedly increase in the course of time, problems of a mechanical or hydraulic nature may be suspected. Measuring vibrations is therefore a very good way of checking conditions of a pump. On smaller pumps such checking is mostly performed with portable instruments. On large installations, especially feed pumps, water transport pumps and injection pumps, vibration measuring devices are often fitted permanently, in conjunction with alarm, tripping and/or continuous recording in the control room.

The causes of aggravated vibration are manifold. Detailed measurements and analyses carried out by a specialist are often needed for a reliable diagnosis. Frequency analysis and evaluation is an important step in this diagnostic work. A distinction must be drawn between system-related and pure pump problems.

Typical system-related problems:

- Unfavorable dynamic behavior of foundations, supporting structures or pipelines (e.g. resonance excitation by forces at rotational frequency);
- Excitation from a component in the pipeline (valve, filter, etc.);
- Excitations from the coupling, especially due to misalignment;
- Excitations from the drive (motor, steam turbine, gearing);
- Unfavorable approach flow conditions like insufficient suction pressure (cavitation), swirling (intake vortex, suction pipe with bends in more than one plane);
- High-pressure pulsations due to hydraulic instability of the entire system.

Typical problems of the pump itself:

- Mechanical imbalance of the rotating parts due to inexpert balancing, careless assembly or operational influences (cavitation erosion, deposits, corrosion, damaged impellers, jammed parts, abrasion);
- Unfavorable dynamic behavior of the rotor due to excessive seal or bearing clearances;
- Increased hydraulic forces when the pump is operated outside of the admissible operation range (some increase in vibration is normal when departing from the optimum flow rate);
- Mechanically defective bearings.

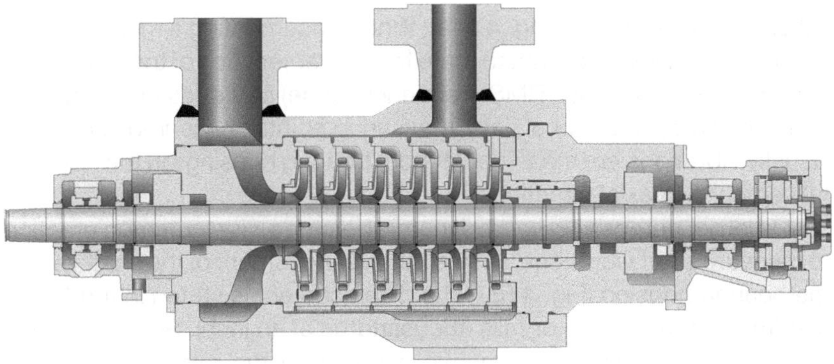

Figure 4.53 Injection pump with hydrodynamic bearings

4.6.2 Measuring bearing housing and shaft vibration

The *vibration velocity at the bearing housing v* is commonly measured to assess running behavior (Fig. 4.53). ISO 13709 specifies bearing housing velocity limits for all pumps regardless of bearing type. It is usual to determine the root mean square value (RMS) of the vibration velocity expressed in mm/s or in/s. The relations for a sine-wave signal are given in Fig. 4.54.

Many instruments display the RMS vibration velocity directly, but there are also instruments showing the peak value (v_p), though its determination is not standardized. Conversion from v_p to v_{RMS} is possible only

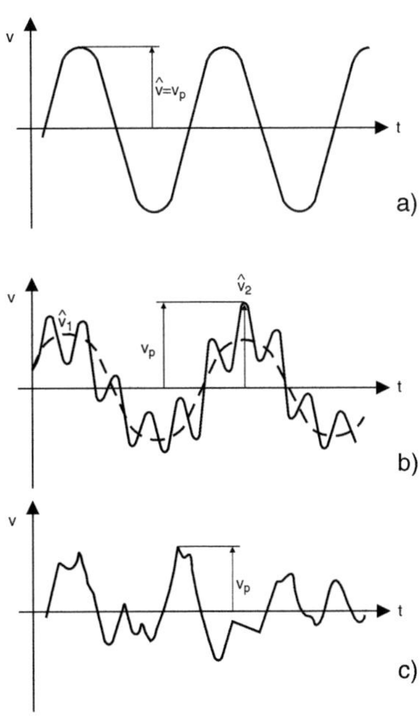

$$v_{RMS} = 0.707 \cdot \hat{v}$$

v_{RMS} = RMS value of vibration velocity (mm/s, in/s)
$\hat{v}$ = peak value of vibration velocity for harmonic signals (mm/s, in/s)
v_p = peak value (mm/s, in/s)

Where the vibration consists of several harmonic components (Fig. 4.54b), then:

$$v_{RMS} = \sqrt{\hat{v}_1^2 + \hat{v}_2^2 + \ldots \hat{v}_2^2}$$

For non-periodic or random signals, by definition:

$$v_{RMS} = \sqrt{\frac{1}{t_0}\int_0^{t_0} v^2(t)\,dt}$$

Figure 4.54 Vibration velocity as a function of time

with a sine-wave signal (Fig. 4.54a). When documenting a measurement it is therefore very important to state the characteristics of the measuring instrument (e.g. velocity, RMS, frequency range) as well as the usual data (machine type, operational data, measuring points, measured variables). Usually vibrations are measured at both bearing housings in the horizontal vertical and axial directions.

On larger pumps with hydrodynamic bearings it is a common practice to measure shaft vibrations, i.e. the relative movement of the shaft versus the bearing housing (Fig. 4.53). The displacement is stated in μm (thousandths of a millimeter) or mils (thousandths of an inch). Such measurements usually enable the dynamic behavior of the rotor to be judged more directly than by measuring on the bearing housing. However, shaft vibration measurements can be falsified by shaft non-cylindricality as well as material inhomogeneities. For correct assessment of the measured results, these defaults (called mechanical or electrical runout) must be known or have only small values. Often, each bearing is fitted with two transducers at right angles to each other. The orbit of the shaft center can then be plotted (Fig. 4.55).

For assessment purposes, therefore, a distinction must be drawn between measured values for individual directions (s_1, s_2) and measured values for the orbit (s_{max}). The magnitude s_{max} cannot be calculated from s_1 and s_2 alone. It is difficult to measure s_{max} directly but a suitable approximation can be made using one of the methods described in ISO 7919. The method most commonly used is to take the maximum value of peak-to-peak displacement measured in two orthogonal directions, *X* and *Y*.

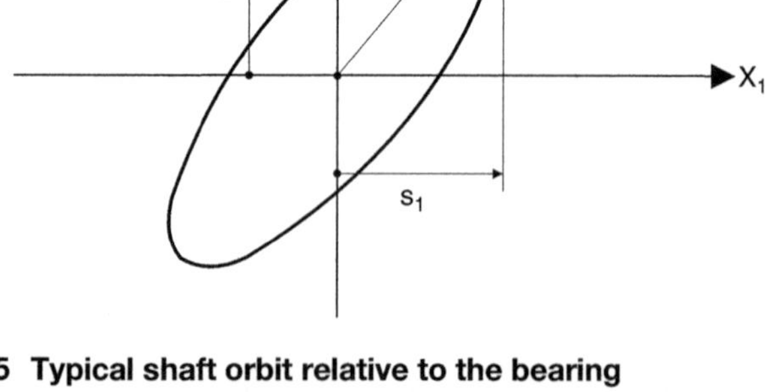

Figure 4.55 Typical shaft orbit relative to the bearing

4.6.3 Vibration limits

Admissible vibration limits depend primarily on the machine design and should therefore be fixed by the pump manufacturer. However, usually vibration standards are applied. The ISO 10816 series of standards are widely used to evaluate mechanical vibrations of machines. Vibration limits for pumps with multi-vane impellers and separate drives (i.e. not close coupled) are proposed in Table A.1 of ISO 10816-7 reproduced below.

Table A.1 Zone limits for vibration of non-rotating parts of rotodynamic pumps with power above 1 kW, applicable for impellers with number of blades $z_l \geq 3$

Zone	Description (see section 5.2 for details of zone definitions)	Vibration velocity limited RMS value			
		Category		Category	
		>200 kW	≤200 kW	≤200 kW	>200 kW
A	Newly commissioned machines in preferred operating range	2.5	3.5	3.2	4.2
B	Unrestricted long-term operation in allowance operating range	4.0	5.0	5.1	6.1
C	Limited operation	6.0	7.6	8.5	9.5
D	Risk of damage	>6.6	>7.6	>8.5	>9.5
Maximum ALARM limit (=1.25 times the upper limit of zone C)[b]		5.0	6.3	6.4	7.6
Maximum TRIP limit (=1.25 times the upper limit of zone C)[b]		8.3	9.5	10.6	11.9
In situ acceptance test	Preferred operating range	2.5	3.5	3.2	4.2
	Allowable operating range	3.4	4.4	4.2	5.2
Factory acceptance test	Preferred operating range	3.3	4.3	4.2	5.2
	Allowable operating range	4.0	5.0	5.1	6.1
		≤2	≤2	≤3	≤3

[b]*Recommended values. The vibration magnitudes should be above these limits for about 10 s before an ALARM or TRIP is released to avoid false alarms and trips.*

[c]*For acceptance tests in the allowable but outside the preferred operating range, the filtered values for rotational frequency* (f_n z_l) *may be expected to be higher (1.3 times) than the values for the preferred operating range.*

Table A.2 Additional criteria for vibration limits on non-rotating parts of rotodynamic pumps with running speed below 600 rpm valid for filtered displacement values (0.5 times, 1 times and 2 times the running speed)

Zone	Description (see section 5.2 for details of zone definitions)	Vibration displacement limit peak to peak value μm
A	Newly commissioned machines in preferred operating range	50
B	Unrestricted long-term operation in allowable operating range	80
C	Limited operation	130
D	Risk of damage	>130
Maximum ALARM[a]		100
Maximum TRIP[a]		160
In situ acceptance test	Preferred operating range	50
	Allowable operating range	65
Factory acceptance test	Preferred operating range	65
	Allowable operating range	80

[a]Recommended values. The vibration magnitudes should be above these limits for about 10 s before an ALARM or TRIP is released to avoid false alarms and trips.

NOTE: The limits are applicable for each discrete frequency mentioned.

Table B.1 Recommended values for maximum relative displacement of the shaft as a function of the nominal diametrical clearance for rotodynamic pumps with hydrodynamic bearings

Zone	Description (see section 5.2 for details of zone definitions)	Limit for peak-to-peak shaft vibration displacement in relation to the diametrical clearance of oil lubricated hydrodynamic bearing[a]
A	Newly commissioned machines in preferred operating range	0.33 × bearing clearance in new state
B	Unrestricted long-term operation in allowable operating range	0.5 × bearing clearance in new state
C	Limited operation	0.7 × bearing clearance in new state
D	Risk of damage	>0.7 × bearing clearance in new state
Maximum ALARM[b]		0.6 × bearing clearance in new state

(Continued)

Table B.1 Recommended values for maximum relative displacement of the shaft as a function of the nominal diametrical clearance for rotodynamic pumps with hydrodynamic bearings *(continued)*

Zone	Description (see section 5.2 for details of zone definitions)	Limit for peak-to-peak shaft vibration displacement in relation to the diametrical clearance of oil lubricated hydrodynamic bearing[a]
Maximum TRIP[b]		0.9 × bearing clearance in new state
In situ acceptance test	Preferred operating range	0.33 × bearing clearance in new state
	Allowable operating range	0.5 × bearing clearance in new state
Factory acceptance test	Preferred operating range	0.33 × bearing clearance in new state
	Allowable operating range	0.5 × bearing clearance in new state

[a] *The pump manufacturer shall specify the nominal value of the hydrodynamic bearing clearance.*

[b] *Recommended values. The vibration magnitudes should be above these limits for about 10 s before an ALARM or TRIP is released to avoid false alarms and trips.*

Ratings according to ISO 10816-7. The vibration values are stated for pumps with multi-vane impeller and with separate driver (centrifugal, mixed flow or axial flow) with rated power above 1 kW.

Sometimes the vibration velocity at the bearing housing is dominated by high-frequency vibration, often corresponding to the number of impeller vanes multiplied by the rotation frequency. In order to judge the rotor behavior it is helpful to filter out these high frequencies. Such high-frequency vibrations may also be falsified by transducer resonances.

For oil and gas as well as hydrocarbon processing purposes ISO 13709 (API 610) is the standard of choice.

Table 4.2 ISO 13709 (API 610) Vibration Limits for Overhung and Between Bearings Pumps

Criteria	Location of vibration measurement	
	Bearing housing	Pump shaft adjacent to bearing
	Pump bearing type	
	All	Hydrodynamic journal bearings
	Vibration at any flow rate within the pump's preferred operating range	
Overall	For pumps running at up to 3600 rpm and absorbing up to 300 kW per stage[a] $v_u < 3.0$ mm/s RMS	$A_u < (5.2 \times 10^6/n)^{0.5}$ µm peak to peak [$(8000/n)^{0.5}$ mils peak to peak] Not to exceed: $A_u < 50$ µm peak to peak[a]

(Continued)

Table 4.2 ISO 13709 (API 610) Vibration Limits for Overhung and Between Bearings Pumps *(continued)*

Discrete frequencies	For pumps running above 3600 rpm and absorbing more than 300 kW per stage[a]. See Fig. 4.56 $v_f < 0.67v_u$	(2.0 mils peak to peak) For $f < n$: A_f, $0.33A_u$
Allowable increase in vibration at flows outside the preferred operating region but within the allowable operating region	30%	30%

[a]*Calculated for BEP of rated impeller with liquid relative density (specific gravity) = 1.0.*
[b]*Values calculated from the basic limits shall be rounded off to two significant digits*

where:
v_u is the unfiltered velocity, as measured
v_f is the filtered velocity
A_u is the amplitude of unfiltered displacement, as measured
A_f is the amplitude of filtered displacement
f is the frequency
n is the rotational speed expressed in RPM

Table 4.3 ISO 13709 (API 610) Vibration Limits for Vertically Suspended Pumps

	Location of vibration measurement	
	Pump thrust bearing or motor mounting flange	Pump shaft adjacent to bearing
	Pump bearing type	
	All	Hydrodynamic guide bearing adjacent to accessible region of shaft
	Vibration at any flow rate within the pump's preferred operating range[a]	
Overall	$v_u < 5.0$ mm/s RMS	$A_u < (6.2 \times 10^6/n)^{0.5}$ µm peak to peak [$(10\,000/n)^{0.5}$ mils peak to peak] Not to exceed:
Discrete frequencies	$v_f < 0.67v_u$	$A_u < 100$ µM peak to peak[a] (4.0 mils peak to peak) $A_f < n$: A_f, $0.33A_u$
Allowable increase in vibration at flows outside the preferred operating region but within the allowable operating region	30%	30%

[a]*Values calculated from the basic limits shall be rounded off to two significant digits*

where:
v_u is the unfiltered velocity, as measured
v_f is the filtered velocity
A_u is the amplitude of unfiltered displacement, as measured
A_f is the amplitude of filtered displacement
n is the rotational speed expressed in RPM

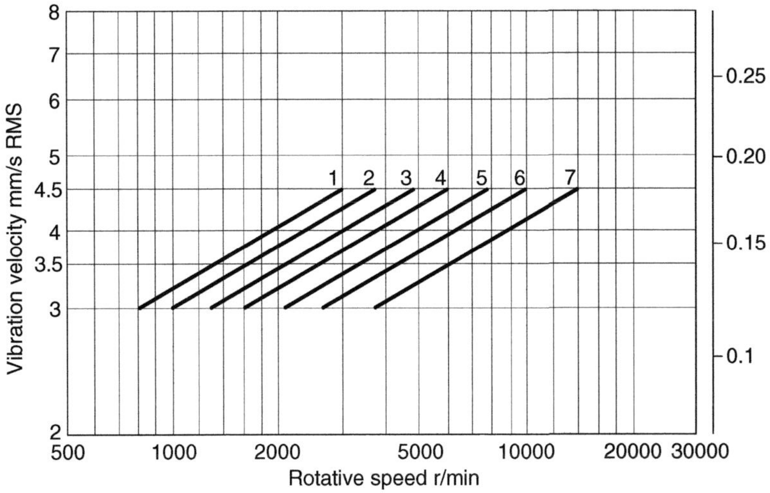

Key – powers in kW per stage

1. $P = 3{,}000$ 2. $P = 2{,}000$ 3. $P = 1{,}500$ 4. $P = 1{,}000$
5. $P = 700$ 6. $P = 500$ 7. $P = 300$

Figure 4.56 For pumps running above 3600 rpm and absorbing more than 300 kW per stage

Chapter | five

Mechanical Components

5.1 SHAFT COUPLINGS

The principal function of the shaft coupling is the transmission of torque from the driving unit (e.g. electric motor with or without gear, steam or gas turbine) to the pump shaft. There are:

- rigid couplings
- flexible couplings
- couplings with torsional elasticity.

Rigid couplings are of the shell or flange type. These are used chiefly where no journal or thrust bearing is provided for the pump or the motor shafts, as is often the case with vertical pumps (Fig. 5.1). The motor bearing at the coupling end then takes over radial guidance of the pump shaft. The axial thrust is also transmitted to the motor shaft, and must be accommodated by the motor bearings.

Flexible couplings are employed in all cases where the drive and pump shafts are supported independently by journal and thrust bearings. These couplings must be capable of accommodating axial, radial and angular misalignments (Fig. 5.2).

For simple cases, i.e. low to medium speeds and power ratings, couplings with flexible plastic or rubber elements are often used (Fig. 5.3). They also possess a certain torsional elasticity and damping.

On pumps with higher speed and power ratings such as boiler feed pumps, gear, flexible membrane and diaphragm couplings are mainly employed (Figs 5.4 and 5.5).

Gear type couplings require lubrication, either by a permanent grease filling or by oil through continuous spray lubrication. The coupling type depends on a large number of factors and needs careful consideration.

It must be noted that all couplings transmit axial forces from one shaft to the other if there is relative axial displacement between the two.

Axial forces are transmitted:

1. By forces dependent on torque:
   a. With couplings having elastic elements of plastic or rubber (Fig. 5.3a and b), by the friction between these and the metallic mating parts.

Centrifugal Pump Handbook. DOI: 10.1016/B978-0-7506-8612-9.00005-X

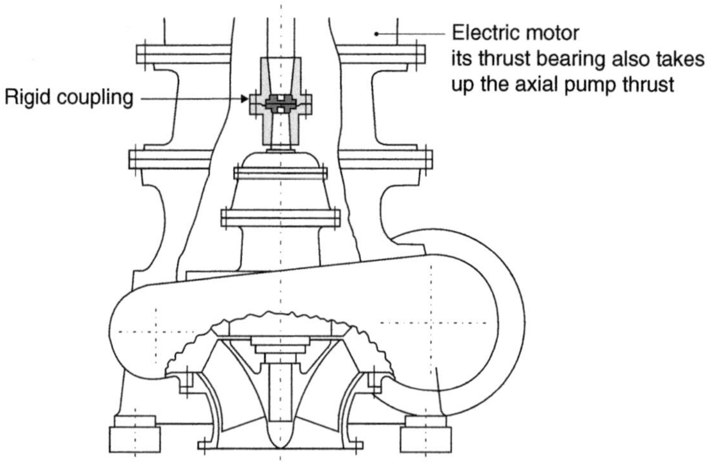

Figure 5.1 Type SPV mixed-flow cooling water pump with rigid shaft coupling

   b. With gear couplings, by the friction forces on the tooth flanks.
   c. With couplings as in Fig. 5.3d, the deformation of the toroidal transmission element sets up an axial force. Additional mutual displacement of the shafts alters this axial preload as a function of the axial stiffness of the transmission element.

2. By the axial stiffness of the transmission elements:
   a. With diaphragm couplings, by the stiffness of the diaphragm assemblies.
   b. On couplings as in Fig. 5.3c and e, by the axial stiffness of the annular or toroidal transmission elements.

Where there is misalignment as in Fig. 5.2b and c, radial forces also act on both shafts.

Couplings fitted with spacer sleeves are installed to allow for disassembly of the shaft seal of the pump, or in case of single-stage pumps to remove the pull-out unit, without removing the driver.

Elastic couplings with appropriate damping elements are used above all where cyclic torsional loading of the shafts due to heavily fluctuating

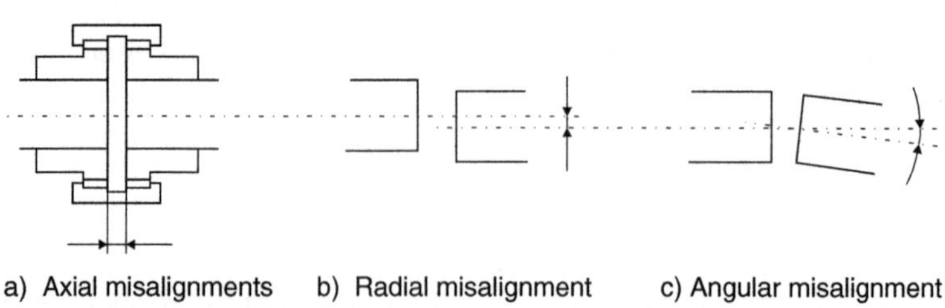

Figure 5.2 Misalignments

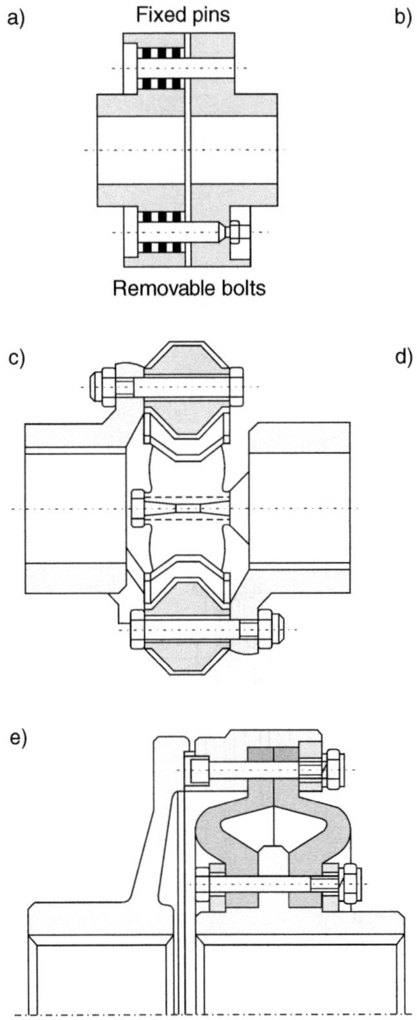

Figure 5.3 Coupling types with elastic elements of plastic or rubber (examples)

torque must be reduced. In such cases careful matching of the coupling characteristic is necessary, entailing calculation of the forced torsional vibration of the entire shaft string.

Owing to their mass, shaft couplings affect the rotor dynamic behavior. Careful selection of the coupling type and size is therefore extremely important. Coupling size is normally selected with the maximum torque at normal operation:

$$T_{B} = K^{P}/W$$

The service factor K is governed by the type of driver and is stated as an empirical value by the coupling supplier or international standards.

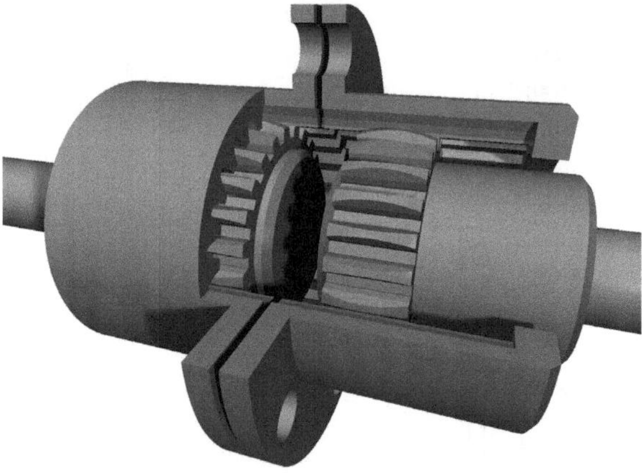

Figure 5.4 Typical gear coupling

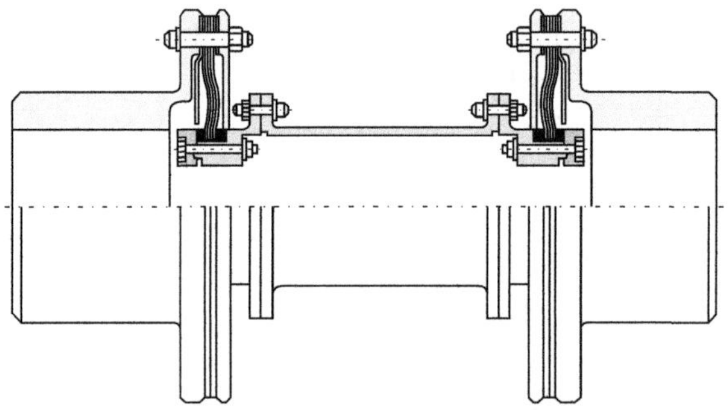

Figure 5.5 Typical diaphragm coupling

In critical cases the fluctuating operating torque and the maximum transient torque (e.g. during startup) are determined by calculating the forced torsional vibration of the shaft string. The choice of coupling is then based on these data.

Shaft alignment is vital to the reliability of the coupling. Accuracy is commensurate with the speed and power rating of the drive and with the type of coupling. In simple cases, alignment with rule and feeler gauge is sufficient (Fig. 5.6). With high-speed machines optical alignment instruments must be used to check the mating surfaces and centers for true radial and axial alignment (Fig. 5.7).

On pumps operating at elevated temperatures, allowance for horizontal and vertical misalignment is made due to differential thermal expansion of the coupled components (e.g. gear and pump). It is advisable to check alignment during commissioning, especially, again, for pumps with higher operating temperatures.

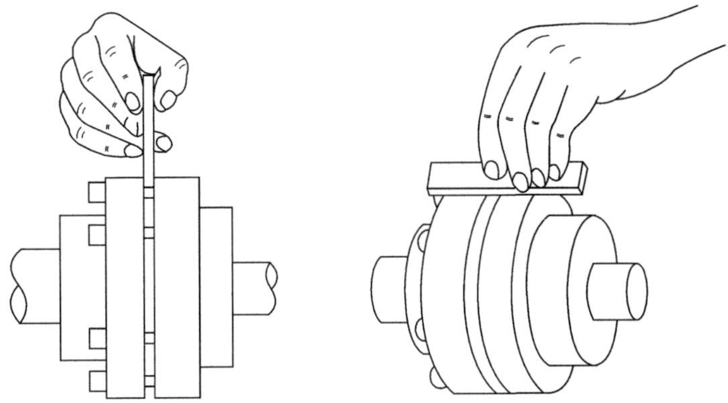

Figure 5.6 Aligning a coupling with feeler gauge and straight edge for simple applications

Figure 5.7 Coupling laser alignment equipment

5.2 BEARINGS

5.2.1 Rolling contact bearings

Where the bearing load and speed permit, as is mostly the case with standard pumps, rolling contact bearings of the ball or roller type are used frequently. Figure 5.8 shows the shaft bearings of a single-stage process pump. Such bearings provide both radial and axial guidance of the shaft. To accommodate larger axial forces, self-aligning roller thrust bearings are often used (Fig. 5.9). Lubrication is with oil or grease, depending on service conditions.

5.2.2 Sliding contact bearings

Sliding contact bearings are employed where the service conditions (e.g. speed, radial and axial loads) no longer allow the use of rolling contact

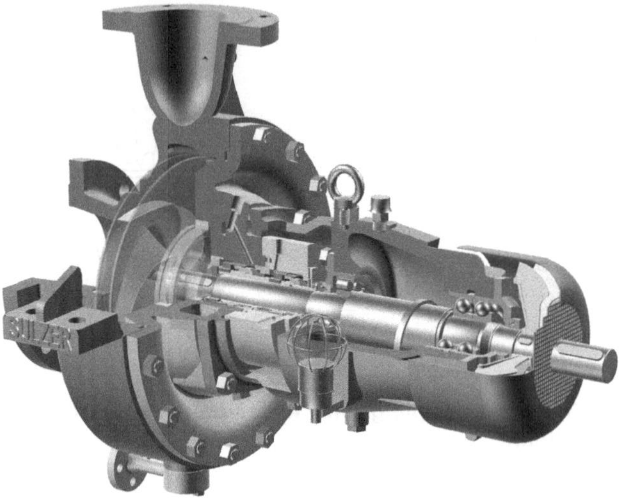

Figure 5.8 Shaft bearings of a single-stage process pump

bearings, and also wherever it is necessary to use the pumped medium to lubricate the bearings.

All these sliding contact bearings are hydrodynamic types, deriving their load-bearing capacity from the viscosity of the lubricant. In normal operation oil-lubricated bearings run with pure fluid friction, i.e.

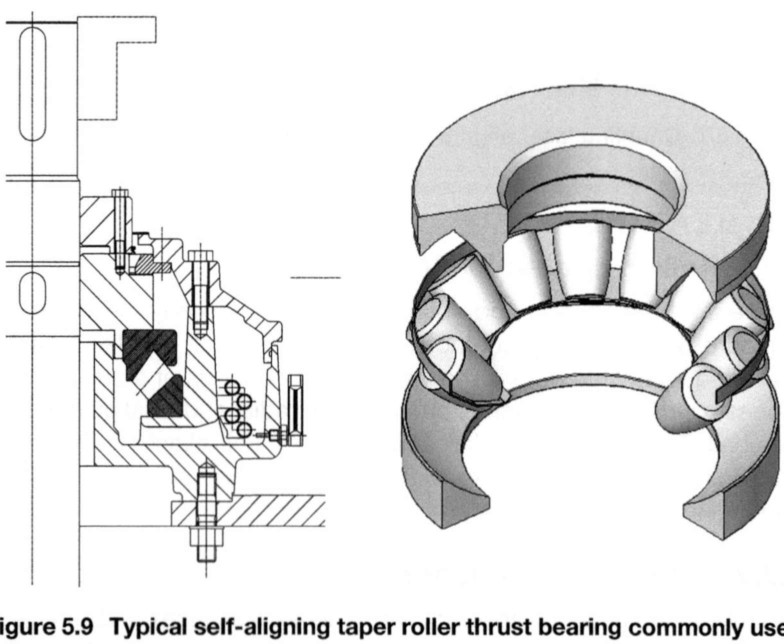

Figure 5.9 Typical self-aligning taper roller thrust bearing commonly used in vertical pumps

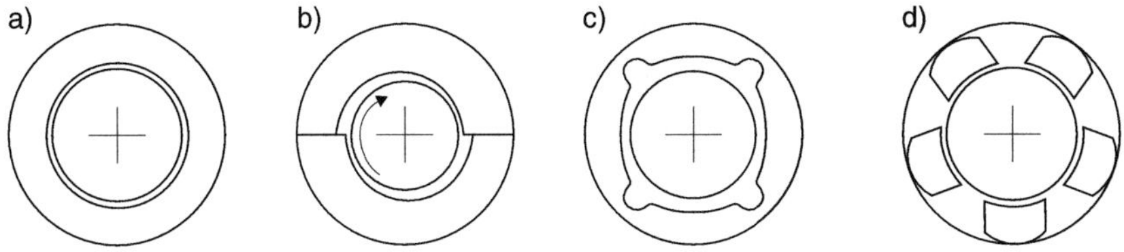

a) Cylindrical bearing
b) Offset halves bearing
c) Multilobe bearing
d) Tilting-pad bearing

Figure 5.10 Journal bearing types

shaft and bearing shell are completely separated by the oil film. Sliding contact bearings provide stiffness and damping to the rotor and therefore are important elements for the rotor dynamic behavior of the pump.

On the other hand, bearings lubricated with the pumped medium frequently operate in the mixed friction range, involving partial contact between the two surfaces. This calls for a judicious pairing of materials if seizure or excessive wear in service is to be avoided. The corrosive properties of the pumped medium must also be taken into account.

Oil-lubricated journal bearings are mainly employed in the following types (see Fig. 5.10):

1. *Cylindrical bearings* are the simplest type of journal bearings and are employed often in multistage pumps running at lower speeds and low to high bearing loads.
2. *Offset halves bearings* are cylindrical bearings with shifted bearing halves in the horizontal plane.
3. *Multilobe bearings* are typically used in high-speed pumps with low bearing loads.
4. *Tilting-pad bearings* are partial arc types. The pad can tilt on the pivot point to conform with the dynamic loads from the lubricant and shaft. These bearings are typically used in high-speed pumps with low bearing loads.

Journal bearings for lubrication by the pumped medium are made of rubber, carbon, metal (often bronze) or ceramics with several longitudinal or helical grooves (Fig. 5.11).

Hydrodynamic thrust bearings are mainly of the tilting-pad type (Fig. 5.12).

5.2.3 Hydrostatic radial and thrust bearings

In these bearings the load is borne by forcing lubricant into separate chambers of the bearing by means of a pressure source. Hitherto they have been used somewhat rarely in pump engineering.

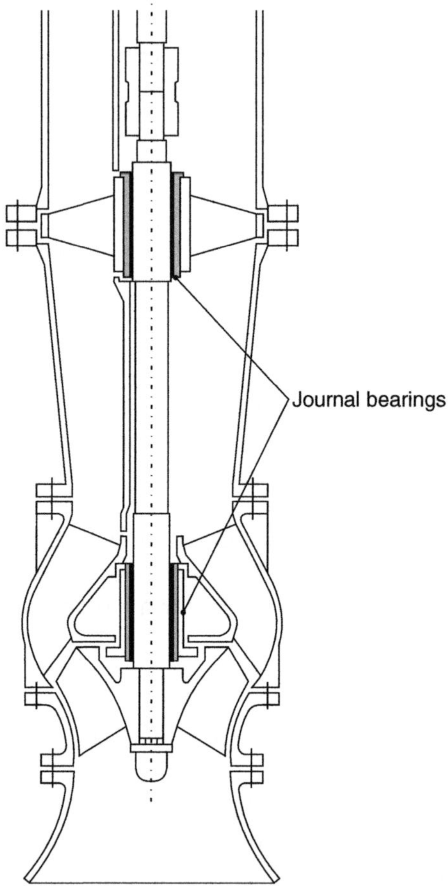

Figure 5.11 Typical (rubber) radial bearing lubricated by the pumped medium (right) or clean liquid via the shaft covering (left) on a vertical pump

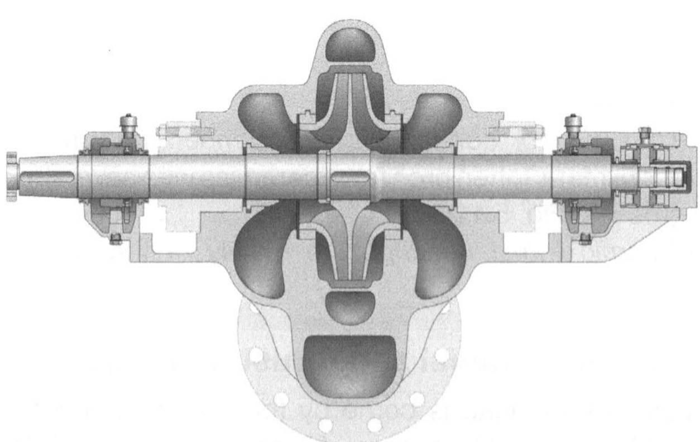

Figure 5.12 Sulzer HSB pipeline pump with tilting-pad thrust bearings

A combination of hydrostatic and hydrodynamic tilting-pad thrust bearings is employed if the bearing has to be started against a high static thrust. In this case the hydrostatic pressure source is used only during startup and rundown. During normal operation non-return valves prevent the lubricant from being forced back by hydrodynamically generated pressure.

5.3 SEALS

Seals serve to provide leakproof segregation – or if this is impossible or uneconomic for any reason, segregation with minimum leakage – between spaces at different pressures. In centrifugal pumps two kinds of seals are mainly employed.

1. Static seals provide tightness between two sealing faces which are motionless or perform only very limited axial or radial movements in relation to each other. Such movements may result, for example, from differential thermal expansion. Pressure-retaining parts are typically sealed with the following seal types depending on service conditions and design factors:
   a. O-rings
   b. Flat gaskets, often with wire mesh inlay
   c. Spiral graphite seals
   d. Metallic seals.

   Seals (b), (c) and (d) can seal only axially and require considerable preload to perform their function. This in turn affects the sizing of the connecting elements (e.g. screws).

   O-rings capable of sealing axially and radially require only small preload, but like U-packings they too can be employed within limited service temperatures only. Where absolute sealing is required, sealing welds are provided. In certain cases a barrier seal is sufficient. For this two single seals are fitted in series, and a sealing fluid is forced into the space between them at a pressure higher than that in the spaces to be sealed.
2. Shaft seals provide sealing between rotating and stationary faces or bores and seal from inside of the pump to the atmosphere. Shaft seals may be divided into non-contact and contacting seals.
   a. Non-contact shaft seals: the simplest kinds are throttle bush seals which form a small clearance between the stationary bush and the rotating sleeve. The clearances are kept as narrow as possible to minimize leakage.

      Reduction of the clearance and hence of the leak rate is possible by using floating ring seals. Casing and shaft deformation are taken up by radial displacement of the individual floating rings when starting, allowing the clearance to be matched to the maximum shaft vibration amplitudes.

b. Contacting shaft seals: the simplest type, long used in centrifugal pump design, is the *packed stuffing box* (Fig. 5.13). Where these are no longer adequate, on account of excessive pressure difference or circumferential speed or other reasons (e.g. high leakage rate), rotating *mechanical seals* are employed. Both single- (Fig. 5.14) and double-acting (Fig. 5.15) types are used in centrifugal pumps.

Mechanical seals have to be cooled and lubricated with clean liquid in order to achieve long seal life. Small leakage rates can be expected from mechanical seals.

Single mechanical seals are normally used if the pumped liquid is clean, does not crystallize when exposed to the atmosphere and if the pumped fluid is compatible with the environment. These seals are normally flushed with the pumped liquid. Coolers can be installed in the flushing piping to cool hot pumped liquid and cyclone separator to clean it. Jacket cooling located inboard of the mechanical seal is applied if the mechanical seal has to be protected from hot liquid during standstill.

Double mechanical seals are normally used if the condition for the single mechanical seals as explained above is not acceptable. Double mechanical seals need to be flushed with clean and cool barrier liquid that is compatible with the pumped liquid and the environment. The barrier liquid is provided from an external seal system at a higher pressure than the pressure to be sealed so as not to contaminate the barrier liquid.

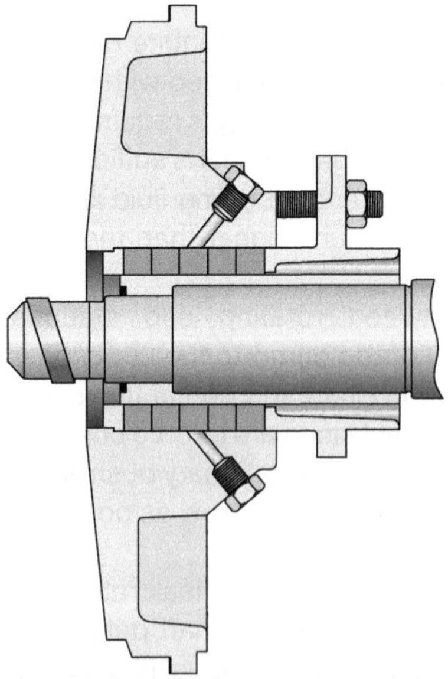

Figure 5.13a Standard packed stuffing box

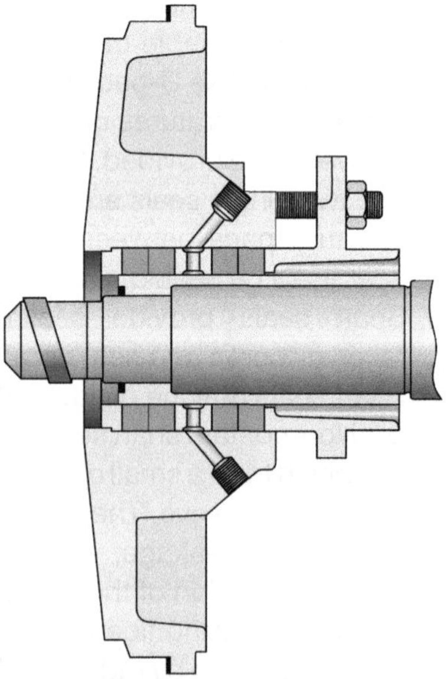

Figure 5.13b Packed stuffing box lantern ring flush

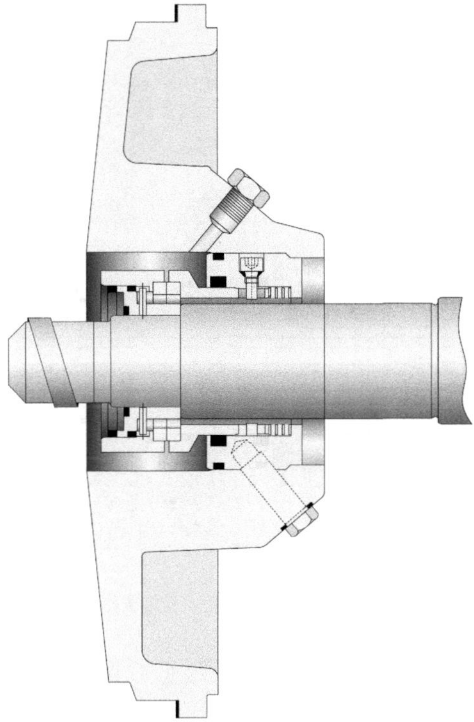

Figure 5.14 Single mechanical seal

5.4 PIPE FORCES

The pipes connected to a pump always impose forces and moments on the pump nozzles. Principal causes of these pipe loads are as follows:

- the deadweight of the pipe
- weight of the pumped liquid
- steady-state internal pressure
- pressure surges (e.g. generated by slamming of non-return valve flap or fast closing valves)
- pressure pulsations
- thermal expansion
- seismic forces
- constraints imposed by erection.

Steady-state or transient internal pressure always makes its full impact on the pump support when axial compensators or expansion sleeves are installed in the line (see Fig. 5.16).

A force amounting to $F = \pi D^2/4 \cdot p$ must be taken by the pump and pipe supports. Such arrangements should be avoided as a general rule unless the pressure p is very low.

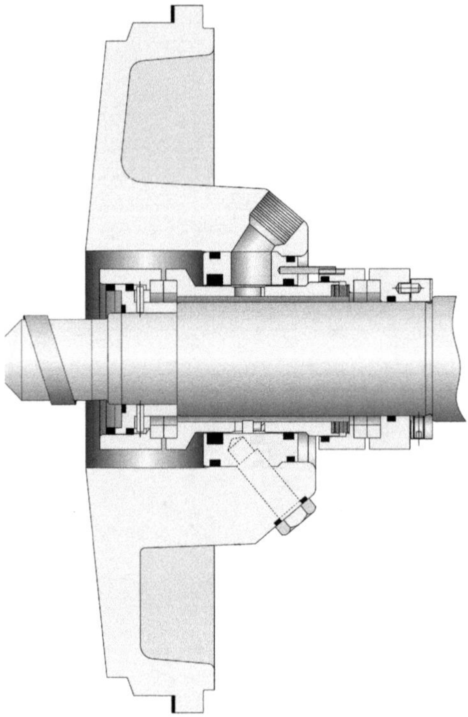

Figure 5.15 Double mechanical seal

Remedy is offered by jointed compensators or hydrostatically balanced expansion sleeves, unless preference is given to a sufficiently flexible pipeline layout.

The effect of pipe loads on a pump is shown in Fig. 5.17, taking a process pump as an example. *Stresses* in the force-transmitting parts *due to nozzle loads* may cause elastic or even permanent *deformation*. The critical zones may be seen from Fig. 5.17. Depending on the material employed and the sizing of the supports, limits are set for the pipe loads

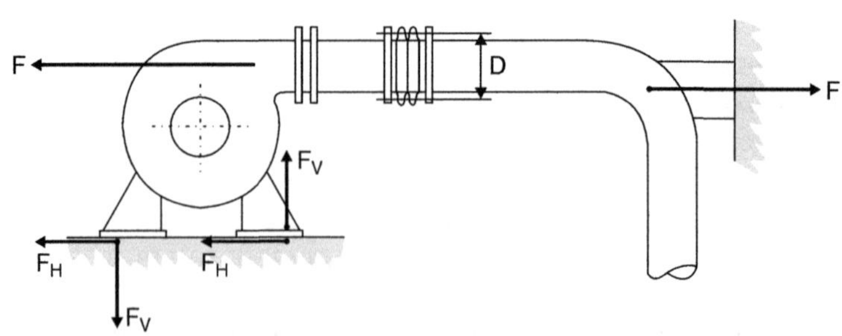

Figure 5.16 Forces to be taken up by the pump when using unbalanced compensators in the delivery line

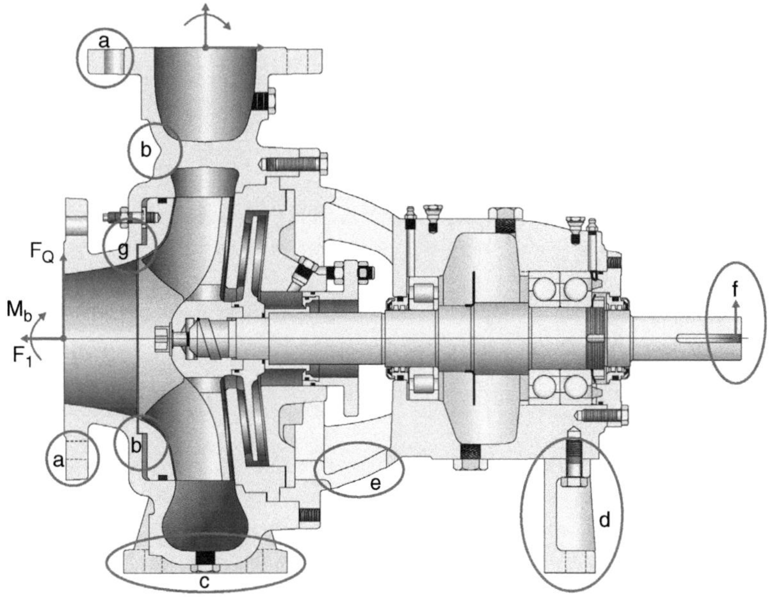

Figure 5.17 Effect of pipe loads on a pump

either by the stresses on the highly loaded parts or more often by the admissible coupling displacement.

Stress caused by nozzle loads can occur in:

a. nozzle flange
b. highly loaded areas
c. casing feet with hold-down bolts
d. support foot with hold-down bolts
e. bearing housing.

Deformation:

f. radial coupling displacement
g. radial impeller displacement (danger of contact).

International pump standards for process pumps provide guidance with respect to allowable forces and moments action on pump nozzles.

Chapter | six

Pipelines, Valves and Flanges

6.1 PRESSURE LOSSES

6.1.1 Pipeline networks

In order to determine the overall pressure losses in a pipeline network, the pressure losses of each individual pipeline are first determined. In the case of pipelines connected in series the losses for given flow rates and in the case of parallel pipelines the partial flow rates for given heads are added together. The diagrams shown in Figs 6.1 and 6.2 are a well-tried graphical method of determining these losses.

6.1.1.1 BRANCHED NETWORK, COMPRISING PIPELINES CONNECTED IN SERIES AND PARALLEL

Water flows into the system through pipeline 1, is divided at point 2, flows together again at point 7 and leaves the system through pipeline 8.

The individual pipelines are replaced for calculation purposes by an equivalent pipeline from point 2 to point 7. The pressure losses H_{v24} and H_{v45} are added together, as are H_{v23} and H_{v35}. These two pipelines are now combined in parallel, giving $H_{v245/235}$. The losses from points 5 to 7 are added, and the resultant losses are $H_{vR23457}$. The combined losses H_{v267} are then taken in parallel with $H_{vR23457}$ to give the total loss H_{v27} in the pipeline network from points 2 to 7.

In the pressure loss diagram (Fig. 6.3) the partial flows in each pipeline for a given flow rate Q can be read off by drawing a parallel to the abscissa.

6.1.2 Series and parallel operation of pumps

The individual characteristics of various elements of the pumping plant, such as pipelines, valves and pumps, can be combined into a resultant characteristic depending on whether they are connected in series or parallel. If, for example, the curves A, B and C are known for three centrifugal pumps and the resultant characteristic is required, the individual abscissae are added in the case of parallel connection, or the individual ordinates in the case of series connection (see Fig. 6.4).

Centrifugal Pump Handbook. DOI: 10.1016/B978-0-7506-8612-9.00006-1

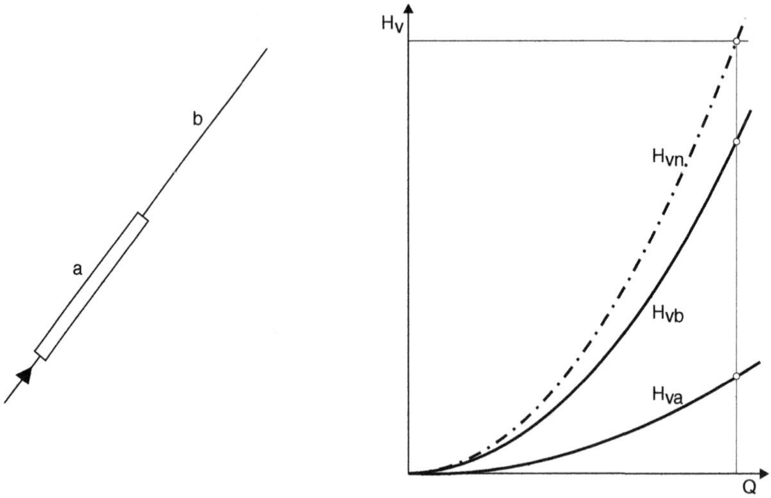

Figure 6.1 Series connection: $H_{vn} = H_{va} + H_{vb}$ for a defined flow rate Q

6.1.3 Establishing pressure loss curves

The pressure drop in a straight pipeline of diameter D and length L is expressed by the following equation:

$$H_v = \lambda \cdot \frac{L \cdot c^2}{D \cdot 2g}$$

Since $c = \frac{Q}{A}$ with $c^2 = \frac{Q^2}{A^2}$, the above equation can be extended as follows:

$$H_v = \lambda \frac{L \cdot Q^2}{D \cdot A^2 \cdot 2g} = C \cdot Q = f(Q^2)$$

where C is a constant.

The equation $H_v = C \cdot Q^2$ can also be written in the form:

$$y = C \times X^2$$

which represents a parabola.

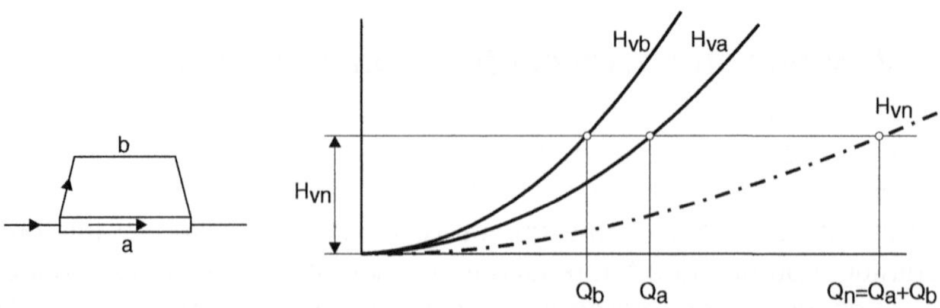

Figure 6.2 Parallel connection: $Q_n = Q_a + Q_b$ for a defined head loss H_{vn}

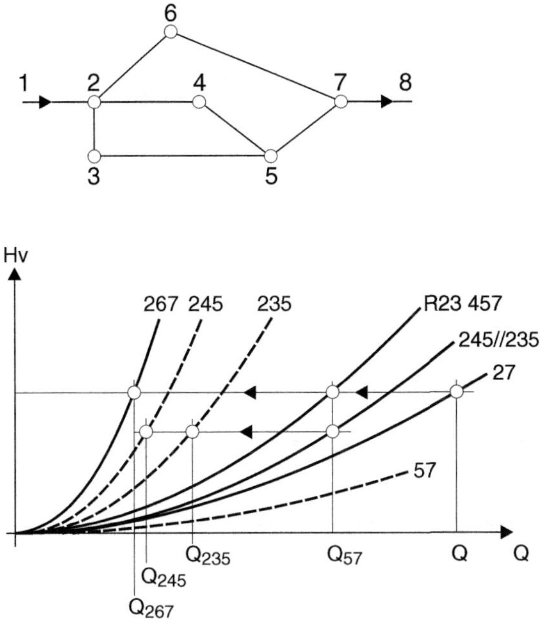

Figure 6.3 Branched network

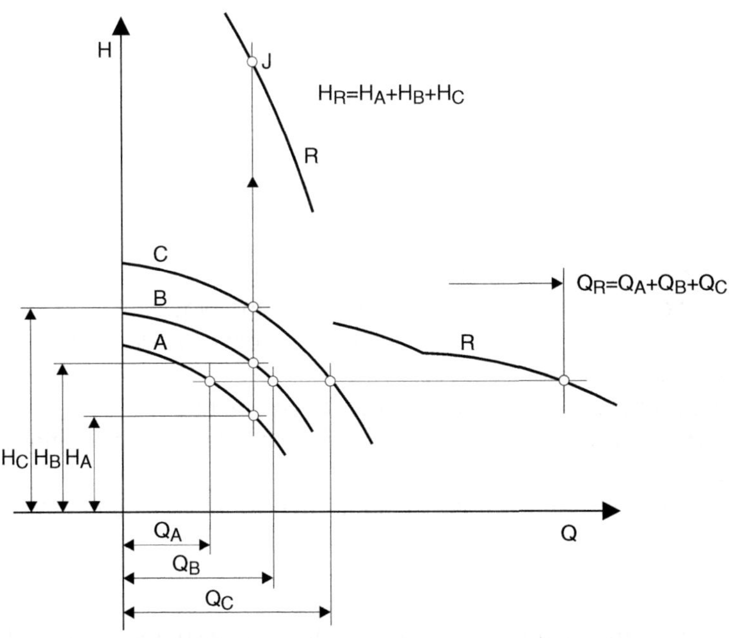

Figure 6.4 Series and parallel operation of pumps

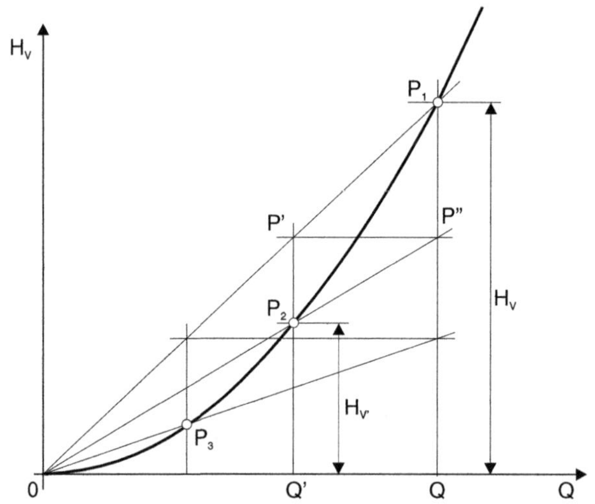

Figure 6.5 Establishing the pressure loss curve

6.1.3.1 ESTABLISHING PRESSURE LOSS CURVES

The pressure loss H_v is known for a given medium flow rate Q, designated by point P_1.

If any point P′ on the straight line OP_1 with abscissa Q' is projected onto the given ordinate H_v, and P″ is connected with the origin 0, the line OP″ cuts the ordinate of point P′ at the point P_2 on the parabola and thus represents the new pressure loss H_v' at the new medium flow rate Q'. The other points on the parabola are determined in the same way (see Fig. 6.5).

6.1.4 Pipelines in general

Pipelines are used for the conveyance and distribution of fluid, gaseous and solid materials.

The following main points should be observed:

1. selection of the most suitable material for the pipe walls and seals;
2. correct sizing of the pipe cross-section, avoiding sudden changes in flow velocity;
3. correct choice and arrangement of anchor points;
4. good accessibility to all parts of the pipeline to facilitate operation and replacement;
5. compensation for changes in length due to alterations in temperature.

6.1.4.1 SELECTION OF THE PIPE INSIDE DIAMETER

For a given mean flow velocity c, which is assumed as constant over the entire cross-section A, the pipe inside diameter can be determined as follows:

$$d_i = \sqrt{\frac{Q}{900 \cdot c \cdot \pi}} \text{ (m)}$$

Q = flow rate (m^3/h)
c = mean velocity (m/s)

6.1.4.2 APPROXIMATE VALUES FOR FLOW VELOCITIES IN PIPELINES UPSTREAM AND DOWNSTREAM OF THE PUMP

Suction pipelines

Generally: $c = 1$ to 2 m/s.
The permissible flow velocity c_{per}:

- rises with increasing flow;
- depends on the permissible head loss H_{vs} in the pipeline.

For pumps with positive suction head:

$$\frac{P_{c_{ans}} - P_D}{p \cdot g} + Z_c - Z_s + \frac{c_e^2}{2g} - H_{vs} \geq \text{NPSH}_{req}$$

(see section 1.5.3, equation (16) and Fig. 1.2) with intake water level at atmospheric pressure and negligible c_e.

For pumps with negative suction head:

$$\frac{P_b - P_D}{p \cdot g} - (Z_e - Z_s) - H_{vs} \geq \text{NPSH}_{req}$$

In general the suction pipeline diameter should never be smaller than the pump intake nozzle.

Inlet pipelines

Generally: $c = 1.5$ to 2.5 m/s.

- Same remarks as for suction pipelines;
- Additionally in the following cases: booster pumps for boiler feed pumps and heater drain pumps; the flow velocity still depends on the given rate of pressure drop in the intake vessel (drop in suction head in the case of sudden and drastic load fluctuations), and can rise to 5 m/s.

Delivery pipelines

Generally: $c = 1.5$ to 3 m/s.

- Particularly in the case of long pipelines the most economical value of c, i.e. of the nominal diameter, can be determined in each individual case by means of only one calculation involving:
  - capital investment cost of the pipeline (interest and depreciation);
  - annual operating costs (power consumption of the pump);
  - water hammer conditions.

6.1.4.3 CALCULATING PRESSURE LOSSES IN STRAIGHT PIPELINES

When conveying liquids various kinds of resistance in the pipeline must be overcome, and these depend on the pipe diameter selected. To determine the pressure losses it must be known whether the flow through the pipeline is laminar or turbulent. The boundary between laminar and turbulent flow is given by the critical Reynolds number $R_e = 2320$.

Pipe friction loss coefficients λ

The pressure or head loss in a straight pipeline of internal diameter d_i and length L is defined as follows:

Laminar flow	Turbulent flow		
	Hydraulically smooth pipes	Hydraulically rough pipes	Pipes in transient range
	Limits: $R_e \cdot \frac{k}{d_i} < 65$	Limits: $R_e \cdot \frac{k}{d_i} > 1300$	Limits: $65 < R_e \cdot \frac{k}{d_i} < 1300$
Formula for λ $\lambda = \frac{64}{R_e}$ $\lambda = \frac{64 \cdot \nu}{c \cdot d_i}$	Formula for λ: (a) Blasius formula for the range $2320 < R_e < 10^5$ $\lambda = 0.3164 \cdot R_e^{-0.25}$ (b) Nikuradse formula for the range $10^5 < R_e < 5 \cdot 10^6$ $\lambda = 0.0032 + 0.221 \cdot R_e^{-0.237}$ (c) Prandtl and von Kármán formula for the range $R_e > 10^6$ $\frac{1}{\sqrt{\lambda}} = 2\lg\left(R_e \cdot \sqrt{\lambda}\right) - 0.8$	Formula for λ: Nikuradse formula $\frac{1}{\sqrt{\lambda}} = 2\lg\frac{d_i}{k} + 1.14$	Formula for λ: Prândtl-Colebrook formula $\frac{1}{\sqrt{\lambda}} = 2\lg\left[\frac{2.51}{R_e\sqrt{\lambda}} + \frac{k}{d_i} \cdot 0.269\right]$

$$\Delta p = \lambda \cdot \frac{L}{d_i} \cdot c^2 \left(\mathrm{N/m^2}\right) \text{ or } \mathrm{H_v} = \lambda \cdot \frac{L}{d_i} \cdot \frac{c^2}{2g} (\mathrm{m})$$

where:

Δp = pressure loss (N/m²)
H_v = head loss (m)
λ = pipe friction loss coefficient (–)
L = pipe length (m)
d_i = inside diameter of pipe (m)
ρ = density (kg/m²)
g = acceleration due to gravity (m/s²)
c = mean flow velocity (m/s)
k = pipe roughness value (mm)

Table 6.1 Kinematic Viscosity ν of Water as a Function of Pressure and Temperature in 10^{-6} m^2/s

Pressure	Temperature in °C													
bar	0	20	50	100	150	200	250	300	350	400	450	500	600	700
1	1.75	1.00	0.551	20.5	27.4	35.2	43.8	53.4	64.0	75.4	88.0	101.0	131.0	164.0
10	1.75	1.00	0.550	0.291	0.197	3.26	4.20	5.22	6.30	7.48	8.75	10.1	13.1	16.4
50	1.75	1.00	0.550	0.292	0.198	0.156	0.134	0.909	1.18	1.45	1.70	2.02	2.59	3.27
100	1.74	0.998	0.549	0.292	0.198	0.156	0.135	0.126	0.529	0.681	0.821	0.967	1.28	1.63
150	1.73	0.995	0.549	0.292	0.199	0.157	0.136	0.126	0.292	0.421	0.526	0.630	0.846	1.08
200	1.72	0.992	0.548	0.293	0.199	0.157	0.136	0.127	0.122	0.285	0.376	0.459	0.629	0.811
250	1.72	0.990	0.548	0.293	0.201	0.158	0.136	0.127	0.121	0.193	0.284	0.357	0.499	0.647
300	1.72	0.987	0.547	0.293	0.202	0.159	0.137	0.127	0.122	0.128	0.215	0.284	0.408	0.535
350	1.70	0.984	0.547	0.294	0.202	0.160	0.138	0.128	0.122	0.121	0.180	0.242	0.351	0.462
400	1.70	0.981	0.545	0.294	0.203	0.160	0.139	0.128	0.122	0.120	0.152	0.207	0.306	0.406
450	1.69	0.978	0.545	0.294	0.203	0.161	0.139	0.129	0.122	0.120	0.137	0.182	0.271	0.361
500	1.68	0.977	0.544	0.295	0.204	0.162	0.140	0.130	0.122	0.120	0.130	0.164	0.245	0.327

(From VDI Heat Manual, 2nd edition, 1974.)

The dimensionless pipe friction loss coefficient λ is a function of the Reynolds number and, depending on flow velocity, of the surface roughness of the pipe walls:

$$R_e = \frac{c \cdot D_H}{\nu}$$

where:
D_H = hydraulic diameter (m)
ν = kinematic viscosity (m^2/s)

$$D_H = \frac{4 \cdot A}{U} = \frac{4 \cdot \text{cross-sectional area}}{\text{wetted perimeter}}$$

for circular-section pipes: $D_H \triangleq d_i$.

In summary:

For *laminar flow*, i.e. for $R_e < 2320$, independent of the interior pipe wall roughness:

$$\lambda = \frac{64}{R_e}$$

i.e. dependent on R_e only.

For *turbulent flow*, i.e. for $2320 < R_e < 10^5$:

- smooth pipes: corresponding to the three formulas (depending on the R_e range in the above table), λ is dependent on R_e only.
- rough pipes:

$$\frac{1}{\sqrt{\lambda}} = 2 \lg\left(\frac{d_i}{k}\right) + 1.14$$

- i.e. dependent only on roughness and the pipe diameter.
- transitional range:

$$\frac{1}{\sqrt{\lambda}} = -2 \lg\left[\frac{2.51}{R_e \cdot \sqrt{\lambda}} + \frac{k}{d_i} \cdot 0.269\right]$$

- i.e. in this range the frictional loss coefficient λ is influenced by both the Reynolds number R_e and the roughness k/d_i (see Fig. 6.6).

  Roughness values k (mm) for various materials:
  Smooth pipes of plastic, glass, copper, brass; drawn, extruded, ground finished – up to 0.002
  Seamless steel pipes, asbestos cement pipes – up to 0.05
  Welded steel pipes, new – 0.05–0.1
  Welded steel pipes, corroded – 0.15–0.2
  Spun concrete pipes, stoneware pipes, new cast-iron pipes – 0.2–0.2.

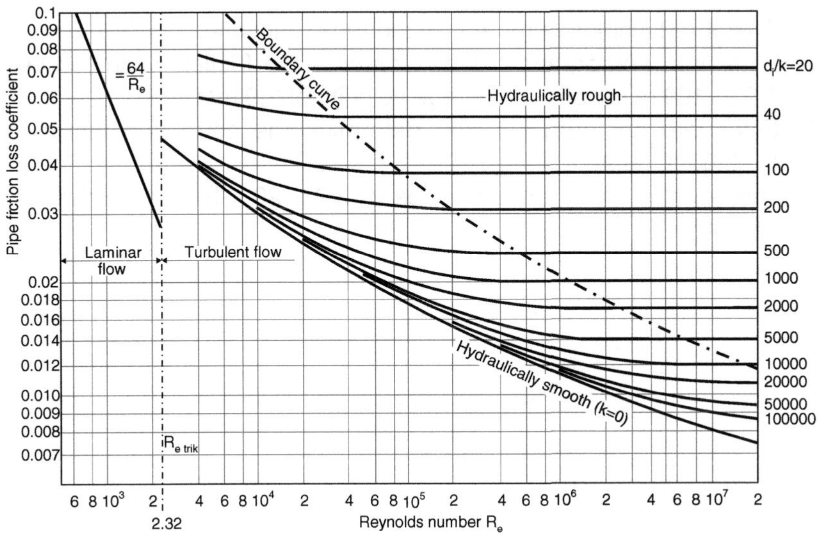

Figure 6.6 Dependence of pipe friction loss coefficient λ on R_e

The pipe friction losses, also called as head losses, H_v at turbulent flow for 2320 $< R_e < 10^5$

The calculation of λ using the Prandtl–Colebrook formula is time consuming. Therefore, it is more common to use a simplified diagram such as nomograms 1 and 2 to derive the head losses for a straight pipe.

Nomogram 2 has been developed for the following applications:

1. Pipe surface roughness $k = 0.1$ mm at pipe wetted surface (e.g. new cast pipe with bituminized lining).
2. Clean water at 12°C with a viscosity ν of 1.236 mm²/s. The nomogram can be applied for similar liquids of nearly identical viscosity and will lead to adequate results.
3. Length of pipe taken as 100 m.

In case the absolute surface roughness k of the used pipe will deviate from the $k = 0.1$ mm please take nomogram 1 to determine the correction factor f for the multiplication:

$$H_v = H_{v(k=0.1)} \times f$$

Nomogram 2 for $H_{v(k=0.1)}$

Nomogram 1 for f (correction factor). The correction factor in nomogam 1 can be referenced for the increase of friction head losses during long operating time, with some effects of increasing roughness due to deposits.

Example

Clean water at $t = 20°C$ ($\nu = 1.0$ mm²/s); $Q = 3000$ m³/h. Asbestos cement pipe DN = 600 mm; length 2000 m; $k = 0.05$.

1. Check if Reynolds number is in the turbulent zone.
2. Enter into nomogram 2 and determine the friction head losses = 1.0 m/100 m (length) at average velocity of 2.9 m/s.
3. Enter into nomogram 1 and determine the correction factor for $k = 0.05$; $f = 0.92$.

The result of friction head losses = 0.92 × 20 = 18.4 m for the total pipe length!

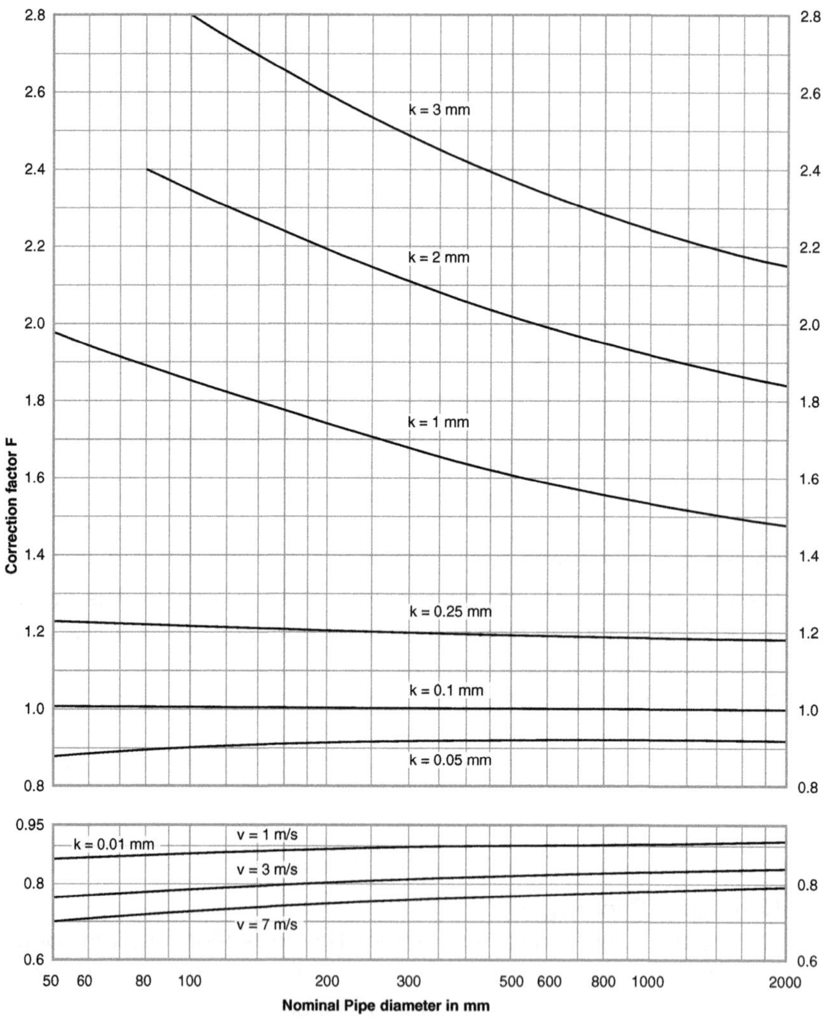

Nomogram 1: correction factors for roughness value $k \neq 1$

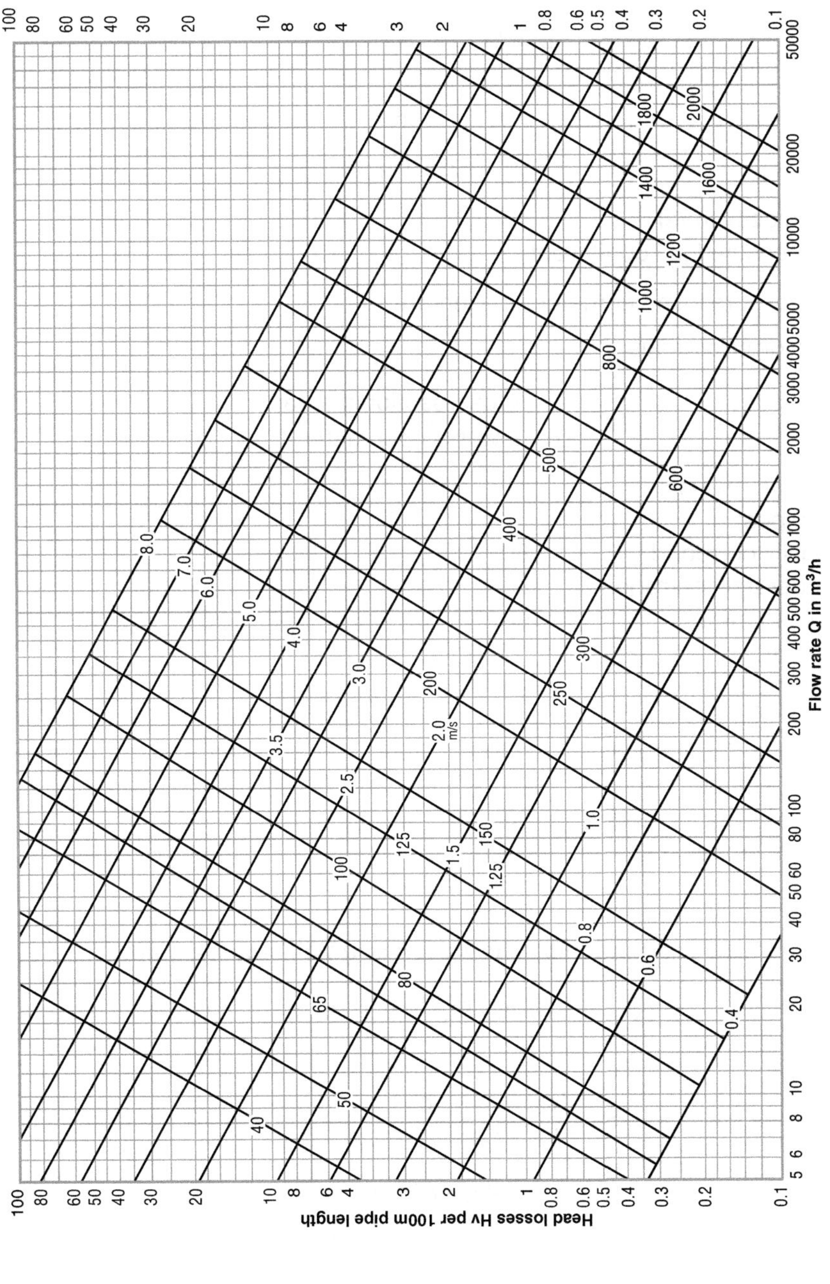

Nomogram 2: friction head losses per 100 m pipe length corresponding to Prandtl–Colebrook (k = 0.1 mm)

Other common methods of computing pressure losses

Pressure loss calculation – the Strickler method. This formula is often used today in structural engineering:

$$c = k_1 \cdot \sqrt{I} \cdot R_H^{2/3}$$

where:

$$I = \frac{H_v}{L} = \frac{c^2}{k_1^2 \cdot R_H^{4/3}}$$

c = velocity (m/s)
k_1 = pressure loss coefficient
I = pressure loss per meter pipe length = $\frac{H_v}{L}$
R_H = hydraulic radius

$$R_H = \frac{A}{U} = \frac{\text{cross-sectional area}}{\text{wetted perimeter}}$$

Strickler k_1 values:

Steel pipes:	
heavily corroded	60
moderately corroded	85
welded, new	95
smooth	≥100
Plastic pipes:	≥100
Pipes with asphalt or cement mortar surfacing:	≥100
Concrete pipes:	
cast in steel shuttering	90–100
cast in timber shuttering	65–70
Penstocks:	(55)–80–(95)

Pressure loss calculation by the Hazen and Williams method.

$$c = 1.319 \cdot C \cdot R_H^{0.63} \cdot S^{0.54} (\text{ft/s})$$

where:

$$S = \frac{H_v}{L} = \frac{c^{1.85}}{1.67 \cdot C^{1.85} \cdot R_H^{1.17}}$$

$$c = \text{velocity (ft/s)}$$

$$R_H = \text{hydraulic radius} = \frac{A}{U}\,\text{ft}$$

$$S = \frac{H_v}{L} = \text{ loss per unit length (ft)}$$

C = surface roughness value

Converted to the metric system:

$$c = \frac{Q}{A} = 0.3548 \cdot C \cdot D_H^{0.63} \cdot S^{0.54} (\mathrm{m/s})$$

$$S = \frac{H_v}{L} = \frac{c^{1.85}}{0.147 \cdot C^{1.85} \cdot D_H^{1.17}}$$

or

$$H_v = 10.64 \cdot L \cdot \left(\frac{Q}{C}\right)^{1.85} \cdot \frac{1}{d_1^{4.87}}$$

where:

H_v = friction losses in the pipeline (m)

L = pipeline length (m)

$D_H = \frac{4A}{U}$ = hydraulic diameter (m)

$D_H = d_i$, for circular-section pipes with full flow

Q = rate of flow through pipe (m^3/s)

C = Hazen and Williams factor

NB The symbols *S* and *C* have been retained to facilitate comparison with the original formula.

Hazen and Williams roughness values *C* for circular-section pipes:

Extremely smooth pipelines	140
Very smooth pipelines	130
Concrete pipes	120
New steel pipes (riveted) and tiled channels	110
Normal cast pipes, 10-year-old steel pipes and old masonry channels	100
Very rough pipelines	60

6.1.4.4 PRESSURE LOSSES IN BRANCH LOOPS (SEE TABLES 6.2–6.4)

Total flow rate = Q ζ_z = resistance coefficient for branch loop.

Flow rate after branching = Q_z ζ_d = resistance coefficient for straight section:

$$\Delta p = \xi \cdot \rho \cdot \frac{c^2}{2}$$

The above values are valid for equal diameters of main and branch pipes, and for sharp edges (as per *VDI Heat Manual*, 2nd edition, 1974).

Table 6.2 Branch Loops (Circular Cross-section)

	Q_Z/Q	0.0	0.2	0.4	0.6	0.8	1.0
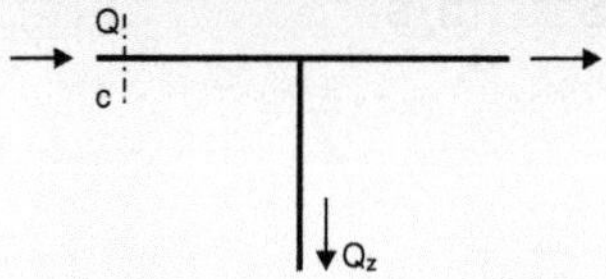	ζ_z	0.96	0.88	0.89	0.96	1.10	1.29
	ζ_d	0.05	−0.08	−0.04	0.07	0.21	0.35
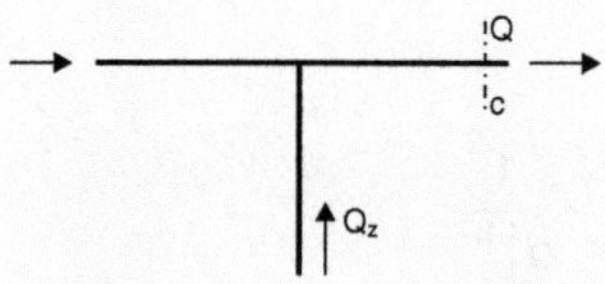	ζ_z	−1.04	−0.40	0.10	0.47	0.73	0.92
	ζ_d	0.06	0.18	0.30	0.40	0.50	0.60
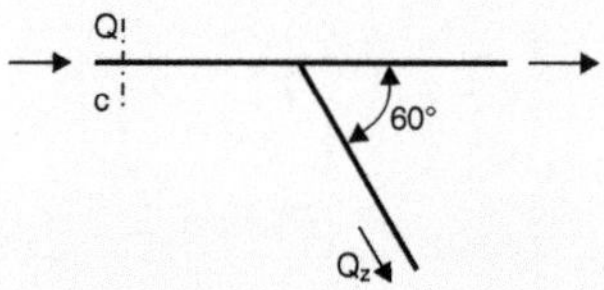	ζ_z	0.98	0.79	0.64	0.57	0.60	0.75
	ζ_d	0.05	−0.0.5	−0.02	0.07	0.20	0.34
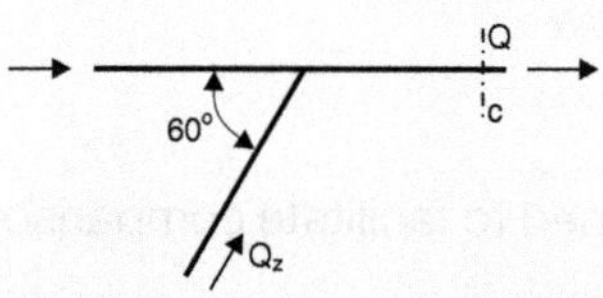	ζ_z	−0.92	−0.30	0.13	0.40	0.57	0.66
	ζ_d	0.04	0.24	0.30	0.25	0.10	−0.19
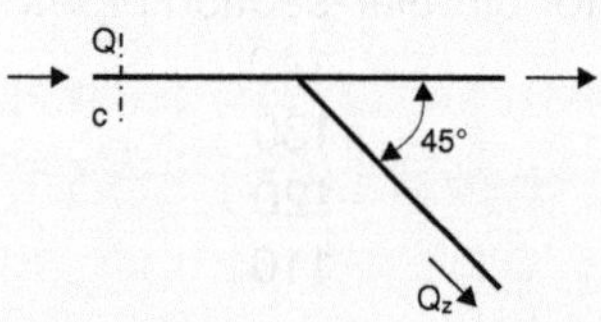	ζ_z	0.90	0.68	0.50	0.38	0.35	0.48
	ζ_d	0.04	−0.06	−0.04	0.07	0.20	0.33
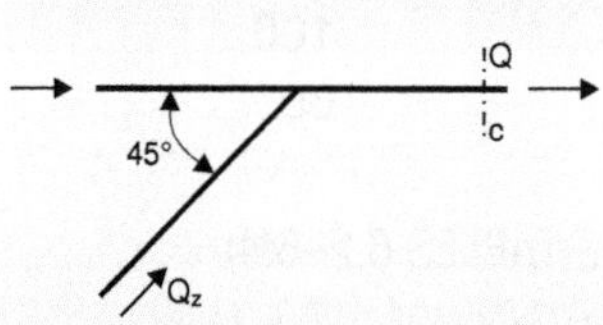	ζ_z	−0.90	−0.73	0.00	0.22	0.37	0.38
	ζ_d	0.05	0.18	0.19	0.06	−0.18	−0.54

6.1.4.5 PRESSURE LOSSES IN BENDS (SEE TABLE 6.5)

Pipe elbow assemblies have lower ζ values than for the sum of the individual elbows (Fig. 6.7). Figure 6.8 shows the values given by Schubart and Kirchbach as presented by Richter. These values clearly indicate minimum losses for defined relative distances *a*/*d* between the elbows.

Table 6.3 T-Pieces, 90° Angle Round or Square Cross-section, Sharp Edges. By rounding off the edges the ζ values are slightly reduced, by 10 to 30% if strongly rounded

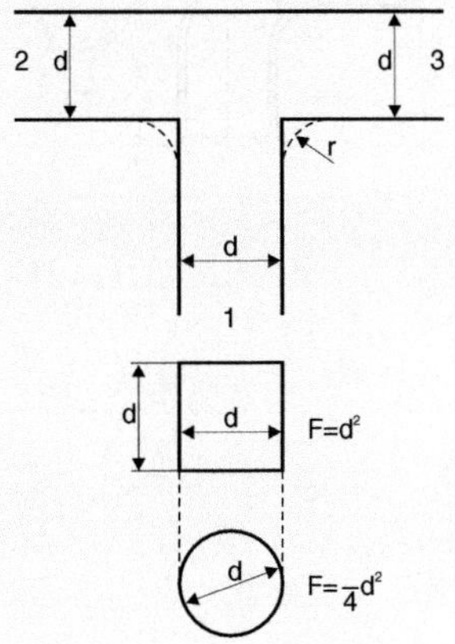

		ζ	
		$A_1 = A_2 = A_3$	$A_2 = A_3 = 0.5A_1$
	Flow direction from 2 to 3; 1 closed	0.50	
	1 to 3 and 1 to 2	1.00	1.90
	1 to 2; 3 closed	1.40	3.70
c b velocity of total flow			

Table 6.4 Y-Pieces Round or Square Cross-section, Sharp Edges

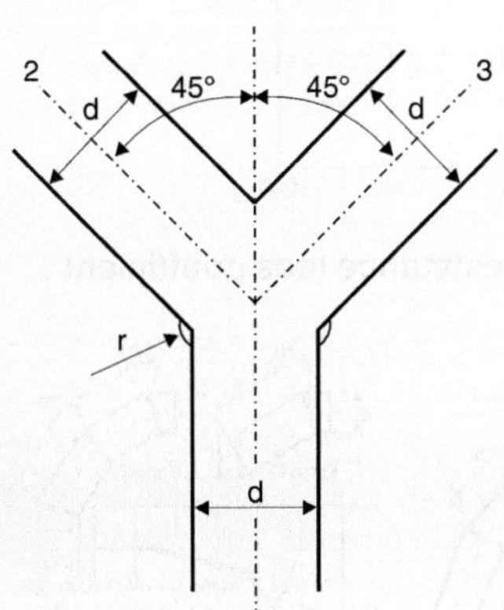

		ζ	
		$A_1 = A_2 = A_3$	$A_2 = A_3 = 0.5A_1$
	Flow direction from 1 to 2 and 1 to 3	0.55	0.75
	1 to 2; 3 closed	0.50	1.35
c at inlet of Y-piece			

Table 6.5 Pipe Bend, Resistance Loss Coefficient ζ

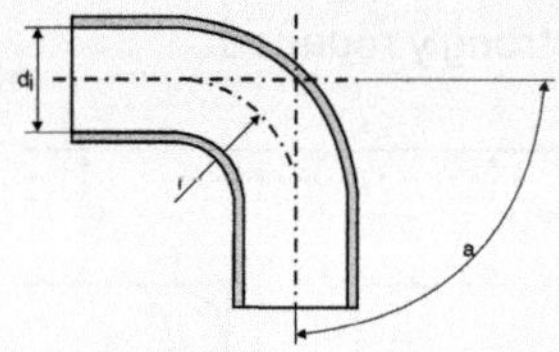

$\frac{r}{d_i}$ α	1	1.5	2	4	6	
15°	0.03	0.03	0.03	0.03	0.03	
30°	0.07	0.07	0.07	0.07	0.07	Interior
45°	0.14	0.11	0.09	0.08	0.075	pipe wall
60°	0.19	0.16	0.12	0.10	0.09	*smooth*
90°	0.21	0.18	0.14	0.11	0.09	
15°	0.10	0.08	0.06	0.05	0.04	
30°	0.23	0.19	0.14	0.11	0.08	Interior
45°	0.34	0.27	0.20	0.15	0.12	pipe wall
60°	0.41	0.33	0.24	0.19	0.15	*rough*
90°	0.51	0.41	0.30	0.23	0.18	

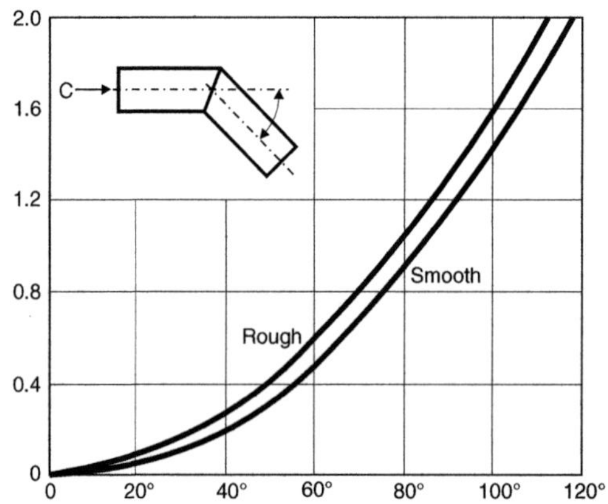

Figure 6.7 Pipe elbows for various angles, resistance loss coefficient ζ

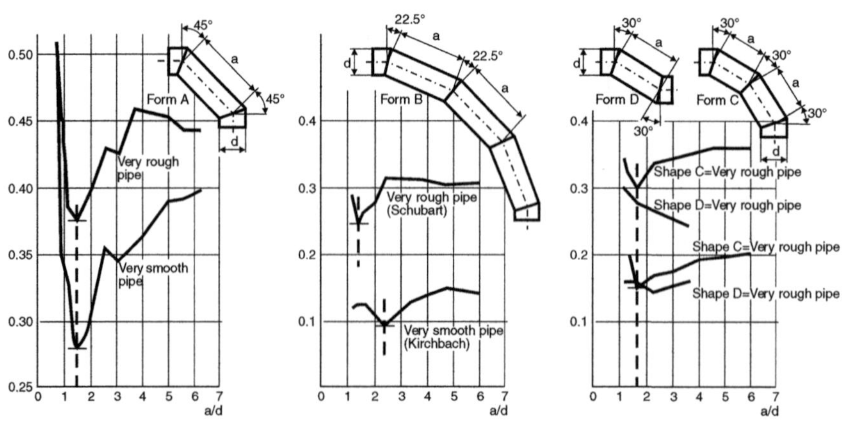

Figure 6.8 Pipe elbow assemblies, resistance loss coefficients ζ

6.1.4.6 PRESSURE LOSSES FOR PIPE INTAKES (SEE TABLE 6.6)

Table 6.6

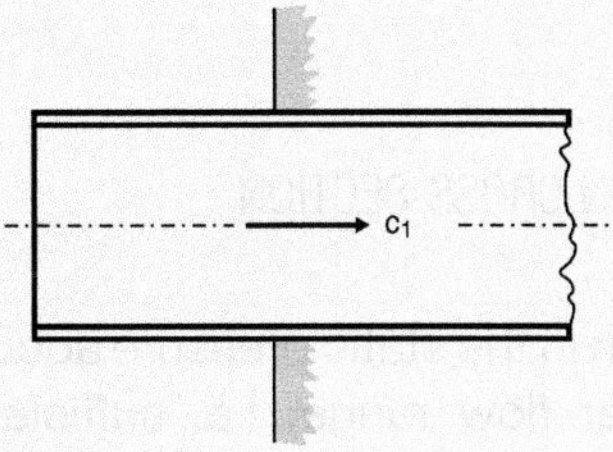

(a)

Simple pipe intake:

– projecting:

sharp-edged: 2.5 ÷ 3

broken: 0.6 ÷ 1

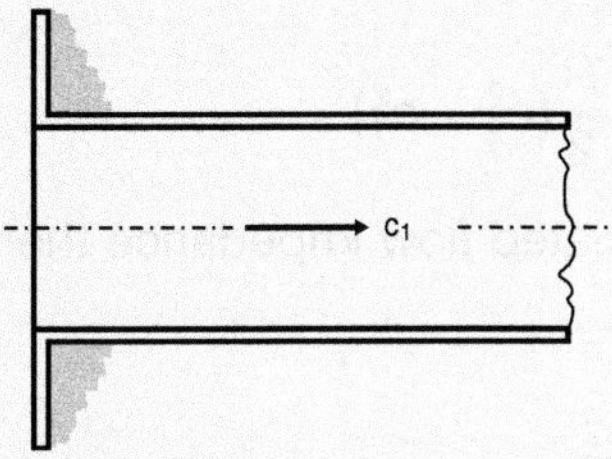

(b)

– even with wall:

sharp-edged: $\zeta = 0.5$

rounded: $\zeta = 0.25$

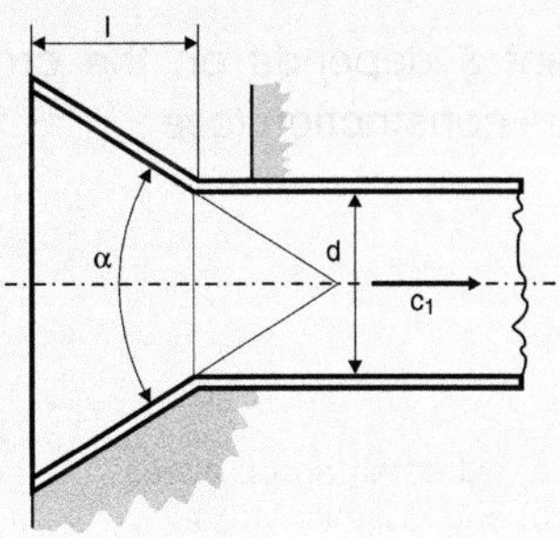

(c)

Enlarged pipe intakes:

– conical shape:

ζ depends on l/d and α

Optimum for $\alpha = 40°$ to $60°$

	Projecting		Even with wall	
l/d	0.1	1.0	0.1	0.6
α	60	40 ÷ 60	40 ÷ 60	40
ζ	0.4	0.1	0.2	0.1

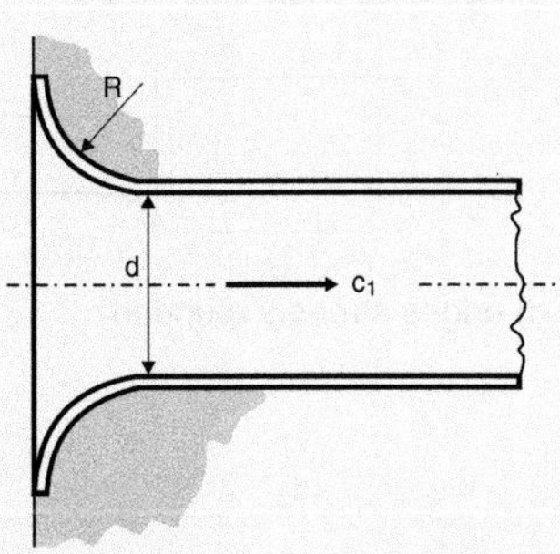

(d)

– rounded shape:

For $R/d \geq 0.2$

$\zeta = 0.03$

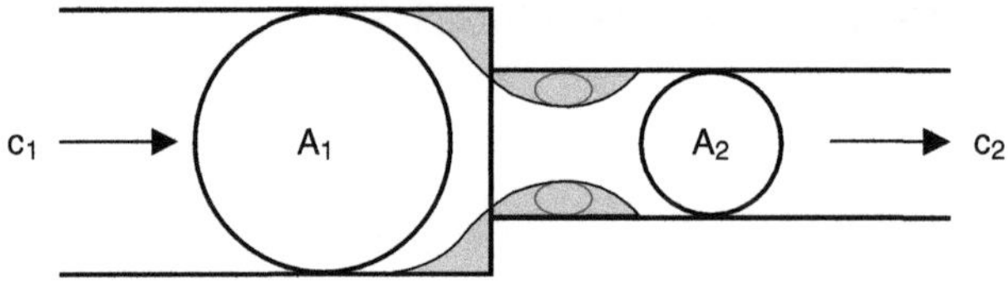

Figure 6.9 Constriction

6.1.4.7 PRESSURE LOSSES DUE TO CHANGES IN CROSS-SECTION

Constriction (see Fig. 6.9)

If the flow were free of losses, the alteration in the static pressure according to Bernoulli would be in the laminar flow range, i.e. sufficiently removed from the constriction:

$$\Delta P_0 = \frac{P}{2} \cdot c_2^2 - \frac{P}{2} \cdot c_1^2 = \frac{P}{2}\left(c_2^2 - c_1^2\right)$$

Due to the constriction and the associated flow impedance there is a pressure drop:

$$\Delta P = \xi \cdot \frac{P}{2} \cdot c_2^2$$

The value of the resistance coefficient ξ depends on the cross-section ratio A_2/A_1 and on the shape of the constriction (see Fig. 6.10).

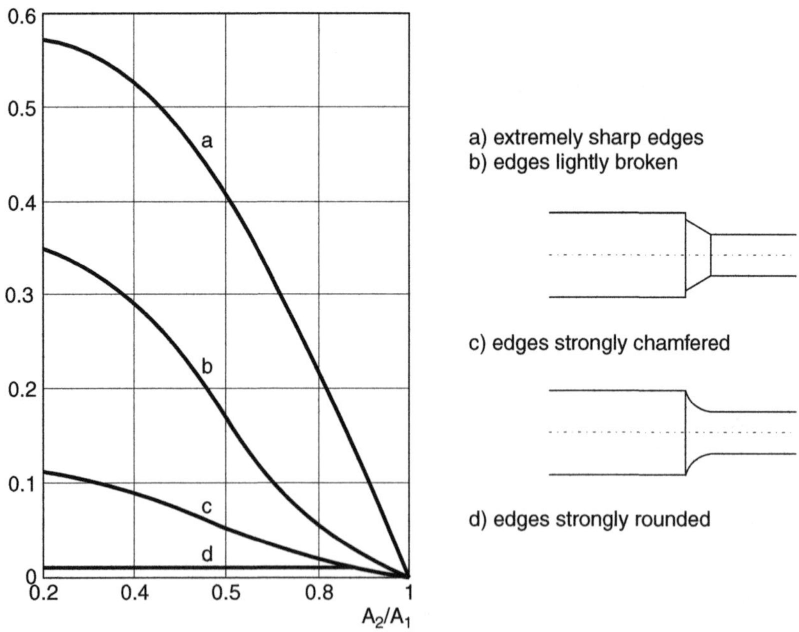

Figure 6.10 Resistance loss coefficient ζ for constrictions

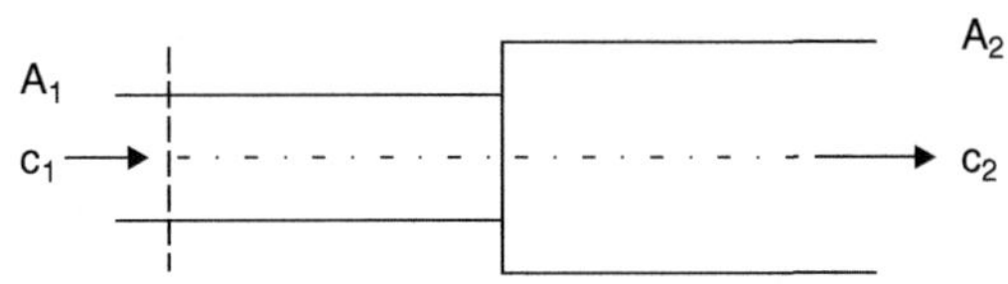

Figure 6.11 Diffuser with sudden divergence

Divergences of taper pipe

Diffuser with sudden divergence (see Fig. 6.11). The losses occurring due to a sudden divergence can be summarized by the theorem of momentum:

$$P_2 - P_1 = \Delta P_1 = \rho \cdot c_2^2 \cdot (c_1 - c_2)$$

If the flow were free of losses, according to Bernoulli:

$$\Delta P_0 = \frac{\rho}{2}\left(c_1^2 - c_2^2\right)$$

From these two relationships the efficiency factor can now be derived as:

$$\eta = \frac{\Delta P_1}{\Delta P_0}$$

and where $n = \dfrac{A_1}{A_2}$

$$\eta = \frac{2n}{1+n}$$

$\eta = \dfrac{A_1}{A_2}$	0.2	0.3	0.4	0.5	0.6	0.7	0.8	1.0
η	0.33	0.46	0.57	0.66	0.75	0.82	0.80	1.0

This means that for small divergences, i.e. for large values of n ($n \geq 0.6$), an acceptable efficiency is obtained even without using diffuser with continuous divergence.

The pressure losses are defined as follows:

$$\Delta\rho = \xi \cdot \frac{\rho}{2} \cdot c_1^2$$

where $\xi = \left(1 - \frac{A_1}{A_2}\right)^2$ (see Fig. 6.12).

The above considerations have led to the development of so-called multi-stage shock diffusers (Fig. 6.13).

Model tests have shown that if such multi-stage diffusers are properly designed (length of each stage about six times the diameter differential

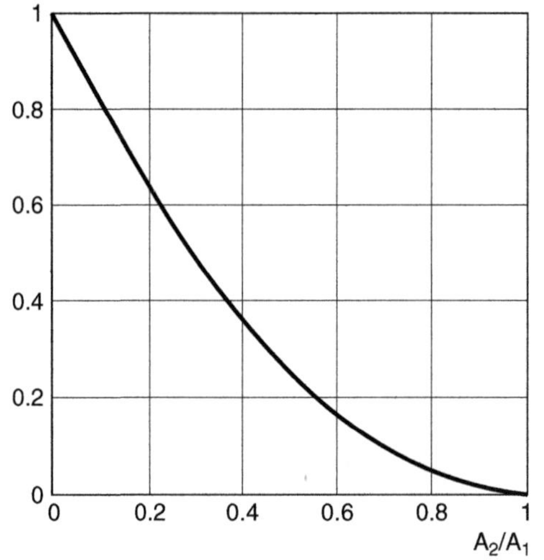

Figure 6.12 Resistance loss coefficient ζ for sudden divergence

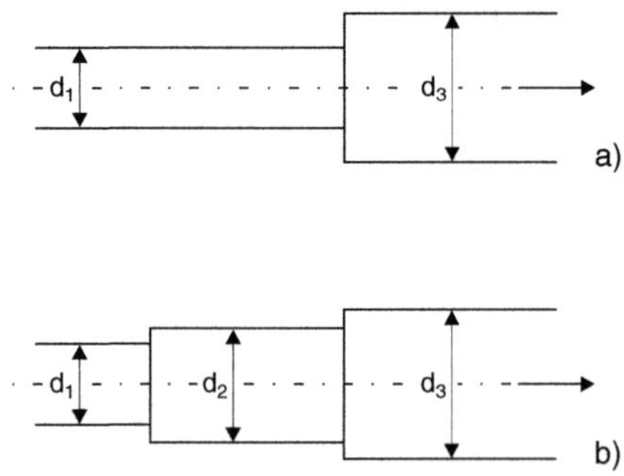

Figure 6.13 Single-stage (a) and two-stage (b) shock diffusers

and equal velocity reduction per stage), the pressure loss is only $1/x$ (for x stages) in comparison with a single-stage shock diffuser.

Diffuser with continuous divergence. Diffusers with continuous divergence (Fig. 6.14) should have a cone angle of $\alpha \leq 8°$. Larger angles

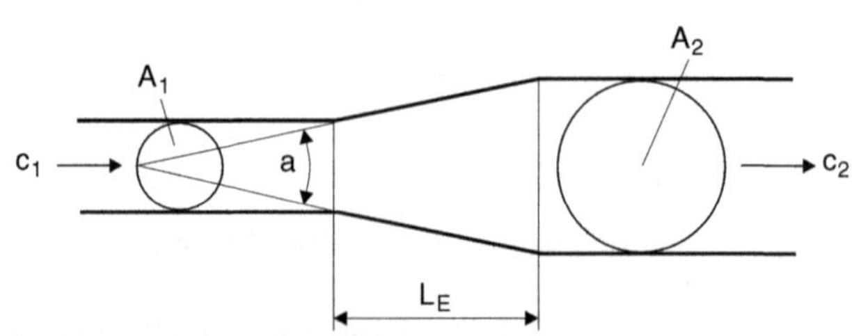

Figure 6.14 Continuous divergence

lead to flow discontinuities and increased pressure losses. The equivalent length L_E of these diffusers is identical to the geometrical length:

$$L_E = \frac{d_2 - d_1}{2} \cdot \cot \alpha/2$$

α	4°	5°	6°	7°	8°
$\frac{L_E}{d_2 - d_1}$	14.64	11.45	9.54	8.17	7.15

Pressure loss: $\Delta p = \xi \cdot \frac{\rho}{2} \cdot c_1^2$
Resistance loss coefficient ξ as a function of α:

A_1/A_2	0.25	0.3	0.4	0.5	0.6
ζ-values $\alpha = 8°$	0.06	0.05	0.04	0.03	0.02
$\alpha = 16°$	0.15	0.13	0.10	0.07	0.05
$\alpha = 24°$	0.26	0.23	0.17	0.12	0.08

Weld seams at the two ends of the enlargement should, if necessary, be taken into account by an additional resistance loss coefficient of $\zeta = 0.02$ to 0.05. Reference value here is always the local flow velocity.

Additional note:

Without any losses, the pressure rise in the diffusor according to Bernoulli would be:

$$\Delta P_0 = \rho/2 \cdot \left(c^2 - c_1^2\right)$$

However, the real pressure rise is:

$$\Delta p = p_2 - p_1$$

It follows that the efficiency is defined as:

$$\eta = \frac{P_2 - P_1}{\rho/2 \cdot \left(c_1^2 - c_2^2\right)} = \frac{P_2 - P_1}{\rho/2 \cdot c_1^2} \cdot \frac{1}{(1 - n^2)}$$

where $n = \frac{A_1}{A_2} = \frac{c_2}{c_1}$

$$P_2 - P_1 = \rho/2 \cdot c_1^2 \cdot \left(1 - n^2\right) \cdot n \text{ and with}$$

$$\left(1 - n^2\right) = \beta \quad P_2 - P_1 = \rho/2 \cdot c_1^2 \cdot \beta \cdot \eta$$

This means that for given values of c_1 and η the pressure rise depends on β: the above calculation shows that for enlargements of

$A_2/A_1 = 3 \div 4$, about 90% of the theoretically feasible figure (given infinite enlargement) is already attained. For this reason divergences of taper pipe should not be greater than $3 \div 4$, since in spite of additional expense only a slight increase in pressure rise can be achieved.

$n = \frac{A_1}{A_2}$	0.67	0.5	0.33	0.25	0.20	0.1	0
$\beta = 1 - n^2$	1.5	2	3	4	5	10	∞
$\beta = 1 - n^2$	0.55	0.75	0.89	0.937	0.96	0.99	1

6.1.4.8 PRESSURE LOSSES IN VALVES

The resistance loss coefficients ζ are approximate values. For exact calculations precise values must be obtained from the valve manufacturers.

- *Straightway valves* (valve completely open) (according to I.E. Idel'Chik)
- *Linear seat valves*
- *Inclined seat valves*

Cast valves, DN 25 up to 200	$\zeta = 2.5$
Wrought iron valves, DN 25 up to 50	$\zeta = 6.5$

- *Angle valves* (valve completely open), DN 25 to 200; $\zeta = 2.0$

DN	25	32	40	50	65	80	100	125 up to 200
ζ	1.7	1.4	1.2	1	0.9	0.8	0.7	0.6

- *Non-return valves* *Linear seat valves*, DN 25 to 200; $\zeta = 3.5$
  Inclined seat valves, DN 50 to 200; $\zeta = 2.0$
- *Foot valves* with suction strainer (according to *VAG valves*)
- *Foot valves* arranged in groups (according to *VAG valves*)

	DN	50 up to 80	100 up to 350
ζ-values	c = 1 m/s	4.1	3.0
	c = 2 m/s	3.0	2.25

- *Metallic seal valves, short* (according to Schilling valves)

DN	400	500	600	700	800	1000	1200
ζ-values	7.0	6.1	5.45	4.95	4.55	4.05	3.9

DN	50	100	150	200	300	400	500	600	800	1000
ζ-values	0.23	0.2	0.18	0.17	0.15	0.14	0.12	0.11	0.10	0.09

- *Metallic seal valves, medium length* (according to Schilling valves)

DN	50	100	150	200	300	400	500	600	800	1000
ζ-values	0.29	0.26	0.23	0.19	0.17	0.16	0.15	0.13	0.12	0.10

- *Metallic seal valves, long* (according to Schilling valves)

DFN	50	100	150	200	300	400	500	600	800	1000
ζ-values	0.32	0.29	0.26	0.23	0.20	0.19	0.18	0.16	0.15	0.13

- Shut-off valves with locking parts open (according to Erhard valves)

	DN	150	200	250	300	400	500	600	700	800	900	1000	1200	1400	1600	2000
ζ-values	PN 6	–	–	–	–	–	–	–	–	–	–	–	0.24	0.24	0.22	0.22
	PN 10	–	0.75	0.65	0.60	0.50	0.45	0.40	0.38	0.36	0.34	0.32	0.32	0.30	0.28	0.28
	PN 16	1.50	1.00	0.80	0.70	0.66	0.64	0.62	0.60	0.55	0.55	0.50	0.45	0.40	0.38	0.38
	PN 25	1.50	1.00	0.80	0.75	0.70	0.70	0.70	0.65	0.60	0.60	0.55	0.52	0.45	0.42	0.42
	PN 40	1.50	1.00	0.85	0.80	0.80	0.75	0.75	0.70	0.65	0.65	0.60	0.60	0.50	0.45	0.45

- Non-return valves without lever and weight (according to Erhard valves)

	DN	50	100	150	200	300	400	500	600	800	1000
ζ-values	$c = 1$ m/s	3.1	3.0	3.0	2.9	2.9	2.9	2.8	2.8	2.6	2.4
	$c = 2$ m/s	1.4	1.4	1.3	1.3	1.3	1.2	1.1	1.0	0.9	0.9
	$c = 3$ m/s	0.9	0.9	0.8	0.8	0.8	0.7	0.6	0.5	0.4	0.4
	$c = 1$ m/s	0.8	0.8	0.7	0.7	0.6	0.6	05	0.4	0.4	0.3

- Non-return valves with lever and weight.

No definite data can be given here on resistance loss coefficients, since these depend on the angle of the weight lever to the direction of low, on the size of the weight and on the length of the lever arm between the weight and the shaft. A rough approximation can be obtained for $c \leq 2.5$ m/s by multiplying the above ζ values (without lever and weight) by the factor 2.5.

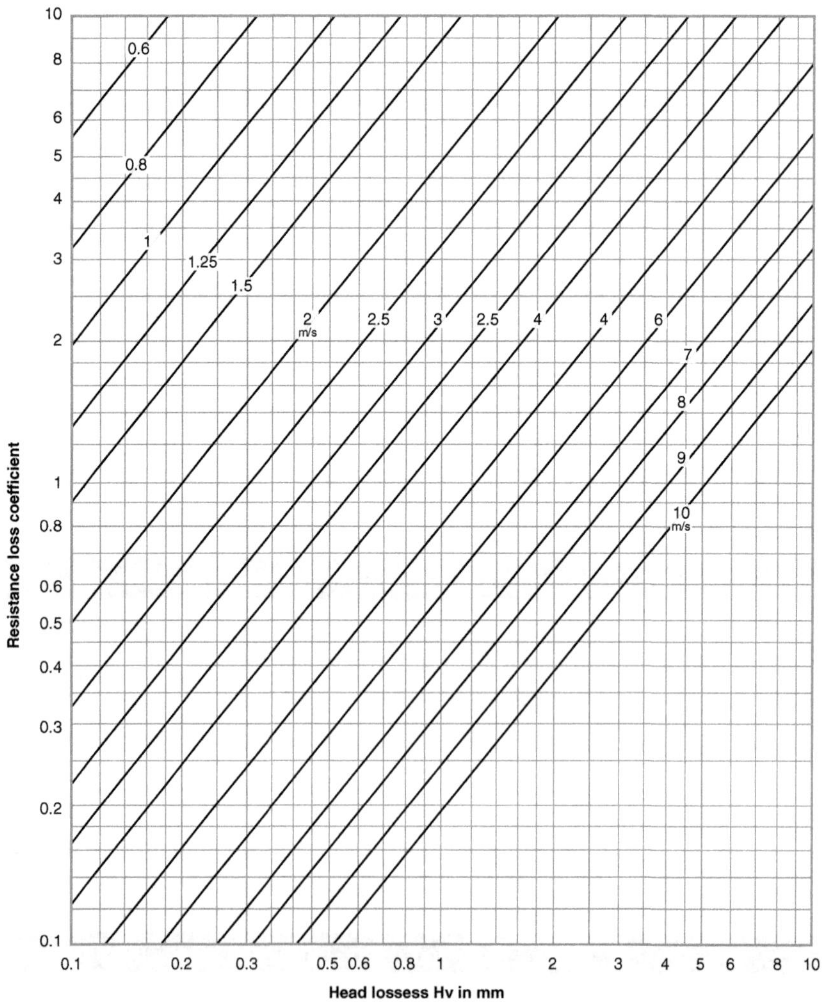

Figure 6.15 Nomogram 3 to determine the friction head losses of bends and valves

6.2 PIPE DIMENSIONS

For a table of pipe dimensions, see ISO 4200–1981 (E) standard.

6.3 BIBLIOGRAPHY

- Technische Strömungslehre, neu bearbeitete 8. Auflage, Springer Verlag, Bruno Eck
- Technische Strömungslehre, Vogel-Fachbuch, Willi Bohl
- Handbook of hydraulic resistance, coefficients of local resistance and of friction. AEC-TR-6630
- U.S. Department of Commerce, National Technical Information Service. By I.E. Idel'Chik 1960
- VDI-Wärmeatlas, 2. Auflage 1974

Chapter | seven

Centrifugal Pump Drives

7.1 ELECTRIC MOTORS

7.1.1 Characteristics of the various electric motors

The functioning of all electric machines is based on the physical laws of induction.

A knowledge of these physical laws and the design of the machines is assumed, so that the following text deals only with *motor characteristics relevant to the operation of pumps.*

7.1.2 Three-phase motors

7.1.2.1 GENERAL

A few guidelines are given below to assist in selecting the correct motor for a particular drive.

The information concerning the principal applications of asynchronous motors, i.e.:

- squirrel-cage motors
- slip-ring induction motors

and of synchronous motors can only be of a general nature, since in many cases a careful investigation will be unavoidable to meet the supplementary requirements specified.

These additional requirements are dependent on the following influencing factors:

- the working machine (a pump)
- the ambient conditions
- the electricity supply grid
- the regulations in force.

Frequently, these requirements involve oversizing or a special design instead of standardized motors.

7.1.2.2 APPLICATIONS FOR THREE-PHASE ASYNCHRONOUS SQUIRREL-CAGE MOTORS

Due to its robust construction, the squirrel-cage motor can be regarded as a general purpose type. It is suitable for a large number of working machines and is therefore utilized in all branches of industry.

Centrifugal Pump Handbook. DOI: 10.1016/B978-0-7506-8612-9.00007-3

In general, a high-speed motor is cheaper for a given torque value. When a gear train is used, the cost of motor plus gear train must be taken into account when determining the appropriate motor speed.

A wide range of special versions facilitate the matching of motors to their applications still further. They include:

- stepped speed designs (pole change motor)
- specially matched torque characteristics
- special types of construction and protection
- infinitely variable speed facility obtained by electronic control via converter-controlled voltage and frequency regulation.

7.1.2.3 APPLICATIONS FOR THREE-PHASE ASYNCHRONOUS MOTORS WITH SLIP-RING ROTORS

The slip-ring rotor type requires greater expenditure both for purchase and in operation (e.g. servicing of slip-rings and brushes). It is therefore used for those applications where the simple squirrel-cage motor is inadequate. Its advantages are to be found in the following properties:

- low starting current
- high starting torque
- easy speed changing (loss-free or dissipative)
- permits a high switching frequency.

7.1.2.4 APPLICATIONS FOR THREE-PHASE SYNCHRONOUS MOTORS

Unlike the asynchronous motor, the synchronous type requires no reactive power for its excitation. Depending on the field current value the synchronous motor can even deliver reactive power to the electricity supply grid.

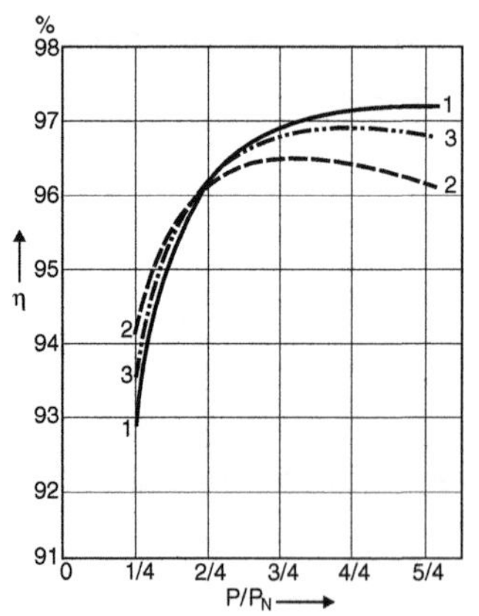

1 = Synchronous motor, 2500 kW, 6 kV, 1500 min^{-1}, cos $\varphi = 1.0$
2 = Synchronous motor, 2000 kW, 6 kV, 1500 min^{-1}, cos $\varphi = 0.8$ over-excited.
3 = Squirrel cage motor, 2500 kW, 6 kV, 1500 min^{-1}.

Figure 7.1 Approximate efficiency curves of electric motors as a function of load P/P_N

Its range of application covers drives of medium and high power, at continuous constant-speed operation.

7.1.3 Boundary conditions affecting motor sizing

Selection of a motor is dependent on the power requirement, speed and drive conditions of the centrifugal pump. Motor size is also influenced by the type of current, the voltage and the frequency of the electricity supply grid.

The following additional information is necessary to assure appropriate sizing:

- the requirements imposed by the centrifugal pump
- the ambient conditions
- the grid connection data
- any special regulations in force.

7.1.3.1 REQUIREMENTS IMPOSED BY THE CENTRIFUGAL PUMP ON THE MOTOR

1. Type of centrifugal pump
2. Power requirement or power demand diagram (as a function of flow rate, speed or time)
3. Function
   Working cycle (switching frequency)
   Gear
4. Speed
   Speed adjustment range
   Direction of rotation
5. Startup with or without load
   Load moment curve
   Mass moment of inertia
6. Control requirements
   Control magnitude
   Accuracy
7. Any mechanical or electrical braking
   Breaking torque curve
8. Installation
   Fastening, foundation
   Physical configuration
   Transmitting torque to centrifugal pump
   Additional axial and radial forces incident on the end of the motor shaft
   Dimensions of shaft end.

7.1.3.2 AMBIENT CONDITIONS

1. Ambient temperature
   Temperature of cooling medium (air, water)
   Contamination of air-acid vapors, moisture, dust
   Analysis of water
2. Special conditions
   Explosive gas mixtures
   Fire-damp (methane)
   Installation in open air, etc.

3. Altitude of installation (meters above sea level)
4. Type of protection
5. Shaking (earthquake)
6. Additional regulations
   Motor noise level
   Motor vibrations, etc.

7.1.3.3 GRID CONNECTION DETAILS

1. Type of current, voltage, voltage fluctuations, frequency
2. Safe operating values: startup current, power factor.

7.1.3.4 SPECIAL SPECIFICATIONS

1. National regulations
2. Climatic conditions
3. Special instructions (operation, storage, transport, servicing, etc.).

7.1.4 Starting characteristics

7.1.4.1 GENERAL

During running up to operating speed, the torque of an asynchronous motor varies in accordance with a characteristic curve which is established by the following values (see Fig. 7.2):

- T_A, starting torque at standstill
- T_K, breakdown torque as maximum value during running-up
- T_S, saddle point torque as minimum value during running-up, where T_S is less than T_A.

The shape of the characteristic curve is dependent on the following:

- in the case of squirrel-cage rotors, on rotor slot shape and cage material; hence it can be matched to operating conditions by appropriate choice of these;
- in the case of slip-ring rotors, on the resistance of the windings and on the external resistance applied to the rotor circuit.

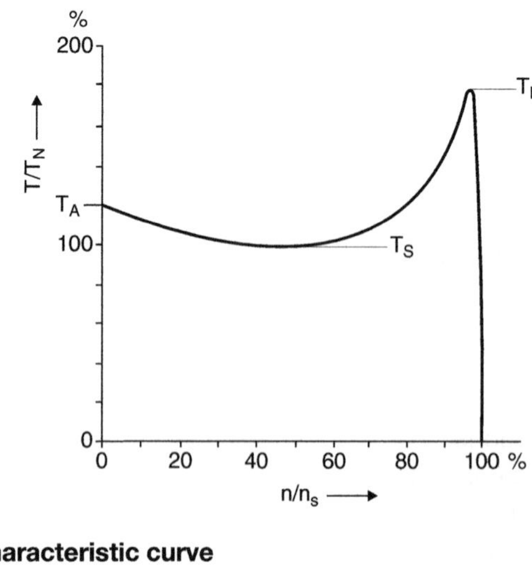

Figure 7.2 Characteristic curve

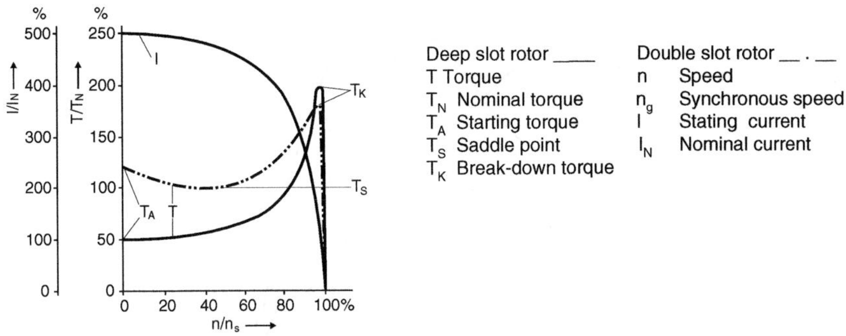

Figure 7.3 Typical starting characteristics of three-phase squirrel-cage motors

7.1.4.2 STARTING CHARACTERISTICS OF THREE-PHASE SQUIRREL-CAGE MOTORS

Typical characteristic curves are shown in Fig. 7.3 for the following types of squirrel-cage rotor:

- deep slot rotor with low starting torque
- double slot rotor with starting torque.

When directly connected at full grid voltage, three-phase squirrel-cage motors take up a transient current which can amount to between four and six times rated current, depending on power and the number of poles. During run-up the current decreases to a value corresponding to the load, e.g. to rated current when rated load is applied. The starting current can be reduced by special startup procedures (when pump startup torque so permits).

7.1.4.3 STARTING CHARACTERISTICS OF THREE-PHASE SLIP-RING INDUCTION MOTORS

In the case of slip-ring induction motors, additional resistances R_v inserted into the rotor current circuit serve to raise the torque during startup and also to increase breakdown slip. In general, the resistances are so chosen that it is possible to start up from standstill with rated torque (which corresponds approximately to rated current). In extreme cases rated breakdown torque is available at standstill (see Fig. 7.4).

At the same time, the starting current of the slip-ring induction motor, by comparison with that of the squirrel-cage motor, is limited by the additional resistances R_v.

7.1.4.4 STARTING CHARACTERISTICS OF THREE-PHASE SYNCHRONOUS MOTORS

Startup of synchronous motors usually takes place asynchronously. The shape of the speed/torque characteristic is dependent on the design of the pole shoes and the damping winding. The characteristic curve of the starting torque produced by a synchronous motor with solid pole shoes is shown in Fig. 7.5.

The design referred to above produces an almost constant torque over the whole speed range.

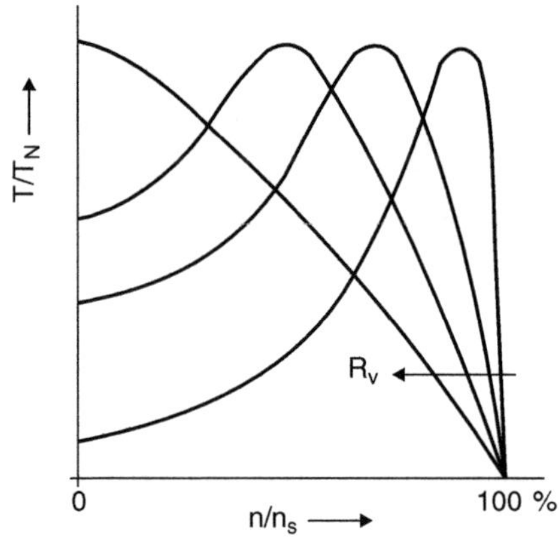

Figure 7.4 Starting characteristics of a three-phase motor with slip-ring rotor

In the case of direct engaging, a transient current occurs as with asynchronous motors. This can be decreased by suitable startup procedures (when pump startup torque so permits).

Changeover to synchronous speed is brought about by connecting the direct current excitation.

7.1.5 Possible starting procedures

7.1.5.1 GENERAL

- Squirrel-cage and synchronous motors are generally designed for direct engaging. When the counter-torque curve permits, these motors can also be supplied for impedance or part voltage startup and for startup via starting transformer.

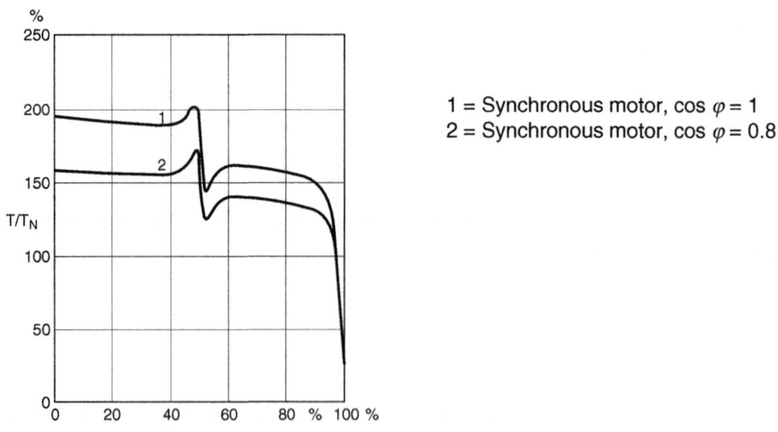

Figure 7.5 Torque characteristics of three-phase synchronous motors, *starting under full grid voltage*

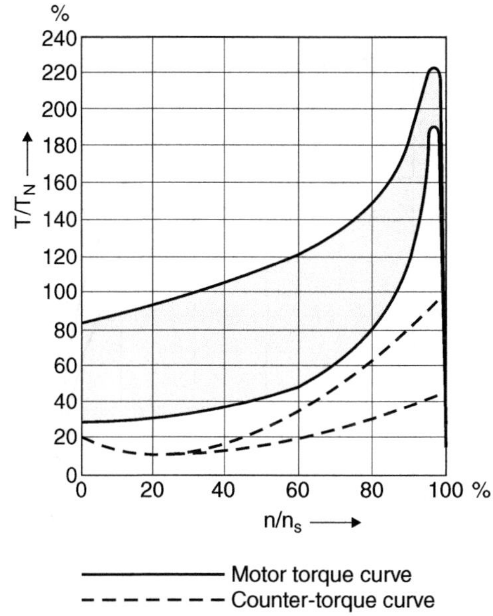

Figure 7.6 Characteristic torque curves

- *Slip-ring induction motors* are run up via starting resistance which is inserted into the rotor current circuit.

Important note

In the case of centrifugal pumps, the starting torque curve of the pump must be taken into account (see section 2.4.1).

7.1.5.2 STARTING THREE-PHASE SQUIRREL-CAGE MOTORS

These motors are designed for direct engaging. The characteristic torque and current curves of squirrel-cage motors (standard values) are shown in Figs 7.6 and 7.7.

The startup characteristics of squirrel-cage motors are shown in the shaded zone as a function of motor size and number of poles.

Current and torque curves are valid for startup at nominal voltage (guide values). When the counter-torque curve so permits, motor suitable for impedance or part voltage starting via block transformer can also be supplied.

NB When starting at reduced voltage, torques decrease approximately in accordance with the square of the voltage current roughly linear to it.

7.1.5.3 STARTING THREE-PHASE SLIP-RING INDUCTION MOTORS

Starting takes place via starting resistances inserted in the rotor current circuit. This increase of the resistance in the rotor current circuit causes a raised starting torque and a reduction of the starting current. The biggest attainable starting torque is approximately equal to breakdown torque.

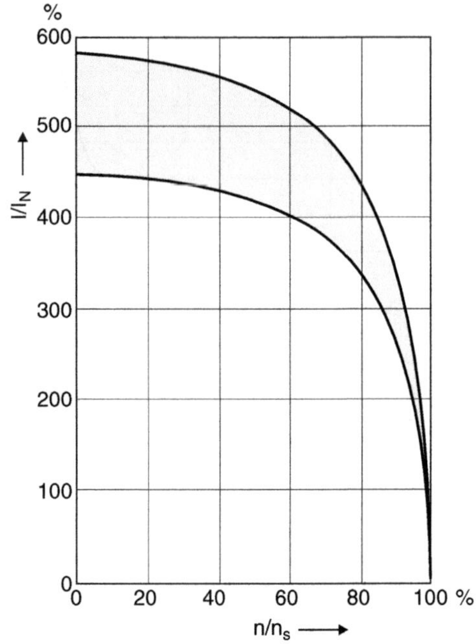

Figure 7.7 Characteristic starting current curves

The shaded areas show the acceleration torque available in each stage, this being a yardstick for starting time (see Fig. 7.8).

The starting heat is dissipated essentially via the starting resistance so that motor stressing during run-up is slight.

7.1.5.4 STARTING THREE-PHASE SYNCHRONOUS MOTORS

The starting of synchronous motors usually takes place asynchronously. The shape of the speed/torque characteristic is determined by the design of the pole shoes.

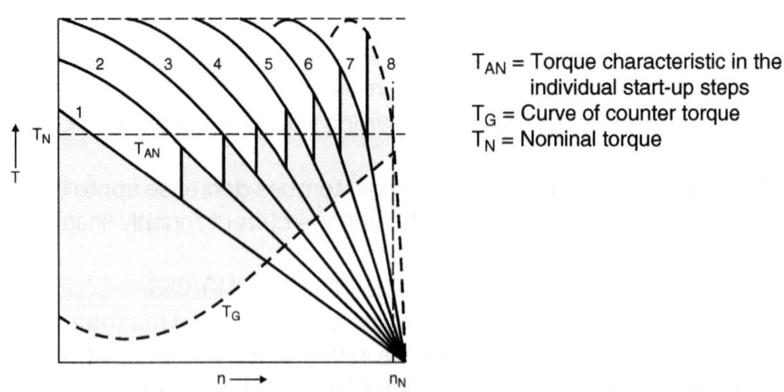

Figure 7.8 Curve of torque obtained using an eight-stage starter (stages 1 to 8)

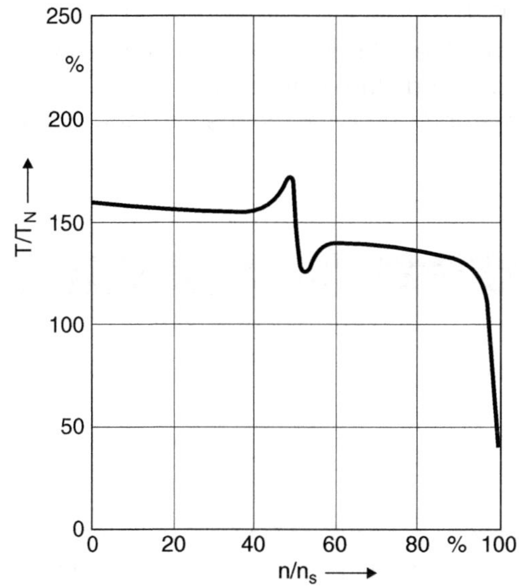

Figure 7.9 Characteristic torque curve of three-phase synchronous motors with solid pole shoes during starting at nominal voltage (guide values)

Figure 7.9 shows the starting torque characteristic curve.

7.1.6 Choice of starting method for three-phase motors

7.1.6.1 GENERAL

When the supply grid characteristics and the drive permit direct engaging, this starting procedure should be used exclusively.

In other cases the most suitable proven starting methods available should be chosen.

7.1.6.2 PROVEN STARTING METHODS FOR SQUIRREL-CAGE AND SYNCHRONOUS MOTORS

	Application with	
	squirrel-cage motor	synchronous motor
a Direct engaging	X	X
Starting reduced voltage		
Star-delta-starting	X	
b Starting via impedance coil:		
– impedance coil in supply line	X	X
– impedance coil in star point	X	X
c Starting via starting transformer	X	X
d Starting via block transformer	X	X

7.1.6.3 COMPARISON OF VARIOUS STARTING METHODS

By way of example, the following table provides a survey of the following:

- Starting currents and starting torques
- The various methods involving reduced voltage, compared with direct engaging (a).

Starting method	a	b	c	d
Terminal voltage U/U_N				
(across motor terminals)	1.0	0.8	0.8	0.8
Starting current $\sim I_A/I_N$	4.8	3.84	3.07	3.84
Starting torque $\sim T_A/T_N$	0.55	0.35	0.3	0.35

The most suitable starting method should be established in consultation with the pump, motor and electricity suppliers.

7.1.7 Speed control/regulation of three-phase motors

7.1.7.1 SPEED CONTROL OF THREE-PHASE SQUIRREL-CAGE MOTORS

The speed of these motors can be changed in steps by *pole switching* or continuously by *frequency changing*.

Pole switching

Squirrel-cage motors can generally be supplied with pole switching for two, three or four speeds. The commonest pole number ratios and the number of stator windings they require are summarized in the following table.

Number of poles	Synchronous speeds in min^{-1} at		Number of stator windings
	50 Hz	60 Hz	
2/4	3000/1500	3600/1800	1
4/6	1500/1000	1800/1200	2
4/8	1500/750	1800/900	1
4/12	1500/500	1800/600	2
6/8	1000/750	1200/900	2
6/12	1000/500	1200/600	1
8/12	750/500	900/600	2
4/6/8	1500/1000/750	1800/1200/900	2
4/6/12	1500/1000/500	1800/1200/600	2

(Continued)

(continued)

Number of poles	Synchronous speeds in min^{-1} at		Number of stator windings
	50 Hz	60 Hz	
4/8/12	1500/750/500	1800/900/600	2
6/8/12	1000/750/500	1200/900/600	2
4/6/8/12	1500/1000/750/500	1800/1200/900/600	2

Other pole number ratios are possible. The required power must be stated for each speed. Pole number ratio, voltage and output determine the limits of practical feasibility.

Frequency change

Frequency change is effected by means of a converter or transformer (up to a few MW output), with a control range 1:5 and above (see Fig. 7.10).

Voltage change

Voltage change is achieved by phase angle control of the working voltage (only applicable for low ratings).

7.1.7.2 SPEED CONTROL OF THREE-PHASE SLIP-RING INDUCTION MOTORS

The following methods are used to speed variation of these motors (see Figs 7.11 and 7.12):

- Control by means of *resistances in the rotor current circuit*. The control causes losses (slip losses) and is therefore usable to a limited extent only (poor efficiency).
- Control by means of *subsynchronous converter cascade* (*or static slip energy recovery system*). This is particularly economical when the speed adjustment range is limited. The slip output is fed back into the three-phase grid. Hence the control is low loss, with good efficiency. Applicable for outputs of up to and more than 10 MW.

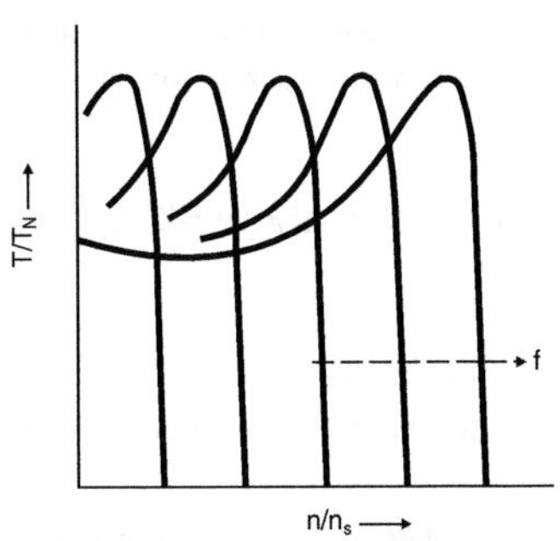

f = frequency of terminal voltage U (voltage U to be changed proportionally to the frequency).

Figure 7.10 Speed control by change of frequency (displacement of characteristic)

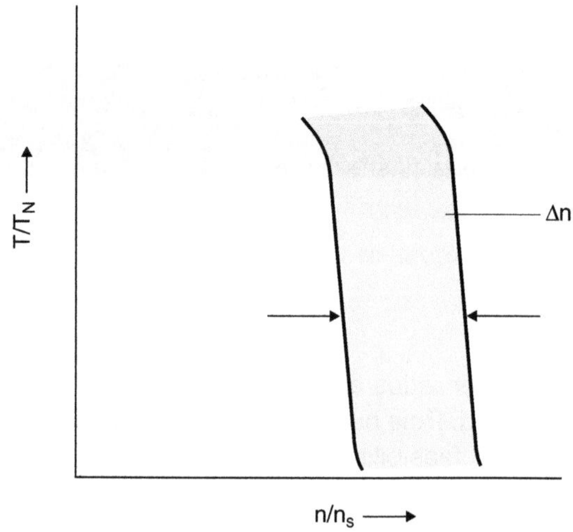

Figure 7.11 Speed control by change of slip, motor with slip-ring and sub-synchronous converter cascade. Δn – usual speed change range

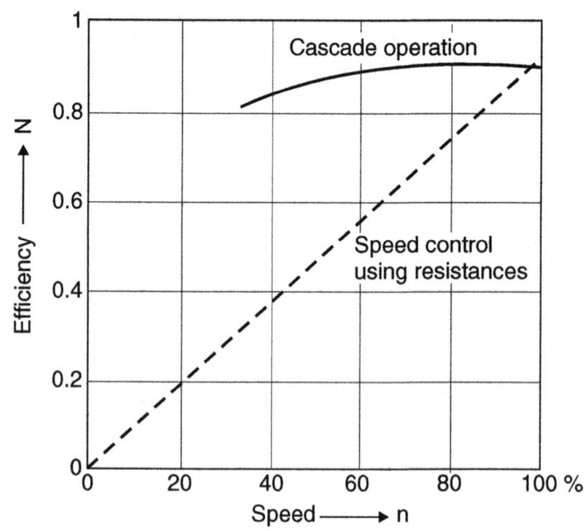

Figure 7.12 Efficiency curve of a three-phase slip-ring induction motor using resistances in the rotor current circuit and with subsynchronous converter cascade (mean output)

The economic speed adjustment range is limited to about 50% of the nominal speed.

7.1.7.3 SPEED CONTROL OF THREE-PHASE SYNCHRONOUS MOTORS

- Normally, synchronous motors are designed to operate at fixed speed, with constant voltage and frequency.
- *In special cases*, speed control is achieved by feeding the motor with variable frequency and voltage by means of static frequency converters. This circuitry, consisting generally of controllable rectifiers, smoothing choke and inverters, is expensive, especially in the higher

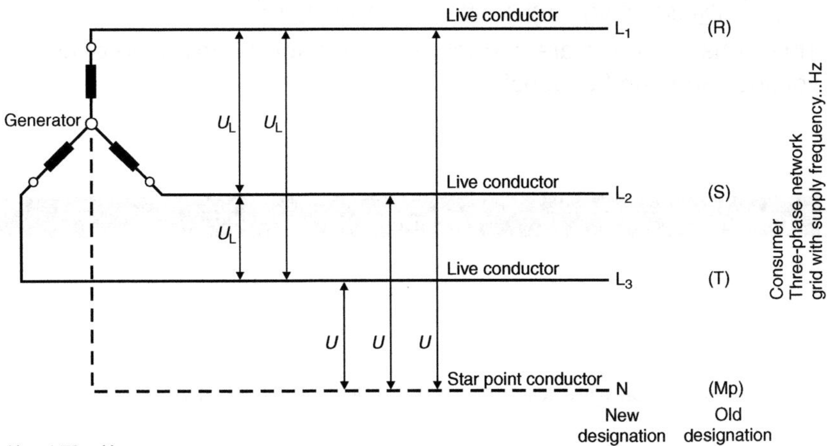

$U_L = 1.73 \times U$

U_L = Voltage between two live conductors (voltage between phases)

U = Voltage between a live conductor and the neutral (star point) conductor (also called star voltage or phase voltage)

L_1 = Live conductor (R) (also called outer conductor)

L_2 = Live conductor (S) (also called outer conductor)

L_3 = Live conductor (T) (also called outer conductor)

N = Star point conductor (Mp middle conductor or middle conductor with protection function, also known as neutral conductor)

Figure 7.13 Symbols for conductors and voltages

MW range. It can be constructed for the highest ratings and for frequencies up to about 120 Hz.

7.1.8 Some features relevant to the choice of induction motors for driving centrifugal pumps

7.1.8.1 THREE-PHASE CURRENT OR THREE-PHASE NETWORK GRID

Symbols for "conductors" and "voltages" are shown in Fig. 7.13.

The grid voltage of a three-phase network = *voltage between two live (outer) conductors* L_1 and L_2 (or L_2 and L_3 or L_3 and L_4).

In Europe the grid frequency is 50 Hz. This frequency can be 50 Hz or 60 Hz in other countries, as quoted below:

Africa	50 Hz	Asia	50 Hz
North America	60 Hz	(*except*:	
(*except Mexico)*	60/50 Hz	Japan	50/60 Hz
		Philippines	60 Hz
South America	50/60 Hz	Taiwan	60 Hz
Oceania	50 Hz	Saudi Arabia)	60/50 Hz

Where doubt exists it is always necessary to ascertain grid frequency definitely.

7.1.8.2 OPERATING VOLTAGES AND GRID FREQUENCIES

Three-phase motors are normally designed for the following operating voltages and grid frequencies:

Frequency	Operating voltage *U* in V			
50 Hz	220	380	500	660
60 Hz	440			

Operating voltages for three-phase, low-tension motors:

Frequency	Operating voltage *U* in kV								
50 Hz	3	3.3	5	5.5	6	6.6	10	11	
60 Hz	2.3		4.16			6.6		11	13.2
									13.8

Operating voltages for three-phase high-tension motors:

Where required, motors can be designed to operate at other, intermediate voltages.

7.1.8.3 VARIATIONS OF GRID VOLTAGE AT CONSTANT GRID FREQUENCY

According to the ICE 34-1/1969 standard, fluctuations of grid voltage of up to ±5% are permissible without a motor output reduction being required. (At boundary voltage fluctuation levels the maximum permissible temperature rise may be slightly exceeded.)

Where a motor designed to function *at a standard voltage has to operate continuously at constant grid frequency* with a voltage variation of ≥5%, motor output must be reduced approximately as shown in Fig. 7.14.

A decrease in torque corresponding to the decrease in output also occurs.

Where changes in voltage of ≥5% occur the motor supplier must be consulted.

7.1.8.4 VARIATIONS OF GRID FREQUENCY AT CONSTANT GRID VOLTAGE

- With variations of ≤5% in rated frequency nominal motor output can still be achieved.
- Where variations of ≥5% in frequency occur the motor supplier must be consulted.

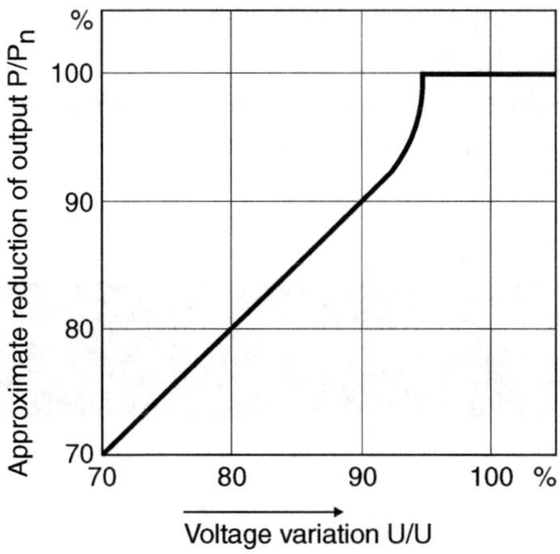

Figure 7.14 Reduction in output as a function of voltage

NB The power input required by the pump must be checked in light of the frequency and/or speed changes that occur.

7.1.8.5 VARIATIONS OF GRID VOLTAGE AND GRID FREQUENCY OCCURRING SIMULTANEOUSLY

Motors can be utilized for other frequencies unmodified provided the change in grid voltage is proportional to the change in grid frequency.

The magnetic conditions are then almost unchanged. Approximately the normal torque is developed by motor.

Speed and output change approximately proportionately to frequency. The approximate conversion factors for various frequencies are given in the table below (guide values):

	(60 Hz)	(55 Hz)	(50 Hz)	(45 Hz)	(40 Hz)
Grid (supply) frequency f	1.2	1.1	1.0	0.9	0.8
Grid (supply) voltage U	1.2	1.1	1.0	0.9	0.8
Nominal output P_N^*	1.15–1.2	1.05–1.1	1.0	0.85–0.88	0.72–0.75
Nominal speed n_N	1.2	1.1	1.0	0.9	0.8
			Basic design 50 Hz		

**The motor manufacturer must be consulted to obtain precise values for motor output and in cases where the nominal voltage does not change proportionally with the frequency.*

NB The power input required by the pump must be checked in the light of the changed speed.

7.1.8.6 SPEEDS OF THREE-PHASE MOTORS AT FREQUENCIES OF 50/60 HZ FOR VARIOUS NUMBERS OF POLES

The following table shows the numbers of poles and associated synchronous speeds for three-phase motors:

Number of poles	2	4	6	8	10	12	(14)	16	(18)
Speed n_s	min^{-1}	min^{-1}	min^{-1}	min^{-1}	min^{-1}	min^{-1}	min^{-1}	min^{-1}	min^{-1}
50 Hz grid	3000	1500	1000	750	600	500	(428)	375	(333)
60 Hz	3600	1800	1200	900	720	600	(514)	450	(400)
Number of poles	20	(22)	24	(26)	(28)	(30)	32	(34)	36
Speed n_s	min^{-1}	min^{-1}	min^{-1}	min^{-1}	min^{-1}	min^{-1}	min^{-1}	min^{-1}	min^{-1}
50 Hz grid	300	(273)	250	(231)	(214)	(200)	188	(176)	167
60 Hz grid	360	(327)	300	(277)	(257)	(240)	225	(212)	200

The figures shown in parentheses should be avoided as far as possible.

However, it should be noted that:

- *Synchronous motors*. Rotor rotates *synchronously* with the rotating stator field (synchronous speed n_s).
- *Asynchronous motors*. Rotor rotates *asynchronously* with the rotating stator field. The relative speed difference between the rotor and the rotating field of the stator (synchronous speed n_s) is referred to as "slip".

Motor output (kW)	1	10	100	1000
Slip at full load (guide values)	5 ÷ 8%	2 ÷ 4%	1 ÷ 2%	0.8 ÷ 1%

The motor supplier must be consulted to obtain precise figures for slip and/or speeds of asynchronous motors operating at full load.

7.1.8.7 INFLUENCE OF AMBIENT TEMPERATURE AND LOCATION ALTITUDE

Nominal motor outputs (as stated by the manufacturer) are normally valid for continuous operation at a coolant temperature (ambient temperature) of 40°C and locations of up to 1000 m above sea level.

Should the actual conditions deviate from the above, a reduction in output as shown in Figs 7.15 and 7.16 (guide values) results.

Should ambient temperature and location altitude deviate simultaneously from the foregoing conditions, the conversion factors must be multiplied together.

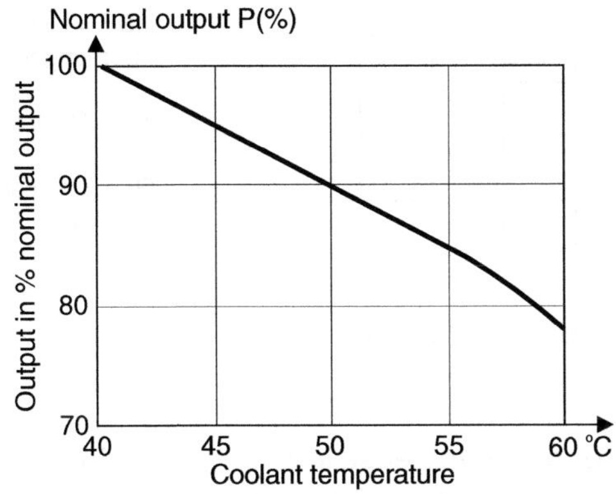

Figure 7.15 Influence of ambient temperature on output

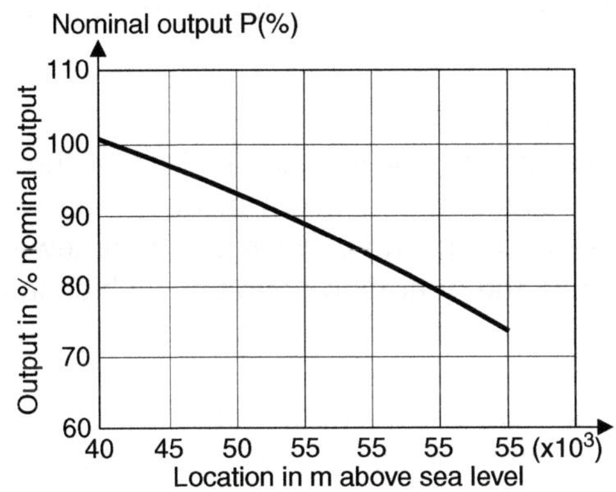

Figure 7.16 Influence of altitude on ambient temperature

7.1.8.8 INSULATION CLASSIFICATION, OPERATING LIFE OF WINDINGS

Insulating materials and their classification

Insulating materials, including impregnants, etc., are divided into classes by VDE standard 0530. Precisely defined operating temperatures are allocated to these classes.

The following table shows the limit of temperature rise in °C for:

Insulation class	B	F	H
Insulated winding (heatup °C)	80	100	125

As illustrated in Fig. 7.17, the highest safe continuous operating temperature of the individual insulating materials is derived from the coolant temperature, the boundary excess temperature value and a safety margin. This safety margin was introduced because the usual measurement

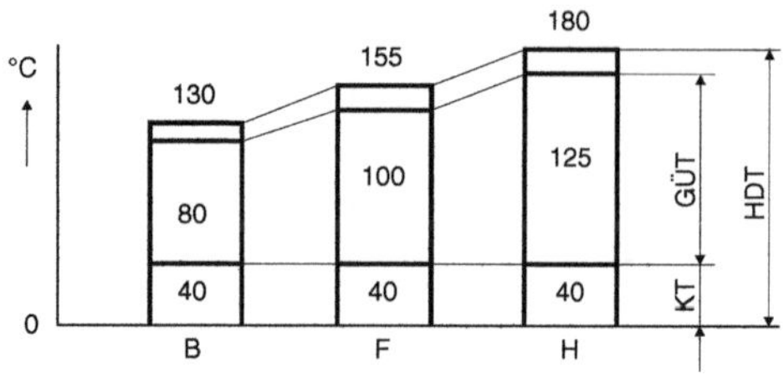

Figure 7.17 Temperature rise classes

method, i.e. determining increase in winding resistance, does not establish the hottest point in the winding, merely a mean temperature rise value. Information concerning motor outputs is based on mean temperature rise value. Information concerning motor outputs is based uniformly for all classes of insulating materials on a coolant temperature of 40°C.

Operating life of windings

The operating life of a motor is equal to that of the insulation of its windings, when no account is taken of operating wear incurred by bearings, brushes and slip-rings, which can be renewed with very little expenditure. Hence the operating conditions which influence temperature rise and thus the insulation must be given particularly careful consideration.

Heat dissipation capability is dependent on the surface area of a motor and on the ventilation conditions. Since the life of winding insulation decreases with increasing temperature, the boundary values for winding temperature defined in VDE 0530 must be complied with. These values correspond to the temperature stability of the materials as stated in the relevant insulating material classification tables. In this connection, a mean life of some 20 years is thereby assumed.

7.1.8.9 STRUCTURAL CONFIGURATIONS OF THREE-PHASE MOTORS

Standard structural configurations (see Table 7.1) are generally manufactured in the following versions:

- IM B3 (IEF code I)/IM1001 (IEC code II).
- IM V1 (IEC code I)/IM3011 (IEC code II) according to IEC Publ. 34–7/1972.

Other structural configurations are also possible (see Table 7.2). Consult motor manufacturer.

7.1.8.10 TYPES OF PROTECTION FOR THREE-PHASE MOTORS ACCORDING TO IEC RECOMMENDATION 34–5 AND DIN 40050 (SEE TABLE 7.3)

The most frequently used types of protection are: IP 23, IP 44 and IP 54. Choice of most appropriate type: consult the motor manufacturer.

Table 7.1 Standard Structural Configurations

Code designation	Schematic diagram	Bearings	Housing	Shaft	General construction	Fastening/ Erection
IM B 3 (IM 1001)	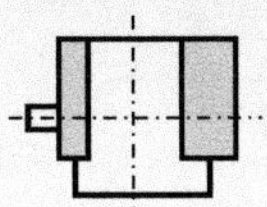	2 end plates	with feet	free shaft end		installed on foundation
IM V 1 (IM3001)	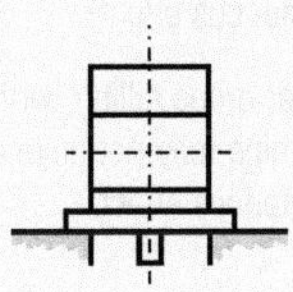	2 end plates	without feed	free shaft end downwards	fixing flange external, near de-bearing	flanged support downwards

Table 7.2 Additional Structural Configurations

D5 (IM 7211)	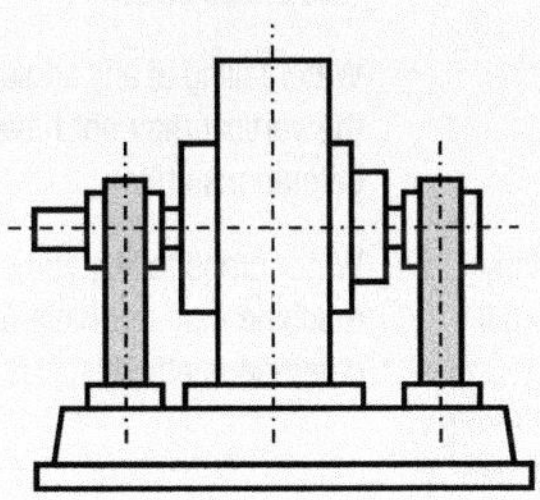	2 pedestal bearings	with feed and baseplate	free shaft end, horizontal
V7 (IM 3811)	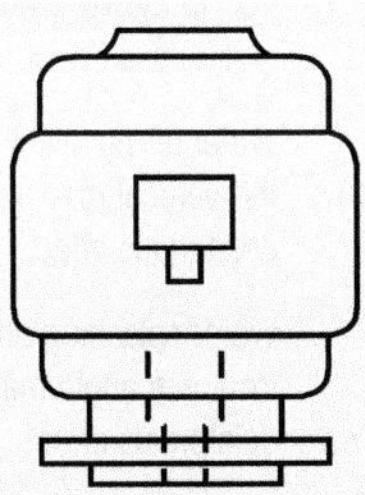	2 end plates	without feet	shaft end, horizontal

7.1.8.11 TYPES OF COOLING FOR THREE-PHASE MOTORS (SEE IEC RECOMMENDATION 34–6) (SEE FIG. 7.18)

The most frequently used types of cooling for centrifugal pump motors are:

- IC A 01 Machine with unrestricted air inlet and outlet (cooling medium is air); self-cooling.

Table 7.3

Type of protection	Contact and foreign body protection	Protection against water
IP 00	No particular protection against random contact with any component under carrying voltage or in motion. No protection of the machine against penetration of solid foreign bodies.	No particular protection.
IP 02		Water drops falling at any angle up to 15° to the vertical may not have any deleterious effect.
IP 11	Protection against random, large-area contact with components under voltage or with internal moving parts, e.g. hand contact, but no protection against intentional access to these parts. Protection against penetration of solid foreign bodies with diameters greater than 50 mm.	Water drops falling vertically on the machine may not have any deleterious effect.
IP 12		Water drops falling at any angle up to 15° to the vertical may not have any deleterious effect.
IP 13		Water falling at any angle up to 60° to the vertical may not have any deleterious effect.
IP 21	Protection against finger contact with any components under voltage or any internal moving parts. Protection against penetration of solid foreign bodies with diameters greater than 12 mm.	Water drops falling vertically onto the machine may not have any deleterious effect.
IP 22		Water drops falling at any angle up to 15° to the vertical may not have any deleterious effect.
IP 23		Water falling at any angle up to 60° to the vertical may not have any deleterious effect.
IP 44	Protection against any components under voltage or any internal moving parts being touched with tools, wires, etc. with a thickness greater than 1 mm. Protection against penetration of solid foreign bodies with diameters greater than 1 mm. Exceptions to the above are cooling air openings (inlets and outlets of external air fans) and condensation water drainage holes from closed machines which may have Grade 2 protection.	Water splashing onto the machine from any angle may not have any deleterious effect.

(Continued)

Table 7.3 *(continued)*

Type of protection	Contact and foreign body protection	Protection against water
IP 54	Complete protection against contact with components under voltage or with internal moving parts. Protection against depositing of deleterious layers of dust. Penetration of dust is not completely prevented, but dust may not penetrate in such quantities that the operation of the machine is impaired.	Water splashing onto the machine from any angle may not have any deleterious effect.
IP 55		A water jet from a nozzle, directed onto the machine from any angle, may not have any deleterious effect.
IP 56		During submersion caused, for example, by heavy seas, water may not penetrate the machine harmfully.

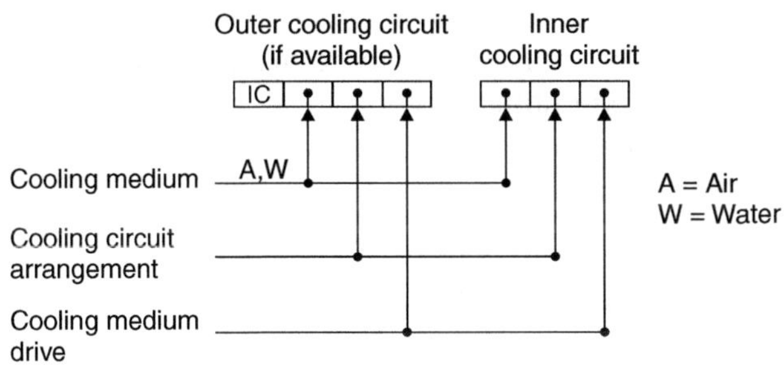

Figure 7.18 Types of cooling for three-phase motors (IEC Recommendation 34–6)

- IC W 37 A 81 Machine with built-in heat exchanger (cooling medium is water, inner cooling circuit – air); self-cooling.
- IC W 37 A 85 As above, but cooling of internal air circuit is by means of built-in ventilation equipment, the drive of which is not dependent on the machine (independently operating fans).

NB Where cooling is solely by air, code letter "A" can be deleted. Other types of cooling are possible. Motor manufacturer to be consulted.

7.1.8.12 POWER UPTAKE BY THREE-PHASE MOTORS

The active power absorbed (P_1) in kW is:

$$P_1 = \frac{P_N}{\eta}\,(\text{kW})$$

The reactive power absorbed (P_s) is expressed by the following equation:

$$P_s = \frac{P_N}{\eta \cdot \cos\varphi}\,(\text{kVA})$$

The nominal current (I_N) is calculated from:

$$I_N = \frac{1000 \cdot P_1}{U_N \cdot \sqrt{3} \cdot \cos\varphi}$$

or

$$I_N = \frac{1000 \cdot P_N}{U_N \cdot \sqrt{3} \cdot \cos\varphi \cdot \eta}\,(\text{A})$$

or

$$I_N = \frac{1000 \cdot P_s}{U_N \cdot \sqrt{3}}\,(\text{A})$$

where:

P_N = nominal output (at motor shaft end) (kW)
P_1 = active power uptake (at motor terminals) (kW)
P_S = reactive power uptake (kVA)
U_N = nominal voltage (V)
η = efficiency (–)
$\cos\varphi$ = power factor (–).

7.2 DIESEL ENGINES

7.2.1 General

Internal combustion engines are generally chosen as the drive for centrifugal pumps where either electrical energy is not available or the pumpset is essential in the event of power failure.

In most cases use is made of diesel engines, which are distinguished by a relatively high efficiency (35–40%) and also by good torque controllability (control of injection quantity) and starting reliability.

Only diesel engine drives are considered hereunder. Gas engines may be employed where adequate supplies are available.

7.2.2 Operating data and operating features

7.2.2.1 ROTATION SPEED

The choice of rotation speed depends on engine availability and on the operating conditions, i.e. for continuous duty a low speed is advantageous, whereas for operation of short duration higher speeds may be allowed by the engine manufacturer. Engine speeds range between 1800 rpm for standby ratings and up to 2 mW and 600 rpm for continuous ratings between 1 mW and 5 mW.

7.2.2.2 POWER

The power ratings of internal combustion engines are related to specific atmospheric reference conditions or standard temperature and pressure.

In accordance with ISO 3046/1 reference conditions are defined as 100 kPa = 1 bar, 27°C air temperature, 60% relative air humidity and 27°C cooling water temperature for the charging air.

If the conditions at a site are different from these, power must be converted from reference conditions to actual conditions. The following data are guide values:

- Engine installation sites at altitudes ≥500 m above sea level involve a power reduction of 1% for every 100 m altitude.
- Similarly, a power reduction of 1% per 2°C temperature must be taken into account for ambient temperatures ≥27°C.
- Continuous rating: ISO 3046/1 defines this as the power at which an engine can be operated continuously at a given speed and under the conditions subsisting locally during the period between overhauls prescribed by its manufacturers.
- Overload capacity: in the absence of other data ISO 3046/1 defines this as 110% of continuous rating for the duration of one hour (with or without interruptions) within an operating period of 12 hours.

7.2.3 Conditions for installation

7.2.3.1 FUEL (DIESEL OIL)

Normal consumption is of the order of magnitude of 200 to 220 grams per kW per hour.

Local regulations must be taken into consideration for the storage of fuels (tanks).

7.2.3.2 DISPOSAL OF EXHAUST GASES

The exhaust gases are ducted to the exterior through the exhaust pipe to which the following conditions apply:

- The limits for back pressure stated by the engine makers must not be exceeded.
- The exhaust pipe must be insulated to reduce heat radiation within the enclosure.
- One or more silencers (depending on ambient conditions) must be mounted before the outlet to meet required noise limits.

7.2.3.3 FRESH AIR REQUIREMENT

A diesel engine requires fresh air for combustion and for cooling the engine room as compensation for the heat radiated by the engine:

- Combustion air requirement at normal atmospheric pressure: about 6–7 m^3/h/kW.

- Air volume for cooling the engine room, if the temperature rise from engine heat radiation needs to be limited to 10°C: about 15–20 m^3/h/kW.

Air inlets and outlets may be equipped with silencers to limit noise levels outside the enclosure. Care must be taken to prevent negative pressure in the engine room.

7.2.3.4 COOLING WATER REQUIREMENT

With the exception of low-power air-cooled engines, engine cooling (cylinder, lubricating oil and combustion air in the case of supercharged engines) is generally effected by a closed cooling water circuit.

This circuit in turn has to be cooled with either air through a radiator or water through a heat exchanger, each being supplied by the engine manufacturer.

The amount of heat to be dissipated for engine powers <100 kW is approximately 3600 kJ/kW/h and for engine powers <500 kW is approximately 3200 kJ/kW/h.

7.2.3.5 ENGINE STARTING METHODS

The engine may be started either by an electric starter motor or by compressed air. Hydraulic systems are also employed for standby units such as fire fighting pumps.

- Electric starter motor:
  For engine powers $P < 350$ kW
  Capacity of starter motor $P = 2$ to 20 kW
  Electric stator motors are usually powered by 24 V DC battery systems up to about 2 mW engine rating. Larger engines may be started by direct air injection into the cylinders or pneumatic motors.
- Compressed air:
  For engine powers $P > 200$ kW
  Required compressed air flow approx. 1.51/kW
  for each start at 10 to 30 bar.

7.2.3.6 PREHEATING THE ENGINE

To ensure full power development in standby sets within the shortest possible time the engine cooling water circuit and lubricating oil must be kept constantly at a temperature of approx. 50°C. This is usually achieved by means of immersion heaters.

7.2.3.7 NOISE EMISSION

Depending on the installation site special arrangements must be made to suppress noise emission from the engine body.

7.2.3.8 TORSIONAL VIBRATIONS

Due to the non-uniformity in the driving torque delivered by a diesel engine the complete shafting must be checked for torsional vibrations.

7.2.4 Applications

Configuration	Advantages	Disadvantages
1. Direct driving of pumps	Possibility of individual speed regulation	Number of diesel engines equal to number of pumps
2a. Drive via diesel/ generating set (auxiliary generating set)	Reasonable price	All the pumps are not in operation simultaneously all the time. When they are not, the diesel engine operates at part load (uneconomical operating mode). No reserve for diesel engine overhauls or failures
2b. Distribution among several diesel/generating sets	Better engine workload, reserve during overhauls	Expensive
3. Pump with electric motor and diesel engine on single shaft	Diesel engine operates during power failures only	– Disengageable coupling between diesel engine and electric motor – Electric motor with two shaft ends and a speed matched to that of the diesel engine, otherwise gear transmission required

7.3 GAS TURBINES

7.3.1 Introduction

During recent decades gas turbines have gained in importance as pump drives.

Their increasing use is based on the following advantages by comparison with other competitive equipment:

- low space requirement and weight per kW
- self-contained system of relatively low capital expenditure
- high reliability
- easy maintenance
- can burn various fuels
- no cooling water required.

7.3.2 Thermodynamic principle and features

7.3.2.1 WORKING PRINCIPLE

The theoretical gas turbine working principle is the Brayton or Joule cycle with adiabatic compression, a temperature rise at constant pressure and adiabatic expansion (see Fig. 7.19).

The compression work is $h_2 - h_1$, the turbine output is $h_3 - h_4$ and the difference is the net work output. The heat input is $h_3 - h_2$. The thermal efficiency can be expressed as follows:

$$\eta - \frac{(h_3 - h_4) - (h_2 - h_1)}{h_3 - h_2}$$

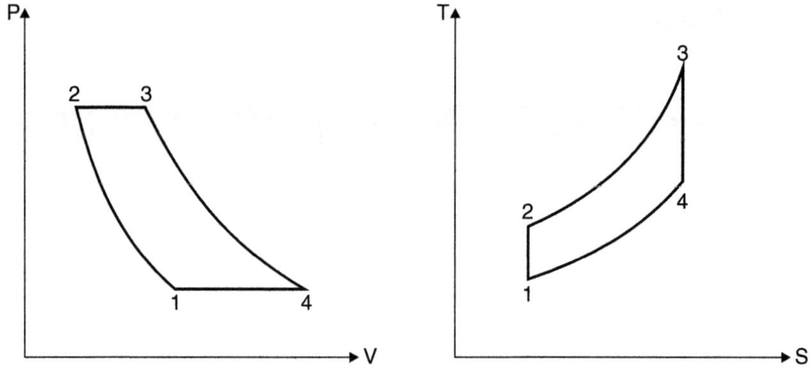

Figure 7.19 P–V and T–S diagrams of the ideal Brayton or Joule cycle

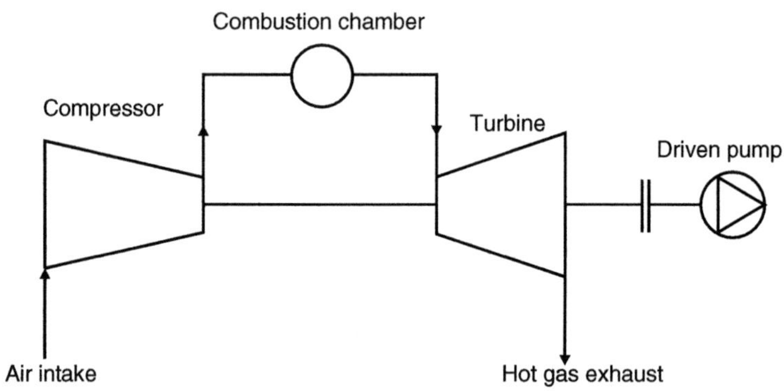

Figure 7.20 Gas turbine (open cycle)

In reality the process is irreversible and the efficiency of compression, combustion and expansion must be taken into consideration. Even so the formula shows that an acceptable turbine output presupposes high compressor and turbine efficiencies. As h_3 is directly proportional to the temperature, it is evident why continuous efforts are made to develop materials and techniques which permit ever higher combustion temperatures and by consequence turbine intake temperatures.

Due to its low investment cost, the open cycle is the most frequently used type for gas turbines driving pumps or compressors (see Fig. 7.20).

The simple open cycle has a relatively low thermal efficiency of approximately 16 to 26%, depending on turbine intake temperature, the exhaust gas heat content representing the major loss.

Various methods of improving efficiency are possible: the most common one is the use of a regenerator in the cycle (see Fig. 7.21).

In this design the air supplied to the combustion chamber is preheated, by the hot exhaust gases, resulting in a lower fuel consumption

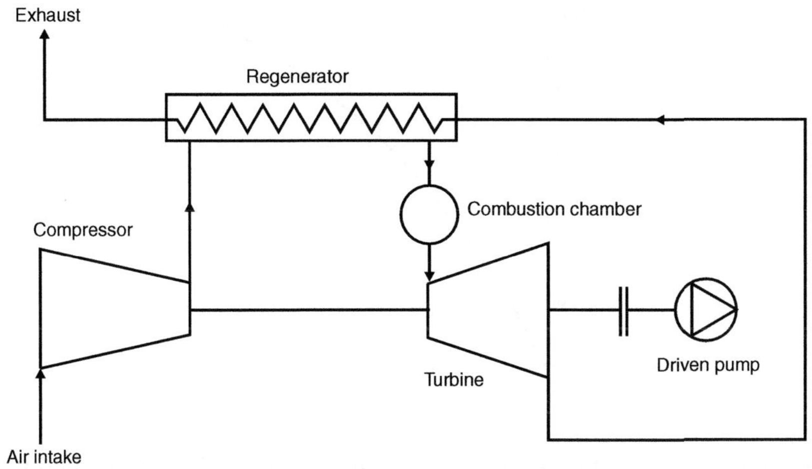

Figure 7.21 Gas turbine (open cycle with regenerator)

to reach the same gas temperature. By this means efficiency can be increased by about 4%. An additional improvement is the use of intercooling during compression, as this reduces the compression work needed for a specific air weight.

7.3.2.2 CONSTRUCTION TYPES

A distinction may be drawn between two different gas turbine types:

1. *The single shaft gas turbine* has been developed mainly for generating electric power and has compressor and power turbine integrated on a common shaft. As such sets are operated continuously at the same speed, compressor and power turbine efficiency can be optimized.
2. *The split shaft gas turbine* is used to drive machines where the speed and therefore the useful power are variable (see Figs 7.22 and 7.23).

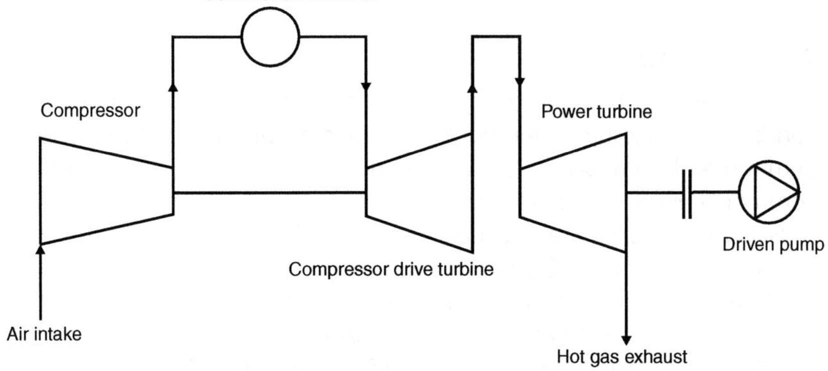

Figure 7.22 Split shaft gas turbine

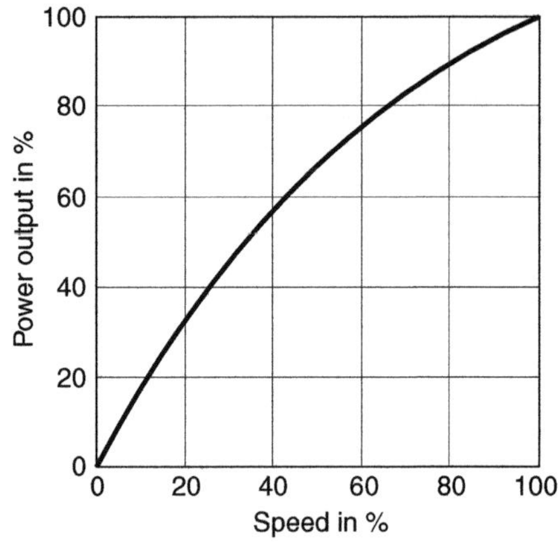

Figure 7.23 Typical power output versus shaft speed for a split shaft gas turbine

Split shaft gas turbines may be of conventional design or of a type derived from aircraft engines. Conventional gas turbines are based on steam turbine technology, feature robustness and can in most cases be overhauled and re-balanced on site.

As its name suggests, the aircraft-derivative version makes use of aircraft engine components, namely a gas generator comprising compressor, compressor drive turbine and combustion chamber. The power turbine is of conventional type. This type of gas turbine is compact and has a low weight-to-power ratio.

When it is recalled that over 90% of the gas turbines produced worldwide are used for aircraft propulsion, it will be clear that the enormous amount of development work done in the field and the rationalized methods made possible by long production runs necessarily lead to important advantages. On the other hand, gas generators have to be returned to the makers for overhauls.

7.3.3 Ratings

In the course of time various organizations have established standard conditions for air intake temperature and installation altitude in order to permit direct comparison between various gas turbines. The four most commonly used definitions are:

- ISO (International Standards Organization):
  - Sea level and 15°C (59°F).
- NEMA (National Electrical Manufacturers Association):
  - 1000 feet (305 m) above sea level and 80°F (27°C)
- CIMAC (Congrés International des Machines a Combustion):
  - Sea level and 15°C (59°F)

- "Site", i.e. the actual altitude of installation and design temperature valid for the installation site of a gas turbine.

With the trend towards ever higher combustion temperatures and in the light of varying operating conditions new specifications have been introduced for design aspects:

- Emergency operation (maximum intermittent)
- Peak load (intermittent)
- Base load.

For installation sites with extreme yearly ambient temperature fluctuations a so-called *flat or straightline rating* (theoretical average annual power output) is established for gas turbines. In this method the lower effective output due to high ambient temperatures is compensated for by calculating the actual higher effective power at lower temperatures. The theoretical average annual effective power output therefore lies considerably above a "site rating" for the design temperature. (Design data at the installation site.) These classifications are based on the number of operating hours per phase of operation and depend on the blading material of the power turbine. When calculating service life materials suitable for 100 000 hours' continuous operation are normally chosen. Higher temperatures at the expense of reduced service life are permitted.

For pumps driven by gas turbines it is advisable to specify the expected operating points with their associated operating duration in order to avoid installing oversized equipment. A typical gas turbine output versus ambient temperature diagram (see Fig. 7.24) demonstrates the necessity for determining the design temperature as accurately as possible. Figure 7.25 shows a typical diagram for elevation correction.

7.3.4 Fuels

In open cycle gas turbines the combustion products come into direct contact with the turbine blading. For this reason fuels are recommended whose combustion products cause no high temperature ash corrosion or

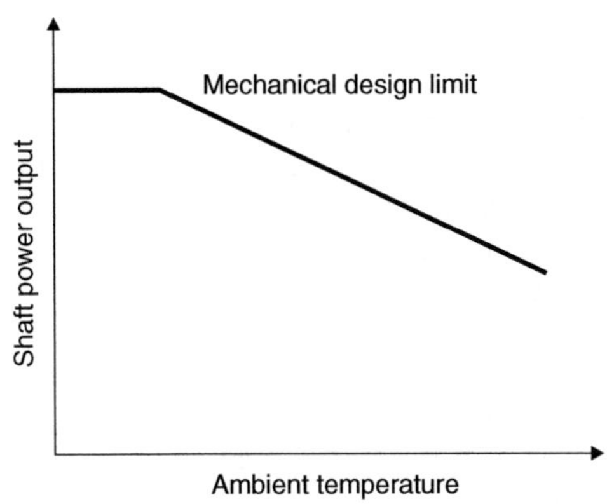

Figure 7.24 Typical gas turbine output/ambient temperature curve

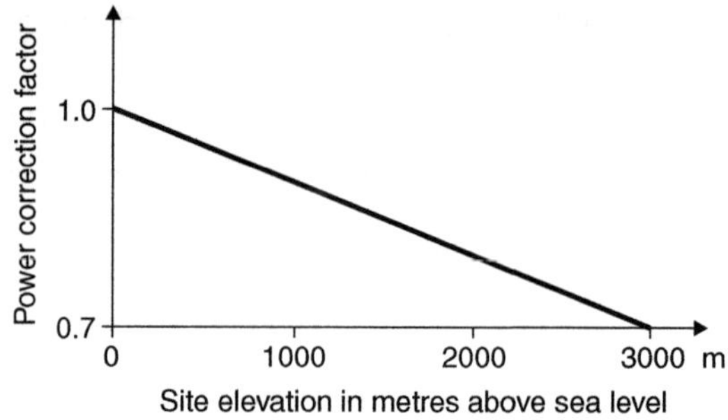

Figure 7.25 Typical gas turbine power correction/altitude curve

depositing and the residual solids are small enough to avoid erosion. Natural gas, refinery gas, blast furnace gas and distillate oils are ideal fuels for gas turbines. Residual fuel oils contain impurities.

Vanadium pentoxide and sodium sulfate are the chief ash components that cause corrosion and depositing in the temperature range of 670–815°C. The most successful method of overcoming the problem seems to be gas scrubbing to eliminate sodium as completely as possible followed by a fuel treatment with additives of magnesium oxide, magnesium sulfate and aluminum silicate. Such treatment systems are in successful operation and appear economically feasible. A detailed fuel analysis should always be part of the bid invitation specification for gas turbines.

7.3.5 Environmental protection

In most countries legislation has been passed to protect the environment. Projects must therefore comply with the legal regulations in force at the installation site.

7.3.5.1 NOISE EMISSION

The principal sources of noise in gas turbines are the turbine intake and the exhaust plus auxiliaries like blowers for air-cooled oil coolers, fans, starting systems and lubricating oil pumps. It is common practice to provide intake and exhaust silencers and some form of acoustic shielding for the auxiliaries. In order to evaluate the extent of acoustic treatment necessary, the following parameters must be taken into consideration for every plant:

1. Laws limiting the time personnel may spend in a noisy environment;
2. Local regulations and their interpretations;
3. Local conditions and the noise level already existing, any sound-reflecting surfaces;
4. Applicable standards and regulations.

7.3.5.2 EXHAUST GAS EMISSIONS

Environment protection regulations prescribe action schedules to obtain satisfactory air quality levels. In particular, the emission of suspended particles and oxides of sulfur and nitrogen is restricted. The relevant regulations and specifications must be studied and complied with right from the planning stage of a project.

It can be stated basically that gas turbines have low particle emissions. The sulfur oxide level in the exhaust gases is directly proportional to the sulfur content in the relevant fuel. Nowadays the aim is to utilize fuels with as low a sulfur content as possible. The formation of nitric oxides is a direct result of the combustion process. Here too manufacturers are energetically studying the problem and its possible solutions.

7.3.6 Gas turbine support systems

7.3.6.1 STARTING SYSTEM

A powered crank is necessary to accelerate the gas turbine to its self-sustaining speed. Possible systems include electric motors, diesel engines and gas expander turbines.

7.3.6.2 LUBRICATING OIL SYSTEM

Normally the gas turbine manufacturer supplies a combined lubricating system for turbine, pump and possibly a gearbox. This applies especially to conventional (industrial) gas turbines. In the case of "aircraft derivative" turbines it is common practice for the pump manufacturer to supply the lubricating system for the pump (and for the gearbox, where there is one), and in certain cases for the power turbine as well. In addition to the main lubricating oil pump sized for a discharge flow adequate for the combined system an auxiliary lubricating oil pump must be provided as standby in case of main pump failure. Normally a reservoir with minimum 4 minutes' retention time is specified. Filtration to 30 μ is sufficient for most gas turbines and pumps.

Lubricating oil cooling can be achieved in various ways. The choice depends on local site conditions. The simplest and cheapest method is to use a shell and tube heat exchanger with water as cooling medium. In remote regions air-cooled closed water circuits are often installed.

7.3.6.3 INTAKE AIR FILTRATION

The degree of filtration needed depends mainly on the air pollution in the area of the plant. For regions where sandstorms may occur and at Arctic sites with risk of icing special intake air filters are available to prevent compressor damages.

7.3.6.4 MONITORING AND CONTROL SYSTEM

Most manufacturers offer monitoring packages which include the proper sequencing of operation for starting automatically, operating and shutdown, via transmitter signals. During an automatic start various transmitters convey signals to a microprocessor. These signals confirm the normal functioning of the auxiliary systems and the entire installation is brought into operation by means of timer relays in accordance with a predetermined sequence of events.

The main operating parameters such as output speed, gas turbine compressor speed, turbine intake temperature and lubricating oil temperature and pressure are continuously monitored during normal service. A signal from the pump discharge pressure transmitter can be fed to a speed and fuel controller, to match the gas turbine to changing conditions of speed and/or power output.

As with starting, normal and emergency shutdowns are also controlled by the microprocessor ("sequencer"). Most standard monitoring systems can be built for remote control via cables or microwaves.

7.3.7 Application as pump drive

Gas turbines for pump drives are available in a wide range of speeds and for power outputs from 180 kW to over 50 MW.

(A summary of the types mostly used is contained for instance in the annual publication *Gas Turbine World Handbook*.)

7.3.7.1 PIPELINE

Since the mid-1950s gas turbines have been used increasingly in crude oil pipe lines as pump drives. This trend is likely to continue. Advantages of gas turbines for this application include:

1. Low installation costs.
2. Variable operating speed permits maintaining a specific pump discharge pressure across a wide range of flow rates, thus providing maximum flexibility.
3. A standard gas turbine control system is well suited for unattended operation and remote control.
4. Operating experience has demonstrated the high reliability of gas turbines.
5. Gas turbines can be assembled as modules, simplifying transportation and erection.

7.3.7.2 INJECTION

An important aspect of crude oil production is secondary recovery. Its purpose is to raise the productivity of an oilfield whose natural pressure no longer suffices to raise the oil to the surface. For years it has been the practice to flood such fields with water (with seawater, formation water, etc.) under high pressure. The development of high-pressure centrifugal pumps permitted the injection of large quantities of water into oilfields. As most oilfields are located in remote areas, pre-assembled gas turbine/pump packages have become more and more popular.

Figure 7.26 27 MW gas turbine for injection pump drive. The complete set on the test bed in Leeds/UK

Offshore platform installations require driving machinery with minimum vibration and low unbalanced inertia forces. The gas turbine fulfills these two requirements ideally (Fig. 7.26).

7.3.7.3 LOADING STATIONS

Another suitable field of application for gas turbines as pump drives is the loading of oil tankers. With a gas turbine pump set a high degree of flexibility can be attained, particularly towards the final phase of the loading operation.

7.3.7.4 APPLICATION CONSIDERATIONS

The uses of gas turbines can vary from simple pump drives with constant discharge flow and head to complex installations with variable speed, discharge head and flow. The hydraulic characteristics of the pump must be matched and optimized with the gas turbine, as only gas turbines with given power output that are already commercially available can be utilized.

The power output of the gas turbine must be equal to or greater than the maximum load the pump may demand. Based on specific power curves similar to Fig. 7.23, this required power output can be established and corrected for site altitude (Fig. 7.25). Gear losses must be added to the pump load. Partial load points should be checked against the speed power curve of the gas turbine. It is important to size turbine-driven pumps for continuous operation at 105% and brief emergency operation at up to 120% of rated speed. On the other hand, where a future need to increase delivery head is expected there is no need to fit a reduced-size impeller to permit subsequent retrofitting with one being capable of at least a 5% head increase, as would be the case with electric motor drives under the terms of API 610.

A torsional analysis of the combined unit must be carried out (generally by the gas turbine manufacturer) to ensure the complete absence of critical speeds in the operating range.

7.4 SPEED CONVERTERS

7.4.1 Preliminary notes

Apart from the speed regulation possibilities afforded by variable-speed driving machinery such as:

- Electric motors with direct speed control (section 7.1.7)
- Diesel engines (section 7.2)
- Gas turbines (section 7.3)
- Steam turbines

pump speed can also be varied over a wide range by the use of variable-speed gearing between electric motor and centrifugal pump (for the advantages of speed control, see section 2.2.2.4).

7.4.2 Variable-speed gearings/speed converters for centrifugal pumps

The variable-speed gearings mostly used on centrifugal pumps are:

- Mechanical stepless speed converters such as belt drives or friction wheel drives
- Hydraulic speed converters:
  - hydrostatic drives
  - hydrodynamic converters
- Electro-magnetic speed converters.

Some basic information is given on hydrodynamic and electromagnetic speed converters hereunder.

7.4.2.1 HYDRODYNAMIC SPEED CONVERTERS (ALSO CALLED FLUID OR HYDRAULIC COUPLINGS)

Advantages

- Stepless control in a range of 4:1 to 5:1 maximum
- Capable of transmitting very high power outputs
- Load-free motor starting
- Gentle pump acceleration
- Damping of any torsional vibrations
- Simple torque matching by modifying the oil fill
- Assured protection against excessive heating by means of fusible cutout

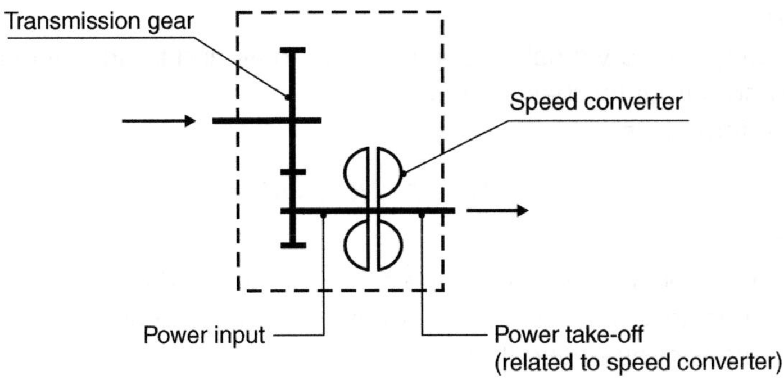

Figure 7.27 Variable-speed coupling schematic

- Wear-free transmission, as the power transmitting parts have no mechanical contact
- High efficiency as rated slip is very low.

Disadvantages

- Price
- Additional space requirement for the converter and its auxiliaries, such as cooler, etc.
- Rated capacity of electric motor slightly increased to compensate for converter losses (rated slip and mechanical loss generators such as bearing and gear train driving power for oil pumps).

Working principle

Put simply, the hydrodynamic speed converter consists of a pump impeller on the driving shaft and a turbine runner on the output shaft. The input torque is transmitted to the output shaft by the mass forces of a fluid (mostly oil) flowing between pump impeller and turbine runner (Fig. 7.27).

Principle: Geared variable-speed device

Power input:	Power take-off:
ω, n_1	ω_2, n_2
T_1	T_2
P_1	P_2

$$\text{Slip: } s = \frac{n_1 - n_2}{n_1} = 1 - \frac{n_2}{n_1}$$

$$\frac{n_2}{n_1} = 1 - s$$

Torque

Neglecting the very small torque loss due to air and bearing friction the input and output torques are equal.

The formula is:

$$T_1 = T_2 = T = K \cdot \rho \cdot n_1^2 \cdot D_p^5$$

where:

K = characteristic comprising the constant factor related to the chosen dimensions and the function of all the other parameters
ρ = density of operating fluid
D_p = converter size

Output

$$P_1 = T \cdot \omega_1$$

$$P_2 = P_1 \cdot \frac{n_2}{n_1} = P_1(1 - s)$$

Transmission efficiency:

$$\eta = \frac{P_2}{P_1} = \frac{n_2}{n_1} = 1 - s$$

Power loss due to heatup:

$$P_v = P_1 - P_2 = P_1 \cdot s = P_1 \cdot \frac{n_1 - n_2}{n_1} = P_2 \cdot \frac{n_1 - n_2}{n_2}$$

i.e. proportional to the slip.

Relation between input power P_1 and power take-off P_2 allowing for all losses:*

$P_1 = \dfrac{P_2\left(1 + a\frac{n_1 - n_2}{n_2}\right) + b}{y}$
P_2 = pumps power uptake at speed n_2
a = 1.12 ÷ 1.18 = coefficient of power lost in the oil circuit and mechanical losses
b = 10 ÷ 50 kW = power required for auxiliary oil pumps (operating and lubricating oil circuits)
y = 0.97 ÷ 0.985 = efficiency of step-up gear (if fitted)
$\left.\begin{matrix} n_1 \\ n_2 \end{matrix}\right\}$ = input and output speeds related to the speed converter.

Power loss (see Fig. 7.28)

From the above equation the power loss:

$$P_v = P_1 - P_2$$

*According to Neyrtec, France (Voith licence).

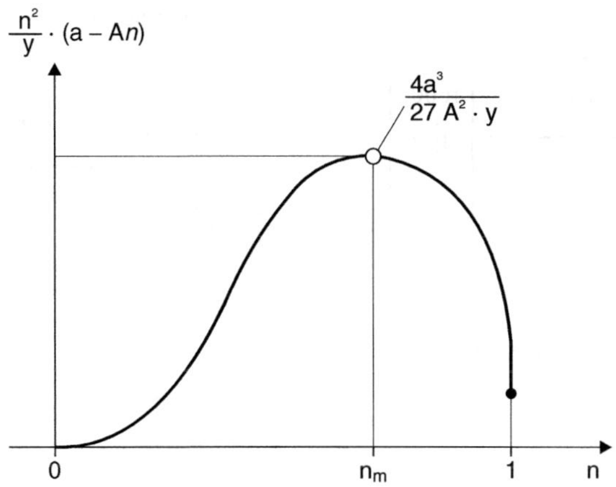

Figure 7.28 Evolution of power loss and of value $\frac{n^2}{y} \cdot (a - An)$ in relation to *n*

can be calculated as follows:

$$P_v = \frac{P \cdot n^2}{y}[a - n(a + y - 1)] + \frac{b}{y}$$

where:
P = rated pump output converted to the speed n_1 (with slip = 0)

$$P = P_2 \cdot \left(\frac{n_1}{n_2}\right)^3 \qquad n = \frac{n_2}{n_1}$$

where: $A = A + y - 1$
and $B = \frac{b}{y}$

the outcome is: $P_v = \dfrac{P \cdot n^2}{y} \qquad [a - An] + B$

This means that for a specific speed converter the lost power P_v becomes a maximum when values are as follows:

$$\frac{n^2}{y} \cdot (a - An) = \text{Maximum}$$

This is fulfilled when: $n = \dfrac{2a}{3A}$

and therefore:

$$\frac{n^2}{y} \cdot (a - An) = \frac{4 \cdot a^3}{27 \cdot A^2 \cdot y}$$

i.e. $P_{v_{max}} = P \cdot \frac{4\ a^3}{27\ A^2 \cdot y} + B$

where: $n_2 - n_m \cdot n_1 = \frac{2a}{3A} \cdot n_1$ (approx 2/3 of n_1).

The above equations correspond to the operation of a pump with origin parabola as resistance curve (pipe resistance characteristic).

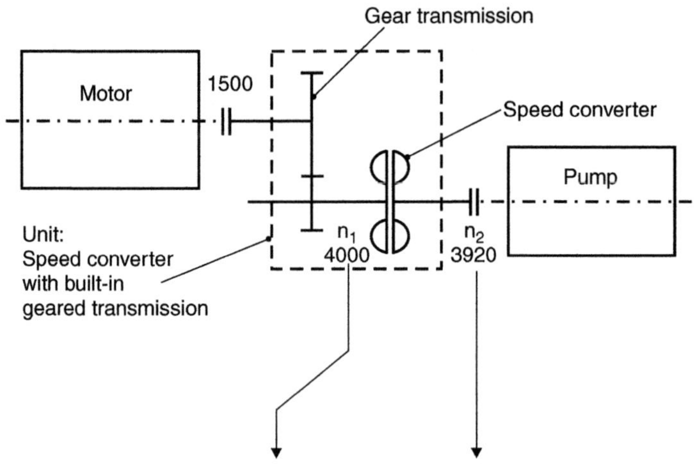

Figure 7.29 Schematic diagram of layout comprising pump, motor and speed converter with built-in geared transmission

When the resistance curve is flat (static and dynamic delivery head) the load variation must be taken into account.

Example of a variable-speed device for a centrifugal pump (see Fig. 7.29)

Given the following:

• Pump:	Rated delivery	$P_N = P_2 = 2588$ kW	
	data:	$n_N = n_2 = 3920\ \text{min}^{-1}$	
• Speed converter:	Rated slip:	$= 2\%$	average
	Characteristic	$a = 1.15$	values
	values:	$b = 15$	according to
		$y = 0.98$	supplier
			specifications

From this it follows that:

$$n_1 = \frac{3920}{0.98} = 4000\ \text{min}^{-1} \qquad P = P_2\left(\frac{4000}{3920}\right)^3 = 2750\ \text{kW}$$

$$P_1 = \frac{P_2\left(1 + a\frac{n_1 - n_2}{n_2}\right) + b}{y} = \frac{2588\left(1 + 1.15\frac{4000-3920}{3920}\right)}{0.98} + 15$$

where:

$P_1 = 2718$ kW (driving power applied to the speed converter)

$P_v = 2718 - 2588 = 130$ kW (power loss at rated speed)

Required to find:

1. Maximum power loss and corresponding speed n_m
2. Pump input and driving power at speed n_m
3. Pump output with throttle control
4. Difference between speed control with speed converter and throttle control (power uptake and savings)

R_e (a) with: $a = 1.15 \quad b = 15 \quad y = 0.98$

$$A = a + y - 1 = 1.15 + 0.98 - 1 = 1.13$$

$$B = \frac{b}{y} = \frac{15}{0.98} = 15.31$$

$$n_m = \frac{2a}{3A} = \frac{2 \cdot 1.15}{3 \cdot 1.13} = 0.678$$

$$P_{v_{max.}} = P \cdot \frac{4 \cdot a^3}{27\, A^2 \cdot y} + B$$

$$P_{v_{max.}} = \text{P}2750 \frac{4 \cdot 1.15^3}{27 \cdot 1.13^2 \cdot 0.98} + 15.31 = 510.5\ \text{kW}$$

i.e. 18.8% of $P_1 = 2718$ kW

$$n_m \cdot n_1 = \frac{2 \cdot a}{3 \cdot A} \cdot n_1 = 0.678 \cdot 4000 = 2712\ \text{min}^{-1} (\sim {}^2/_3 \text{ of } n_1 = 4000\ \text{min}^{-1})$$

R_e (b)

$$P_2 = P \cdot \left(\frac{n_2}{n_1}\right)^3 = 2750 \left(\frac{2712}{4000}\right)^3 = 857\ \text{kW}$$

$$P_1 = P_2 + P_v = 857 + 510.5 = 1367.5\ \text{kW}$$

R_e (c)

$$Q_2 = Q_N \cdot \frac{n_2}{n_1} = Q_N \cdot \frac{2712}{3920} = 0.692 \cdot Q_N$$

$$P_2 = P_1 = \sim 85\% \text{ of } P_N = 0.85 \times 2588 = 2200\ \text{kW}$$

R_e (d)

Difference between throttle control and speed variation control:

$$\Delta P = P_2 - P_1 = 2200 - 1367.5 = 832.5\ \text{kW}$$

i.e. power saving for speed variation control:

$$\frac{832.5}{2200} \cdot 100 = 37.8\%$$

– at rated speed $n_2 = 3920\ \text{min}^{-1}$:

$$\underbrace{\begin{matrix} n_1 = 4000 & n_2 = 3920\ \text{min}^{-1} \\ P_1 & P_2 \quad L \end{matrix}}_{P_v = 1320\ kW}$$

(Power loss corresponding to a rated slip of 2%)

– at rated speed $n_2 = 2712\ \text{min}^{-1}$:

$$\underbrace{\begin{matrix} n_1 = 4000 & n_2 = 2712\ \text{min}^{-1} \\ P_1 = 1367.5 & P_2 = 857\ \text{kW} \end{matrix}}_{P_v = 510.5\ \text{kW (maximum power loss)}}$$

NB Where there are several operating points the speed converter and cooler must be designed for the operating point which when converted to the load curves $H = f(Q)$ and $P = f(Q)$ (without rated slip, i.e. at driving speed nil) results in the largest load and therefore the heaviest slip (see Fig. 7.30).

This is understandable, since operation must be possible across the entire range of this resistance curve (from zero up to the duty point).

7.4.2.2 ELECTROMAGNETIC SPEED CONVERTERS

Electromagnetic speed converters are generally built with a stationary excitation coil or one that turns with the rotor.

The design principle of a speed converter with stationary excitation coil is shown in Fig. 7.31.

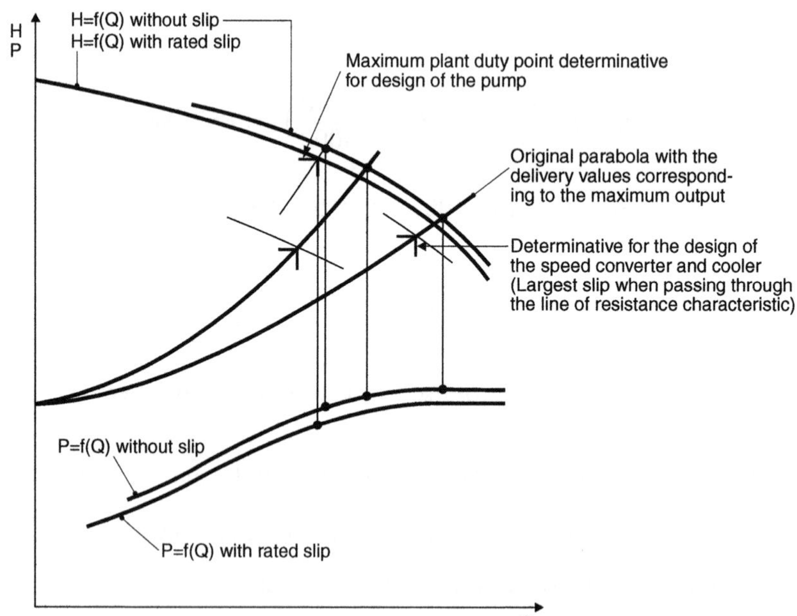

Figure 7.30 Design of a speed converter for several stated operating points

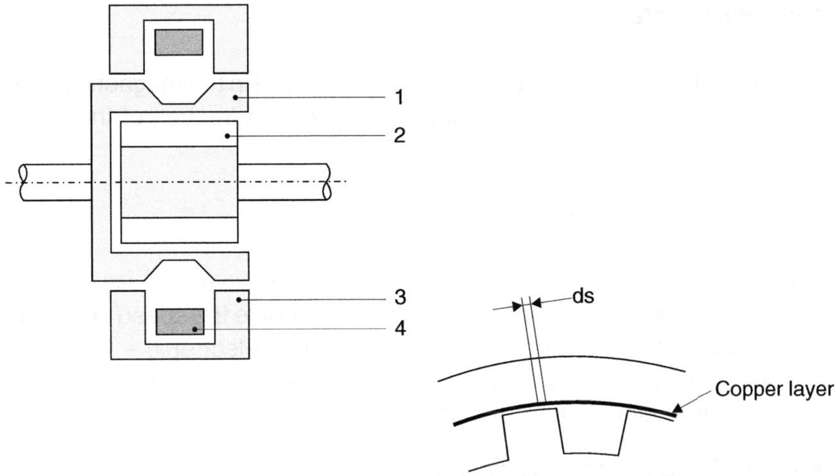

Figure 7.31 (1) Smooth driving rotor with copper layer driven at constant speed. (2) Toothed driven rotor, the speed of which is variable. (3) Stationary electromagnetic circuit excited by a coil. The magnetic field so generated flows through both rotors. (4) Coil

Functional principle

When the driving rotor is rotating and the driven rotor is stationary, one surface of the driving rotor is permeated consecutively by a maximum induction field (opposite a tooth) followed by a minimum induction field (opposite a gap).

As the magnetic flux taken up by this surface is variable, eddy currents are generated here. Like all induced currents these follow Lenz's law, i.e. they attempt to resist the cause of their generation. The driven rotor starts to rotate. The speed differential between the two rotors is reduced.

There is a torque/slip curve corresponding to every excitation value of the electromagnetic circuit (see Fig. 7.32).

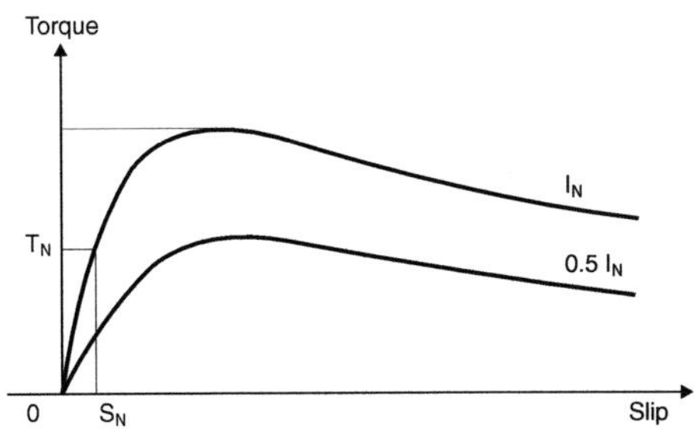

Figure 7.32 Design principle of an electromagnetic speed converter with stationary exciter coil

Operating features

- Good economy
  The electrical power taken up by the exciter coil in the magnetic circuit – power loss for excitation – amounts to only a fraction of the transmitted power.
  - maximum 2% for loads <25 kW
  - approx. 0.5 to 1.0% for loads <500 kW
- Control
  As already mentioned, speed variation is achieved by varying the exciter current. It is therefore possible to combine the speed converter with any kind of control system – preferably electronic – to form a control circuit. Pressure flow, temperature, speed and torque are among the relevant parameters for control.

Range of application and versions

The range of application is limited thermally, i.e. by the need to dissipate the slip power in the form of heat.

Speed converters are built with air or water cooling. In special applications (plants with explosion risk or special environmental conditions) or for large slip power losses water cooling is recommended.

Limits of application:

- max. speed <3600 min^{-1}
- max. torque <540 daNm.

Slip power loss

With respect to slip power loss the considerations set forth in section 7.4.2.1 hold good (but without the inclusion of auxiliary oil pumps and step-up gears).

The minimum rated slip is of the order of magnitude of 3 to 6.5%. The power loss for a specific speed converter corresponds to a maximum of 15 to 18% of rated power at 2/3 of rated speed with an origin parabola as resistance characteristic to slightly less when the resistance characteristic has a flat course.

Chapter | eight

Materials and Corrosion

8.1 INTRODUCTION

DIN 50900 describes *corrosion* as "the reaction of a material with its environment causing a measurable change in the material and possibly leading to corrosive damage". Furthermore, *corrosive damage* is defined as an "Impairment of the function of a component or an entire system through corrosion." This definition clearly distinguishes between the corrosion process and its results, namely the corrosive damage, two terms which are frequently incorrectly equated with each other. Corrosive damage is described as damage to a metallic component originating at its surface through chemical reactions of the metal with constituent elements of the environment. By contrast to mechanical wear, corrosion is therefore fundamentally a *chemical process*, which can be presented by the following reaction equation:

$$\mathrm{Me} \rightleftarrows \mathrm{Me} + Z \cdot \mathrm{e}^- \qquad (1)$$

A metal Me enters into a chemical reaction, as a result of which positively charged metal ions Me^{Z+} and Z electrons e^- are released. The metal ions so produced can remain in a dissolved state in the corrosion medium or settle on the corroding surface as insoluble corrosion products or accumulate as suspended matter in the corrosion medium. If in addition to the chemical corrosion reaction the metal surface is stressed by mechanical wear or by some other mechanical effect the corrosion reaction may be accelerated considerably.

8.2 FACTORS AFFECTING CORROSION

The cause of all corrosion reactions is the thermodynamic instability of metals in relation to air, water or other oxidizing agents. They tend to revert to a compounded state, corrosion products being formed from the pure metals and energy being released in the process. In the case of corrosion a material conversion takes place, the speed of which represents the corrosion rate. This material conversion rate is

Centrifugal Pump Handbook. DOI: 10.1016/B978-0-7506-8612-9.00008-5

governed by a general physical principle, according to which a current or a mass flow is proportional to a motive force and inversely proportional to a resistance:

$$\text{Mass flow} = \frac{\text{motive force}}{\text{resistance}} \tag{2}$$

In corrosion reactions, corrosion or transformation rate replaces the mass flow, the electrochemical potential replaces the motive force and a complex sum of individual resistances (e.g. reaction and diffusion resistances) replaces the resistance. "Motive force" is a thermodynamic quantity, while "resistance" is directly involved with reaction kinetics.

8.2.1 Thermodynamics of the corrosion process

Electrochemical potential is a measure of the ability of a system to function. By way of simplification it may be said that it characterizes the tendency of a metal to pass into the ionic form, i.e. to transfer electrons. Potentials are usually indicated in relation to the hydrogen electrode, which is taken as zero: metals with negative potentials are designated as "base", those with positive potentials as "noble". If metals are listed in decreasing order of potential, the well-known "Electrochemical Series of Metals" (Table 8.1) is obtained. These standard potentials are, however, of secondary importance in practice, as by definition they apply to solutions in which the metal ion concentration amounts to 1 mol/l. Such highly concentrated solutions are rare in practice. Much more use is made of so-called practical electrochemical series, of which the two best known were determined for the phthalate buffer (as model for fresh water) and seawater (Table 8.1). In addition to the absolute potential values, these electrochemical series can in certain cases also provide references to the mutual influencing of two metals in composite construction.

The corrosion resistance of a metal is essentially determined by its potential and the pH value of the solution. These factors affecting use are summarized in the so-called pH/potential diagrams. They contain the equilibrium curves for metal/solution systems and demarcate the conditions under which immunity, corrosion or passivity can be expected.

Figure 8.1 shows the simplified pH/potential diagram for chromium. Diagrams for other metals can be found in the specialized literature. It should be emphasized that passivity is basically different from immunity. While in the case of immunity corrosion is impossible for thermodynamic reasons, with passivity corrosion has practically ceased because of kinetic inhibition (formation of thin and so-called passive films derived from corrosion products). Of technical

Table 8.1 Electrode Potentials of Pure Metals at 25°C (on left) – Potentials in V_H; Potentials of Metals and Alloys in Practical Media (on right) – Potentials in mV_H

Electrode	$E°_{Me/Me^{Z+}}$		Seawater pH 7.5; 25 °C	Phthalate buffer pH 6.0; 25 °C
1	2	Gold	(+243)	(+306)
Cu/Cu^{2+}	+0.337	Silver	+149	+194
Cu/Cu^{+}	+0.521	Titanium	(−111)	+181
Hg/Hg_2^{2+}	+0.789	G-Cu Sn8		+156
Ag/Ag^{+}	+0.7991	L-Ag 60 Cu 27 In		+154
Pd/Pd^{2+}	+0.987	Cu Zn 37 Al	+28	+153
Pt/Pt^{+}	+1.2	Cu Zn 37	+13	+154
Au/Au^{3+}	+1.50	Cu	+10	+140
Au/Au^{+}	+1.7	G-Cu Al 10	−1	(+139)
H_2H^{+}	0.0	Ni 99.6	+46	+118
Fe/Fe^{3+}	−0.036	X22 Cr Ni 17 (800 N/mm²)	−134	+76
Pb/Pb^{2+}	−0.126	X20 Cr 13 (800 N/mm²)	−134	+7
Sn/Sn^{2+}	−0.136	X5 Cr Ni 189	−45	−84
Ni/Ni^{2+}	−0.250	A199.5	−667	−169
Cd/Cd^{2+}	−0.403	Sn	−148	−175
Fe/Fe^{2+}	−0.440	Hard chromium (50 µ) on steel	−291	−249
Cr/Cr^{3+}	−0.74	Tin solder		−258
Zn/Zn^{2+}	−0.763	GG 22 (cupola furnace)	−347	−346
Mn/Mn^{2+}	−1.18	GG 18	−455	−389
Ti/Ti^{2+}	−1.63	GG 12 (electric furnace)	−351	−404
Al/Al^{3+}	−1.66	Cd	−519	−574
Mg/Mg^{2+}	−2.37 (V)	Zn (100 µ) in steel	−806	−794
		Zn 99.995		−827

importance are the conditions necessary for the good corrosion resistance of the passivatable materials (e.g. stainless Cr and CrNi steels, aluminum titanium).

8.2.2 Kinetics of the corrosion process

The most important kinetic factors affecting use are concentration, temperature, mass transfer, potential and time.

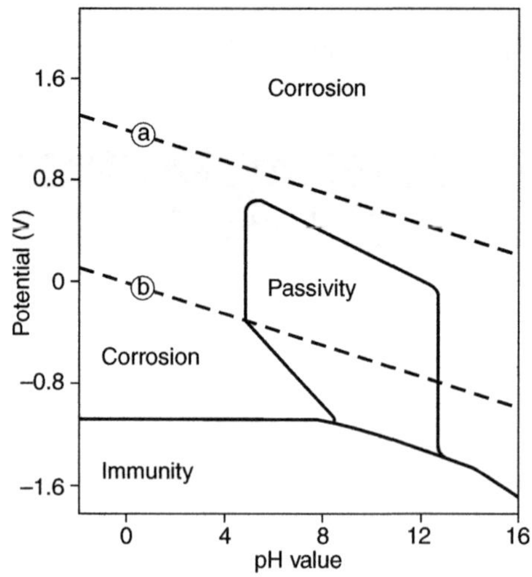

The lines a and b limit the stability range of water.

Figure 8.1 Simplified pH potential diagram for chromium

8.2.2.1 CONCENTRATION

For a given material and medium, the concentration of the aggressive components often plays an important part. An increase in concentration can affect corrosion speed negatively as well as positively. Thus the corrosion of unalloyed steels in saltless water intensifies with increasing oxygen content, and the same behavior is observed with the increases in hydrogen ion concentration (i.e. with a falling pH value). Completely contrary behavior is shown by CrNiMo steel in sulfuric acid, at least concentrations above approximately 50% (Fig. 8.2).

8.2.2.2 TEMPERATURE

It is generally valid that the rate of chemical reactions increases considerably with the temperature. As a rule of thumb, a 10°C temperature increase causes the reaction rate to double. Exceptions to this rule are, however, frequent and depend upon the respective conditions.

8.2.2.3 MASS TRANSFER

Flow usually favors corrosion. Natural and forced convection cause aggressive substances to be carried to the metal surface and corrosion products to be transported away. Moreover, flow-induced mechanical shear stressing of the surface is possible. The consequences are pitting and crevice corrosion in stagnant conditions, intercurrence of mass transfer and phase-boundary reaction at medium flow rates and erosion and cavitation corrosion at high flow rates. In Fig. 8.3 the influence of flow rate on the rate of corrosion is shown schematically. The horizontal

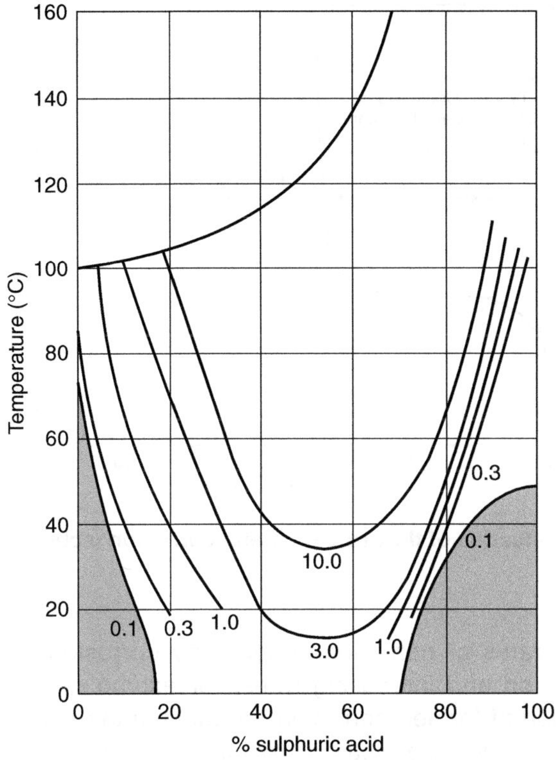

Figure 8.2 Iso-corrosion diagram for steel XSCrNiMo 1 8.10 in sulfuric acid. (Numbers in diagram: mm/year metal losses)

section of the curve corresponds to those conditions under which the phase-boundary reaction alone determines the rate of corrosion. In this case as many or more reactands per unit of time reach the metal surface as are consumed by the electrode reaction.

8.2.2.4 POTENTIAL

Potential displacements at a corroding metal surface usually have a very big effect on the rate of corrosion. It can be shown both mathematically and experimentally that the rates of electrochemical reactions can change by up to a factor of 10^5. Potential displacements can be due to the presence of redox systems, mating with other materials, external current sources, etc.

8.2.2.5 TIME

Time represents an important factor affecting corrosion. The dependence of a corrosion reaction on time can be very varied. Rates of reaction remaining chronologically constant are observed in the case of gaseous corrosion where no covering layer is formed, while

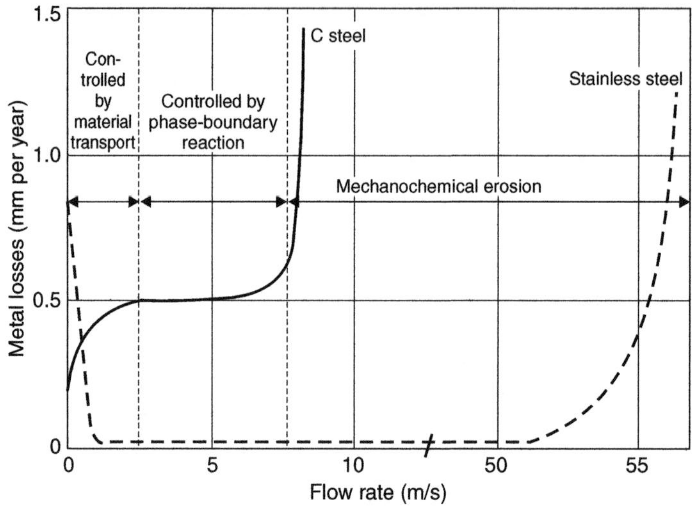

Figure 8.3 Influence of flow rate on metal corrosion (schematic)

decreasing rates of corrosion accompany corrosion with covering-layer formation and increasing rates, for example, are found when concentration of formed corrosion products in the corrosion process takes place (during standstill corrosion). Also typical are incubation times of a few to several thousand hours for the occurrence of certain types of corrosion, such as pitting, crevice and stress corrosion cracking.

8.3 GENERAL CONSIDERATIONS AFFECTING THE CHOICE OF MATERIAL

Reliability and working life of hydraulic machines depend among other things substantially on the chosen materials. In addition to questions of price, mechanical strength and machinability, corrosion resistance must also be considered when choosing materials. The significance of corrosion depends on the aggressiveness of the flow medium. Corrosion resistance is frequently the characteristic that determines the choice of material.

Corrosion resistance is not a simple material characteristic like the yield point of electric conductivity. There is actually no "corrosion resistance" as a material property, since corrosion behavior not only depends on the properties of the surrounding medium, but to a great extent is determined by other factors too (see scheme below).

The material is thus an element of a "stressing system", so that it cannot be considered in isolation, but always in the field of influence of its environment.

8.4 FORMS OF CORROSION

8.4.1 General

Corrosion phenomena on metallic materials in hydrous media (see Fig. 8.4) can be classified in three groups:

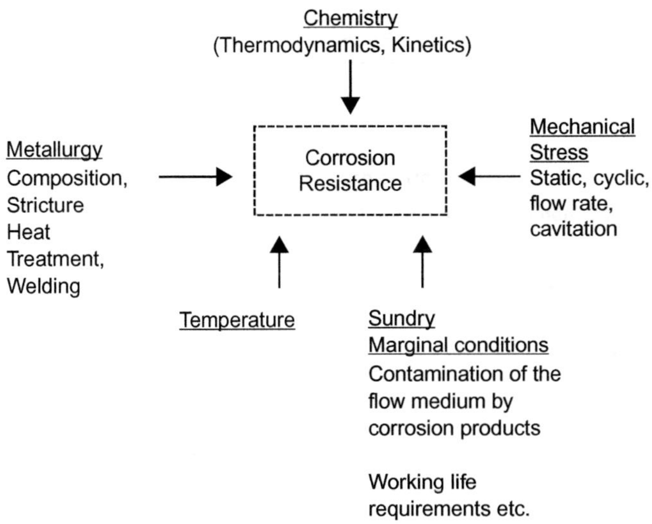

1. *Uniform corrosion*, in which the rate of corrosion is the same at all locations of the metal surface (e.g. metal dissolution in acids, simple rusting processes).
2. *Local corrosion without mechanical stressing,* during which the corrosion is limited to local narrowly defined locations (pitting corrosion, crevice corrosion, intercrystalline corrosion, galvanic corrosion, selective corrosion).
3. *Local corrosion in the presence of mechanical stressing*. It is possible to distinguish additionally between the mechanical stressing acting on the material in depth (stress corrosion cracking, corrosion fatigue) and only on its surface (erosion corrosion, cavitation corrosion).

Numerous forms of corrosion are specific to a given medium and/or material, i.e. they primarily occur in specific media on specific materials.

Uniform corrosion processes are usually much more harmless than local corrosion. In the former case the chronological progress of the material damage can be monitored and suitable measures undertaken such as increasing wall thickness, provision of spare parts, etc. The occurrence and extent of local corrosion, on the other hand, are often undetectable; in addition, the damage generally progresses much more quickly than in the case of uniform corrosion. In especially critical cases sudden fracture of a structural part can occur as a result of stress corrosion or corrosion fatigue.

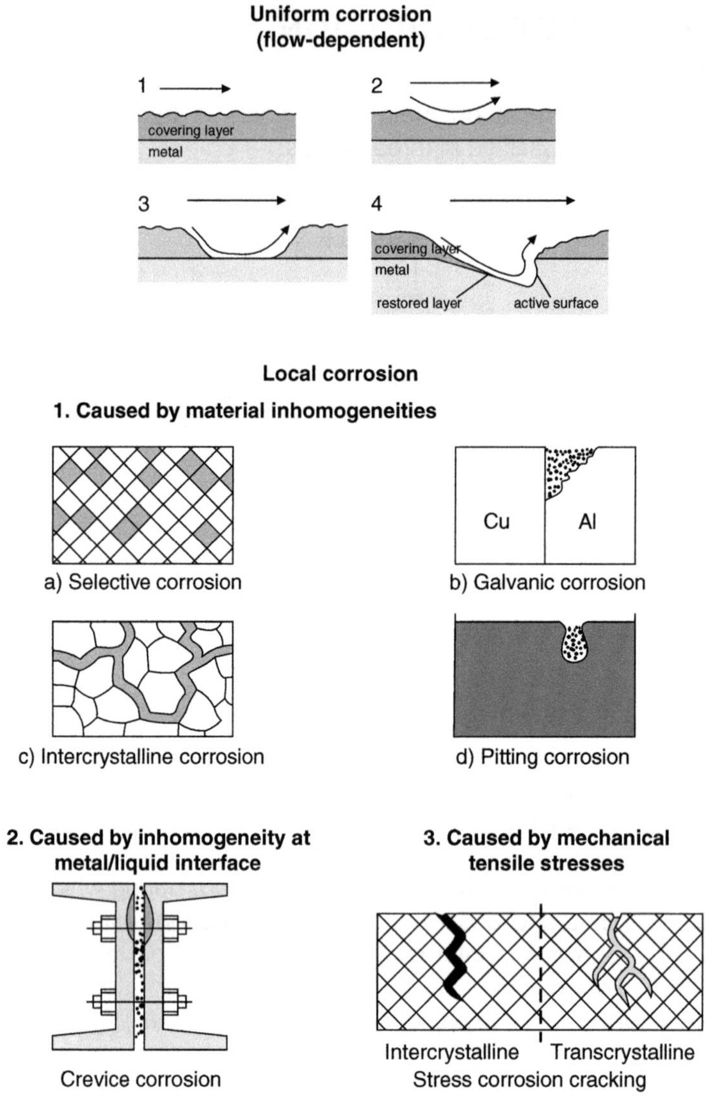

Figure 8.4 Forms of corrosion

Uniform corrosion can be obviated relatively easily by an appropriate choice of material. Frequently a change is made from unalloyed to chromium or chromium–nickel steel grades, usually of high-alloyed type. These display an essentially better resistance against uniform corrosion, but depending on the content of alloying elements and the composition of the flow medium, they can be susceptible to certain kinds of local corrosion. These alloys usually show an apparently bright surface free of covering layer. In fact, within certain potential ranges a very thin ($2–3 \cdot 10^{-6}$ mm) oxidic covering layer, a so-called passive film, is formed, and this is responsible for the excellent corrosion resistance of these materials. This phenomenon is called *passivity*. However, any local damage to this film, whether chemical or mechanical, creates the danger of a local corrosive attack.

8.4.2 Uniform attack

In the case of uniform attack metal removal takes place more or less evenly over the entire metal surface. Certain differences can occur in the microdomain, which lead to roughening. This form of corrosion is characteristic of aggressive media, there being no resultant insoluble corrosion or surface-clinging products to influence the rate of corrosion locally. Uniform attack is observable on numerous materials in acids.

8.4.3 Pitting corrosion

Characteristic of this type of corrosion are individual, mostly hemispherical small cavities which can lead to perforation of pipes and tanks (Fig. 8.5), while most of the metal surface shows no corrosion at all. The cause of pitting corrosion is local destruction of the passive layer with formation of a small corrosion anode, while the oxygen reduction can proceed unhindered on the large passive area. As a result of hydrolysis of the corrosion products, the pH value can fall to very low values in the cavity.

Pitting corrosion is initiated by chlorides in particular. It occurs especially frequently in stagnant solutions, e.g. during machine outages. This type of corrosion is furthered by rising temperatures. All passivatable materials, and especially austenitic stainless steel, are prominent among those susceptible to pitting corrosion. This can be countered by using steels with a molybdenum additive (usually 2.5% Mo), the amount of molybdenum required to ensure protection rising in proportion to chloride content and temperature.

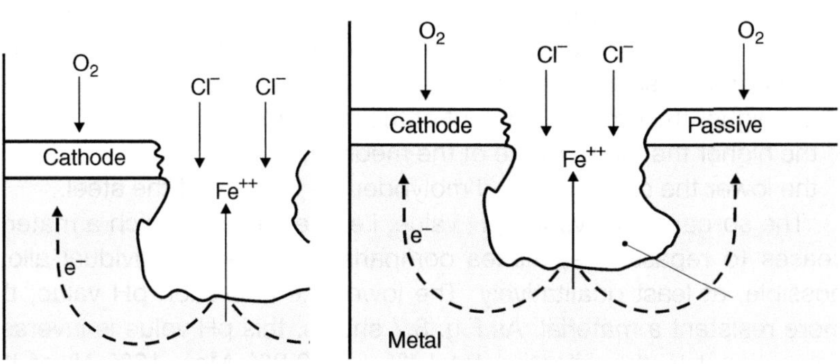

Figure 8.5 Pitting corrosion

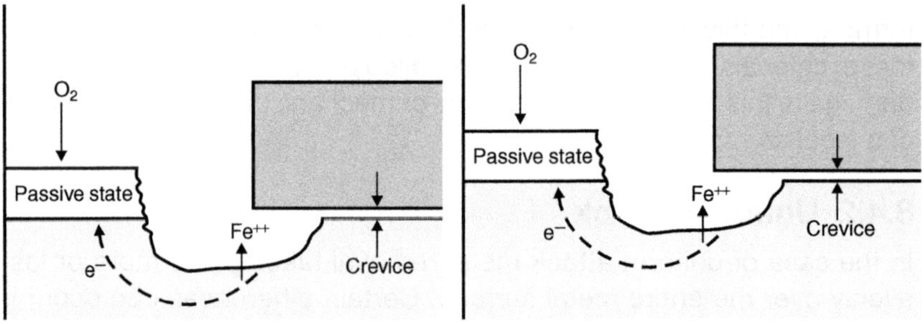

Figure 8.6 Crevice corrosion

8.4.4 Crevice corrosion

This is an attack on surfaces in a crevice filled with liquid (Fig. 8.6). Such crevices can be a consequence of the design but may also result from encrustations and deposits on metal surfaces formed during operation.

Crevice corrosion may be regarded as local breakdown of the passive condition resulting from oxygen deficiency and/or a pH decrease due to hydrolyzing corrosion products in the crevice caused by impediments to emigratory diffusion. Crevice corrosion occurs frequently with stainless steels, in particular when the medium contains chlorides.

The susceptibility of stainless steels to crevice corrosion falls with increasing stability of the passive film, i.e. with increasing chromium and molybdenum content, namely in the sequence 13% chromium steel $<$ 17% chromium steel $<$ 18/8 chromium–nickel steel $<$ 18/8 chromium–nickel–molybdenum steel. In seawater adequate resistance is only guaranteed with an effective total (% Cr + 3.3% Mo + 16% N) $\geq$ 30.

Pitting corrosion and crevice corrosion are very similar, at least in terms of their mechanisms. In both cases the corrosion is brought about by local damage to the passive film, with chlorides playing a significant part. As a result of hydrolysis of the resultant corrosion products, the pH value of the electrolyte within the pits or crevices becomes acid, as a result of which corrosion accelerates and repassivation is impeded. The danger of corrosion is greater

- the higher the chloride content of the medium;
- the higher the temperature of the medium;
- the lower the chromium and molybdenum content of the steel.

The so-called activation pH value, i.e. the value at which a material ceases to repassivate, makes comparison between individual alloys possible, at least qualitatively. The lower the activation pH value, the more resistant a material. As Fig. 8.7 shows, this pH value is inversely proportional to the effective total (% Cr + 3.3% Mo + 16% N) of the alloying elements chromium and molybdenum.

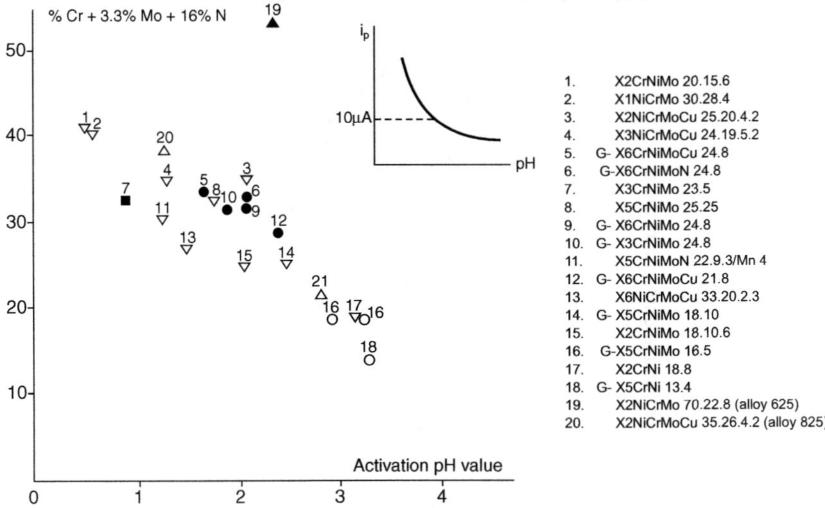

Figure 8.7 Activation pH values for various Fe and Ni alloys in seawater

8.4.5 Intercrystalline corrosion

Intercrystalline corrosion is the term used when corrosion primarily affects the grain boundaries of a polycrystalline metal. Even when total metal loss is very small this form of attack leads to the breakdown of crystallite cohesion and – as a consequence of grain dissociation – grain disintegration (Fig. 8.8). The most important factor affecting the intercrystalline corrosion is the material itself, or rather its composition.

Typical examples of the materials susceptible to this type of corrosion are highly alloyed austenitic chromium–nickel steels and ferritic chromium steels.

Within a temperature range of 550 to 750°C these alloys precipitate chromium carbide with the formula $Cr_{23}C_6$ at the grain boundaries. As a direct result of these carbide precipitations the free chromium content falls to a value which no longer suffices for passivation, namely below approximately 13% (Fig. 8.8). Carbide precipitation can occur at all grain boundaries of a workpiece as a result of heat treatment, but during welding it only takes place at the edges of the weld seam (the so-called heat-affected area). The carbide precipitation and chromium depletion resulting from this are dependent upon the influence of temperature and its duration, and also on the carbon content. Low-carbon steels with C contents below 0.03% (the so-called L grades) are resistant to grain dissociation, if the heat influence does not last too long. Thus heat treatment or welding of these steels is possible without difficulty. However, only stabilized steels (with titanium and/or niobium additions) can be considered for continuous service at higher temperatures, as

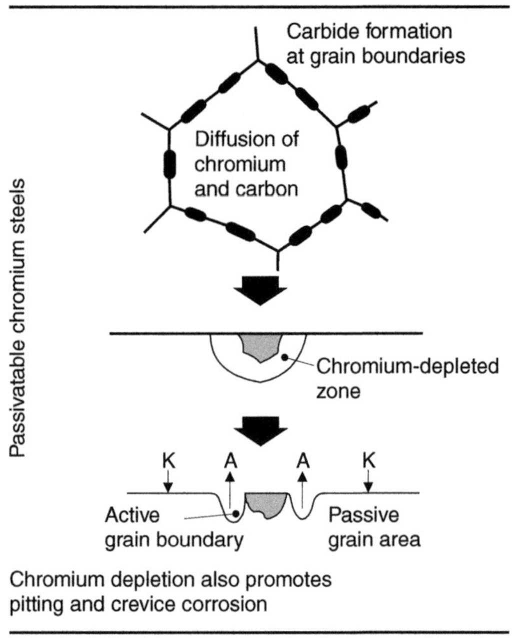

Figure 8.8 Intercrystalline corrosion

these stabilizing additives form more persistent carbides than does chromium. The time, temperature and carbon factors affecting the intercrystalline corrosion are graphically illustrated in the so-called TTS diagrams (time, temperature, sensitization) (Fig. 8.9).

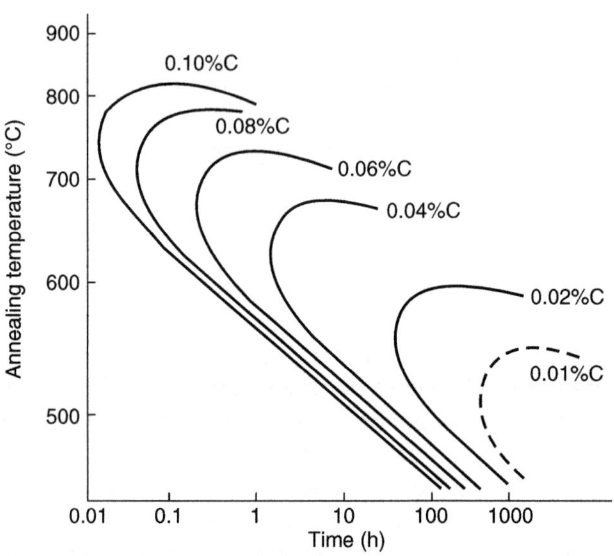

Figure 8.9 Time–temperature sensitization diagram for austenitic CrNi steels as a function of the C content

8.4.6 Galvanic corrosion (contact corrosion)

If two different metals are in electrically conducting contact, galvanic corrosion can occur if their individual potentials are sufficiently far apart. In such cases the base metal corrodes more quickly than if it were isolated from the more noble one. Conversely the contact results in a reduction in corrosion of the more noble metal (cathodic protection). It is especially critical if the surface of the base metal wetted by the medium is small compared with the surface of the more noble one.

The danger of galvanic corrosion is greater the higher the salt concentration (electrical conductivity) of the medium. It is, for example, very small in organic liquids or demineralized water; in seawater, however, even potential differences of 100 mV can become problematical.

8.4.7 Selective corrosion

Selective corrosion can occur in two-phase or multi-phase alloys, in that the base phase is selectively dissolved out while the more noble phase remains undamaged; the form of the component is practically unchanged. The damage is finally manifested by porosity or mechanical failure. The technically most important cases of selective corrosion occur with cast iron (graphitization, spongiosis), brass (dezincification) and certain aluminum bronzes (dealuminification).

Causes of selective corrosion include chlorides, slightly acid pH values, a reduced oxygen supply and insufficient circulation of the medium. The danger of selective corrosion can be prevented by technical alloying measures.

8.4.8 Stress corrosion cracking (SCC) and corrosion fatigue

These types of corrosion are characterized by the occurrence of cracks and/or brittle fractures at stresses far below the yield point of the undamaged material.

Stress corrosion cracking occurs at static tensile load, and in fact only with certain medium–material combinations. The SCC of austenitic CrNi steels in chloride-containing media is of especial technical significance. The danger of damage increases with the chloride content, the temperature and the stress level. Below approximately 70 to 80°C the occurrence of SCC is, however, unlikely.

Since in practice the composition of the medium cannot usually be changed, design measures (e.g. low mechanical stresses) or alloy-technical measures must be taken to avoid SCC. The use of CrNi alloys with increased Ni content (above 25 to 30%) in many cases enables the danger of SCC to be minimized. Duplex steels also show enhanced resistance.

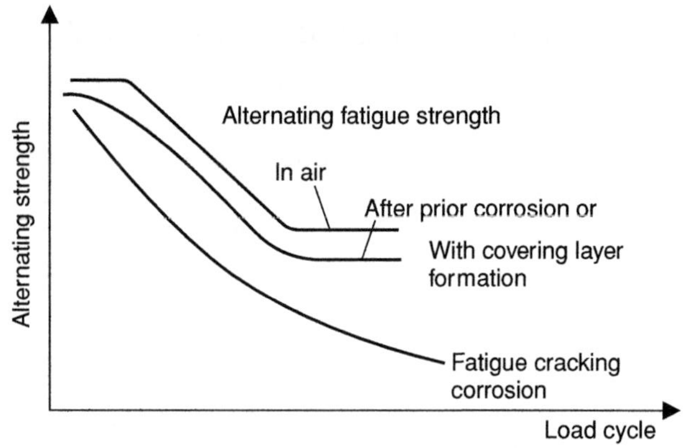

Figure 8.10 Alternating fatigue strength. Schematic for air (Wöhler curve) and in a corrosive medium

Fatigue cracking occurs in the case of a cyclically alternating mechanical load; it is not specific to either materials or media, i.e. it can occur with all metallic materials and in all corrosive media. As Fig. 8.10 shows, endurance limits are no longer reached under cyclic loading when acting *together with* a corrosion influence, in contrast to the fatigue curve measured under exposure to air (Wöhler curve), while for a required fatigue strength area the number of load cycles to fracture is reduced.

Because of the dependence of corrosion on time the attainable number of cycles to fracture becomes exceptionally dependent on frequency. In the fatigue strength area of the Wöhler curve mechanical influences predominate in general when the frequency is not too low, while at low alternating loads and thus a higher number of load cycles the corrosion effect becomes more and more evident, as it has increasingly more time to exert its influence.

8.4.9 Erosion corrosion

Erosion corrosion is the accelerated destruction of metallic materials in fast-flowing corrosive liquids. With increasing flow rate corrosive substances are brought more quickly to the metal surface, while reaction products are completely or partly removed. In Fig. 8.3 an attempt has been made to illustrate schematically the influence of flow rate on the damage rate.

In the velocity ranges controlled by transport as well as those controlled by phase reaction the shear forces generated by the flow at the surface of the metal cause partial or total destruction of the covering film (Fig. 8.11), thus making it possible for corrosion to increase abruptly.

At very high flow velocities and/or with unfavorable hydraulic profiling, high mechanical forces can act on the metal surface (liquid impact,

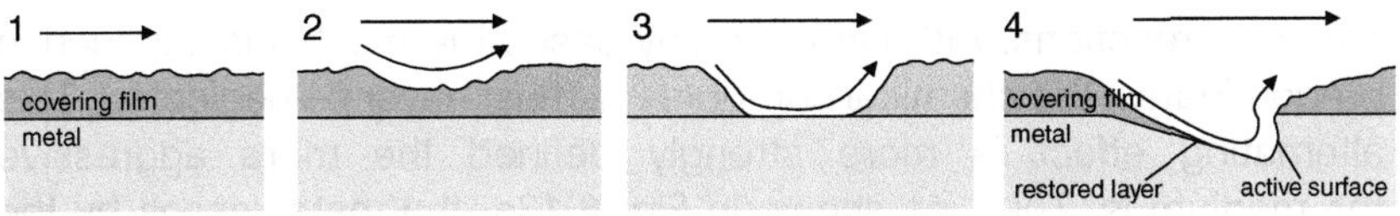

Figure 8.11 Influence of flow rate on the formation of covering film

cavitation) and so cause direct mechanical destruction of the material. In general corrosion is accelerated here as well. The term used for this is mechano-chemical metal damage.

For every material there is a flow rate which must not be exceeded in engineering installations. Furthermore, this flow rate is dependent upon the respective medium, and the more corrosive this medium acts, the lower is the allowable flow rate. Figure 8.12 shows the dependence on corrosive behavior of a number of alloyed steels in chloride brines containing hydrogen sulfide. The limiting flow rates are clearly visible in this presentation.

8.4.10 Cavitation corrosion

If the static pressure in a pump falls below the vapor pressure of the liquid being pumped, vapor bubbles are formed, which on their implosion can cause cavitation damage. As with erosion corrosion, the material cavitation consists of an interplay of mechanical stresses and corrosion–

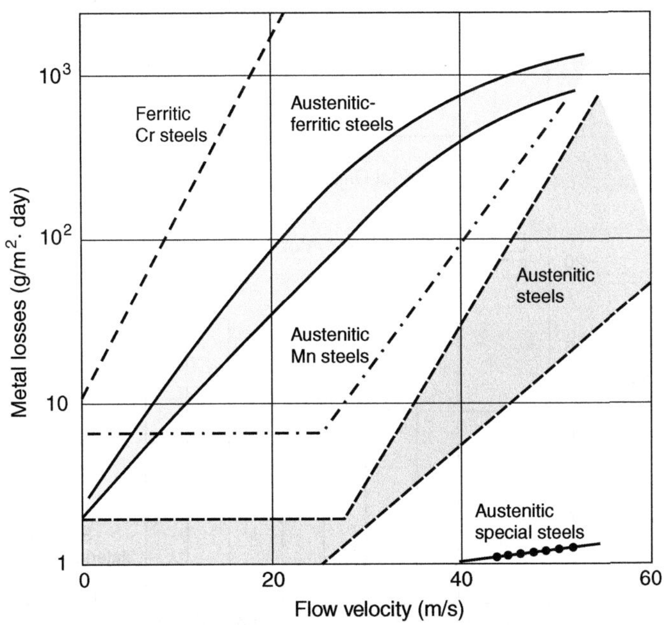

Figure 8.12 Rates of corrosion of stainless steels as function of flow velocity in m/s. Medium: injection water, 60°C, 23 000 mg/l dissolved salts, pH 4.5

chemical reactions. Although in many cases the mechanical influence predominates, the chemical components must never be neglected. This alternating effect is more strongly defined the more aggressive the medium is. Thus, as shown in Fig. 8.13a, the metal losses by the cavitation of an Al-multicomponent bronze in seawater increases with increasing H_2S content. For austenitic CrNi alloys too the damage rate in aggressive seawater is appreciably higher than in less corrosive demineralized water (Fig. 8.13b).

In only slightly corrosive media (e.g. water) the cavitation resistance of a material improves with rising hardness. Thus is illustrated in Fig. 8.14 using Cu alloys as an example. Similar dependence on material hardness has been established for the group of stainless chromium steels. In a corrosive liquid, on the other hand, the choice of a cavitation-proof material may not be based on its hardness alone. On the contrary, the best

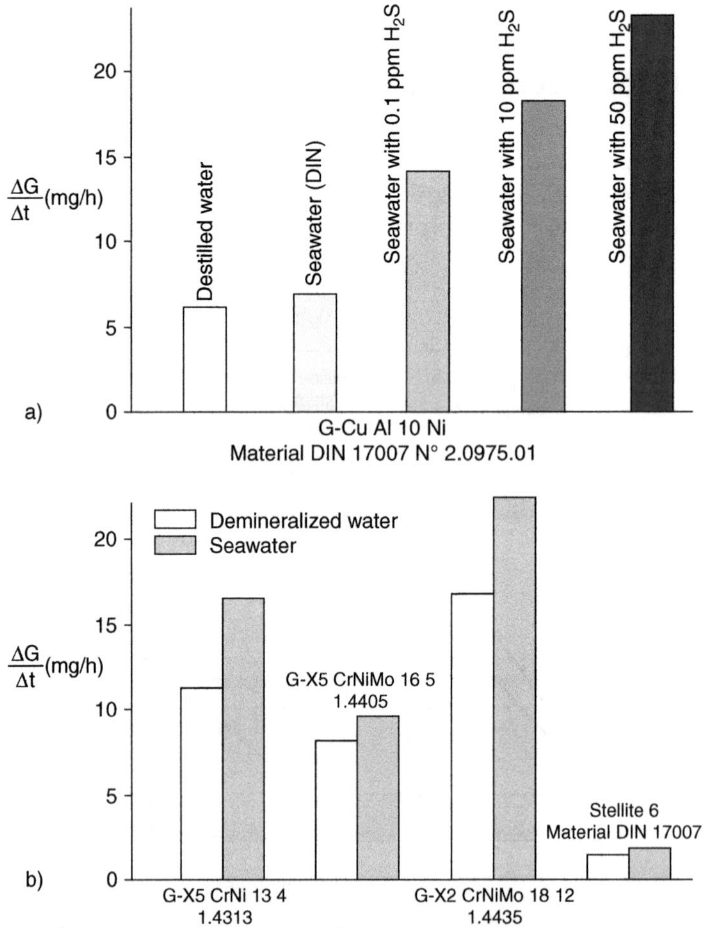

Figure 8.13 Cavitation behavior of metals in synthetic seawater

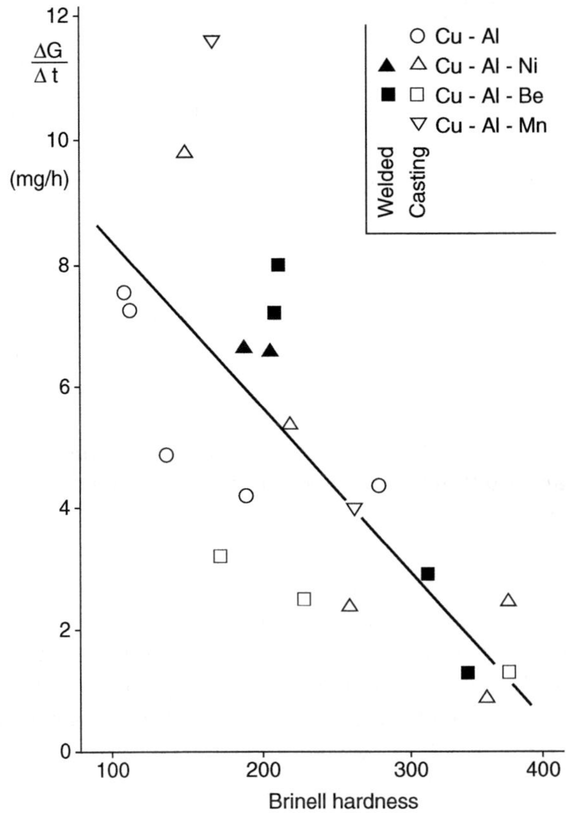

Figure 8.14 Cavitation resistance of Cu alloys in water as a function of hardness

cavitational resistance is displayed by materials with good toughness coupled with a high hardness value and an optimal corrosion resistance.

8.5 CORROSIVE PROPERTIES OF LIQUIDS BEING PUMPED

A distinction is drawn between non-aqueous and aqueous liquids.

Non-aqueous liquids, typically organic substances such as oil, petrol, etc., usually have no corrosive properties at all in relation to metals. If these substances contain water, then an increase in corrosivity will only be observed if the water solubility level is exceeded. Caution is required with organic acids, amines, chlorinated hydrocarbons and the like.

Aqueous liquids often contain dissolved salts and gases; their pH value fluctuates from acid via neutral to alkaline. Depending on the source the following groups of waters can be distinguished:

- natural waters: e.g. groundwater, surface water, seawater;
- treated waters: softened, demineralized and/or degassed water;

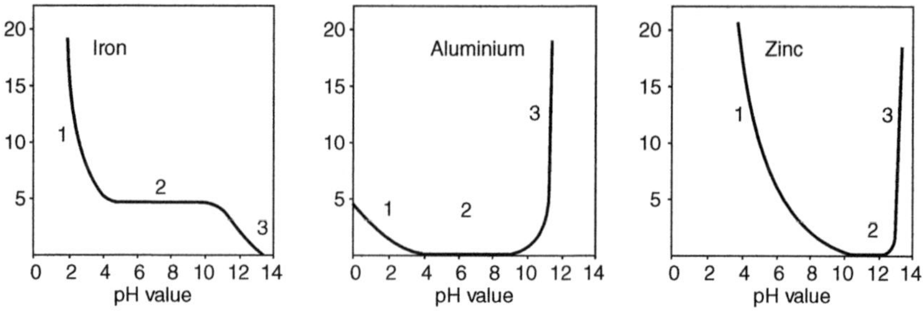

pH-values (above approx. 9–10); non-ferrous metals too, corrode increasingly in the alkaline range.

Iron	Aluminium	Zinc
1 H_2 type corrosion	1 H_2 type corrosion	1 H_2 type corrosion
2 O_2 type corrosion	2 Passivity	2 Passivity
3 Passivity OH	3 H_2 type corrosion	3 H_2 type corrosion

Figure 8.15 Corrosion behavior of iron, aluminum and zinc in solutions containing air at different pH values. Rates of corrosion in g/(cm^2 – day) versus pH values

- special chemicals and salt solutions: e.g. cooling brines, effluent from chemical processes or other industrial sectors.

The corrosion behavior of most metals is determined to a great extent by pH value. This influence can, however, vary considerably, depending on the metal, as is illustrated in Fig. 8.15 for the metals iron, aluminum and zinc. In general – this does not apply only to the metals mentioned – the rate of corrosion at neutral pH values (about 7 to 8) is moderate to slight. At pH values above 9.5, the corrosion rate falls off to very low values, while corrosion of the amphoteric metals (aluminum, zinc) is very strongly accelerated at alkaline pH values (above approx. 9–10); non-ferrous metals, too, corrode increasingly in the alkaline range.

Most metals are unstable in acid media. From case to case highly alloyed Cr or CrNi steels, titanium, Hastelloy alloys, silicon iron and the like can be used, but limits also exist for these materials, depending on type of acid, acid concentration and temperature of application. If necessary, recourse must be made to plastics, ceramics, glass or graphite.

Natural waters contain among other things a varying quantity of calcium carbonate, which is responsible for the "hardness" of the water. The concentration of this substance is stated in degrees of hardness in addition to the customary concentration units. Carbonate hardness is an important criterion when using unalloyed and low-alloyed steels (Cr content <13), as it is favorable to the formation of protective hardness layers. The stability of these layers is dependent on pH value. Equilibrium curves can be determined and based on a given water composition, used to establish whether a water is aggressive or forms covering film, i.e. is non-aggressive (Fig. 8.16). Such curves can help in choosing materials, though additional substances contained in the water must be taken into consideration at the same time. It can also be deduced from this that soft

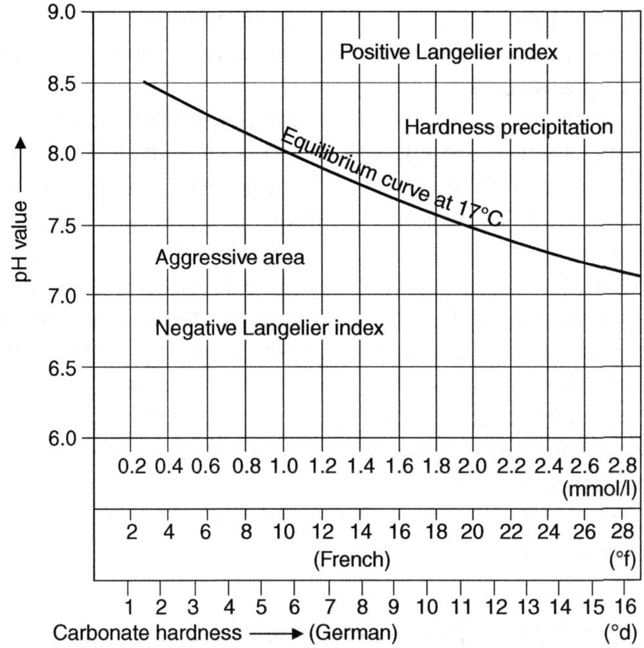

Figure 8.16 Simplified illustration of calcium–carbon dioxide equilibrium in the form of a carbonate/pH diagram

waters, including *softened or demineralized water*, act corrosively on unalloyed and low-alloyed steels and cast iron.

In the range of neutral and weakly alkaline pH values the *oxygen content* of the water plays an essential part in the corrosion process. Without oxygen corrosion the process is normally slow (thermodynamically impossible). The oxygen comes from the air and the absorbed quantity of oxygen is proportional to the partial oxygen pressure and inversely proportional to the water temperature.

Affecting corrosion by removing the dissolved oxygen presupposes that the latter can be reduced to a limit of 30 ppb (parts per billion, number of parts in 10^9 parts; 1 ppb = 1 mg/m^3) or less. This is possible by means of thermal degassing or adding chemicals which remove the oxygen by combination (e.g. hydrazine, sulfite).

Seawater contains on average about 30 000 ppm or 3% dissolved salts, though salt content can vary considerably from one sea to another (e.g. Baltic Sea 0.7%, Dead Sea 21%). About 80% of the salt content of seawater is sodium chloride, which means that seawater is very corrosive. Components which are in contact with seawater must be made of high-grade alloys (e.g. Albronze, CrNi steels containing molybdenum) and/or protected by suitable methods (cathodic corrosion protection, non-porous metallic or organic coatings). Particularly in coastal areas, seawater is often heavily polluted (brackish water). Mention may be made, for example, of hydrogen sulfide and ammonium ions, which even in the smallest concentrations – often in the ppm range – can drastically intensify aggressiveness.

In crude oil production too, it is often necessary to handle salt solutions, some of which are aggressive. These liquids, known as injection water, quite often contain a salt content approaching the solubility limit (about 27 to 30% by weight). Contents of hydrogen sulfide of a few hundred milligrams per liter and slightly acid pH values up to about 4.5 are frequent. For pumping such liquids only special materials (among others high-alloyed austenitic steels, duplex steels) can be considered.

Chemical solutions can have very variable compositions, such as salt solutions of highly diverse kinds and concentrations, mineral acids and alkaline solutions and organic liquids. The suitable material must be determined from case to case, and appropriate literature with resistance tables be consulted. The use of such tables, however, calls for great caution, as the data contained therein apply to *pure chemicals* and furthermore are not relevant to pump applications.

8.6 CORROSION PROBLEMS IN HYDRAULIC MACHINES

The choice of materials for hydraulic machines is influenced to a considerable extent by the aggressiveness of the medium conveyed. In principle each of the types of corrosion described can occur in a pump individually or together with others. Moreover high flow rates obtain, so that near the metal surface a complex interaction of mechanical stresses and chemical reactions takes place.

Modern pumps have an extremely wide *spectrum of flow velocities*, namely from about 0 to almost 100 m/s. Conditions for crevice corrosion (stagnant medium) and for erosion and cavitation corrosion often obtain simultaneously in the same machine. These variable flow conditions must be considered when choosing the material. Metal surfaces submerged in flowing media containing chlorides are endangered by local types of corrosion of local character (pitting and crevice corrosion).

The *temperature* of the flow medium also has a great influence on material behavior. Usually the rate of corrosion doubles with a temperature increase of 10°C. In addition temperature presents a critical factor affecting the occurrence of local corrosion. Thus in seawater temperatures over 60°C must be regarded as critical for CrNi steels containing molybdenum.

In non-aggressive media the hardness of a material is a parameter materially influencing *wear characteristics*. (Figure 8.17 shows the wear-resistance coefficients of various materials in sand/water suspensions.) By contrast, much more complicated conditions often prevail in corrosive liquids. Under corrosive–mechanical stress the mechanical

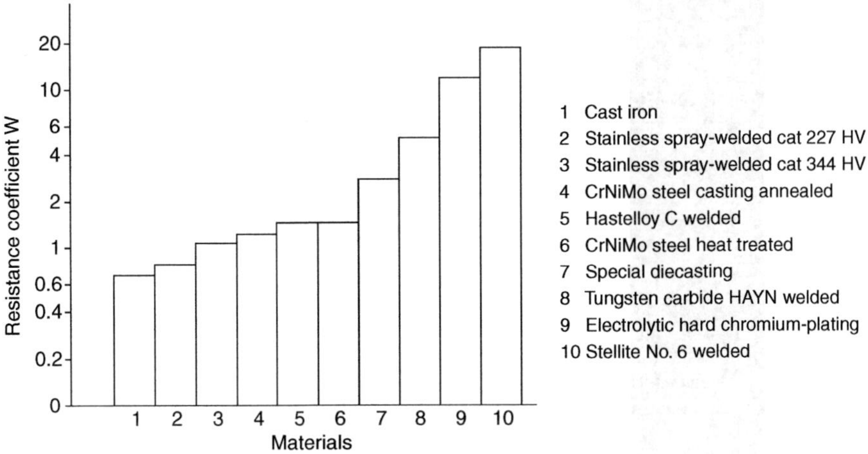

Figure 8.17 Relative coefficients of resistance of materials to abrasive wear

characteristics of the covering layers play a greater part than those of the base material. In addition, for many materials corrosion resistance falls with increasing hardness.

For hydraulic machines, as with other components of an installation, *standstill phases* represent critical periods. Temporary, locally limited changes in the medium enclosed in the pump may cause the formation of macro-corrosion elements which in certain cases lead to corrosion damage. Periodic starting of the pump or flushing with suitable liquid can prevent such damage.

In Table 8.2 an attempt has been made to illustrate the corrosion characteristics of some frequently used materials in various flow media. Such presentations are necessarily highly condensed and must therefore always be treated with great caution by users.

A reliable choice of the construction materials to be used can only be made if the exact *composition of the medium* is known. Components present in very small concentrations only can still be of importance in this connection.

Important parameters include pH value, the content of calcium carbonate, chlorides, sulfates, suspended solids, etc. The total content of dissolved salts is of interest, insofar as it permits deductions as to the completeness of the analysis provided. Information on the content of dissolved gases is also important. Of special significance is the content of dissolved oxygen, carbon dioxide, hydrogen sulfide, etc.

Samples must be taken in such a manner that they supply representative results for the plant (this applies in particular to dissolved gases and suspended matter). Where fluctuating operating conditions occur, care is to be taken that they are covered by a sufficient number of analyses. Because of the danger of a temporary change in the sample,

Table 8.2 Examples of Materials Recommended for Pumps

	Material No. as per DIN (examples)	Natural waters, equil. water	Treated waters: softened, deminer-alized	Water or boiler feedwater O_2-free	Salt water			Sea water	Injection water	Organic liquids, e.g. oil, hydro-carbons
					Neutral	Acid	Alkaline			
Cast iron, incl.	0.6025	+	–	+	%	–	+	–	–	+
Low-alloyed	0.7040									
Steel casting	1.0619	+	–	+	%	–	+	–	–	+
Incl. low-alloyed	1.7706									
Ni-Resist	GGG-NiCr202	+	+	+	+	%	+	+	–	+
Si cast iron	G-X70 Si 15	+	+	+	+	+	+	+	+	+
	1.4027									
13% Cr steel	1.4313	+	+	+	+[1]	%	+	–	–	+
16% Cr steel	1.4405									
CrNi steel	1.4306	+	+	+	+[1]	%	+[1]	–	–	+
Type 18/8	1.4308									
	1.4404	+	+	+	+	%	+	+	%[3]	+
CrNi steel	1.4408									
Type 18/8/2.5	1.4581									

(Continued)

Table 8.2 Examples of Materials Recommended for Pumps *(continued)*

	Material No. as per DIN (examples)	Natural waters, equil. water	Treated waters: softened, demineralized	Water or boiler feedwater O_2-free	Salt water			Sea water	Injection water	Organic liquids, e.g. oil, hydrocarbons
					Neutral	Acid	Alkaline			
	1.4460	+	+	+	+	%	+	+	$+^3$	+
Duplex steels	1.4462									
NiCrMo steel	1.4500	+	+	+	+	%	+	+	+	+
Bronze (Sn)	2.1050.01	+	+	%	$+^2$	–	–	–	–	+
Al bronze	2.0966	+	+	%	$+^2$	–	–	$+^3$	–	+

Symbols: ++ resistant

% conditionally usable
– unstable
[1]in the absence of much chloride
[2]in the absence of ammonia or NH_4^+
[3]according to H_2S content and pH

several analyses must be carried out locally (this applies in particular to the dissolved gases).

8.7 CONCLUSIONS

The corrosive properties of the conveyed medium essentially determine the choice of materials for pumps. In this connection the influence of flow velocity is to be taken into particular consideration. Erosion and cavitation damage can only be avoided by an optimal flow profile, by observing hydraulic requirements and by a correct choice of materials. For this reason the use of so-called material-corrosion tables is very questionable, as the corrosion data they list have usually been determined under stagnant conditions (total immersion tests). A correct choice of material can only be made in the light of many years of experience or practical erosion tests.

Chapter | nine

Principal Features of Centrifugal Pumps for Selected Applications

Sulzer first began making centrifugal pumps in 1860.

Today Sulzer manufactures pumps for the following application areas:

- Oil and gas
- Hydrocarbon processing
- Power generation (fossil and nuclear)
- Water and wastewater
- Pulp and paper
- General industry.

Oil and gas applications include the following:

- Off and onshore production – water injection
- Pipelines
- Floating production storage offloading (FPSO) vessels
- Multiphase.

Hydrocarbon processing applications include:

- Crude oil refining
- Synfuels
- Gas processing
- Petrochemicals
- Nitrogenous fertilizers.

Power generation applications include the following:

- Feedwater circuit
- Condensate circuit
- Cooling water circuit
- Auxiliary circuits.

Water and wastewater applications include:

- Pipeline

Centrifugal Pump Handbook. DOI: 10.1016/B978-0-7506-8612-9.00009-7

- Drinking water supply
- Seawater desalination
- Irrigation
- Drainage
- Mine dewatering
- Sewage.

Pulp and paper applications include:

- Paper production
- Pulp production.

General industrial applications include:

- Biofuels production
- Metals production
- Sugar production.

A few prominent pump types from various application areas are briefly described.

9.1 OIL AND GAS

Liquid hydrocarbons and gas account for nearly 60% of the world's energy consumption. This essential energy source relies on pumps at every stage of production and transport. Typically the pumps specified comply with ISO 13709 (API 610) together with additional stringent customer-specific specifications. Many applications require custom engineered solutions and extensive testing to provide the required performance. Another common feature of many key upstream applications is the requirement for large, high energy machines, with designs of up to 27 MW having already been put into service. Because of the remote nature of many oilfields or pipeline pumping stations reliability and long service life are essential to the economic operation of the field or pipeline. Materials of construction, special design features and surface coatings are used to resist both erosive and corrosive attack, maximizing operating life between maintenance intervals.

9.1.1 Injection pumps (see Fig. 9.1)

As exploitation of an oilfield progresses, the pressure drops in the reservoir. Water injection pumps are used to maintain reservoir pressure and drive the oil to the producing wells. Injection wells are located all round the field and inject sea, brackish or produced water (ρ = 1.03 kg/l) into the oil-bearing formation to boost the yield. Compared with the line losses the delivery head is very high, so that the resistance curve is practically a straight line with slight inclination. Consequently, stable characteristics are crucially important in parallel operation. Depending upon injection pressures needed, horizontally split case or barrel-casing pumps with pull-out cartridge are employed. The pumps must meet strict customer specifications in addition to the requirements of ISO 13709 (API 610).

In order to maintain production rates and reservoir integrity, rapid replacement of the rotor assembly of an injection pump is necessary.

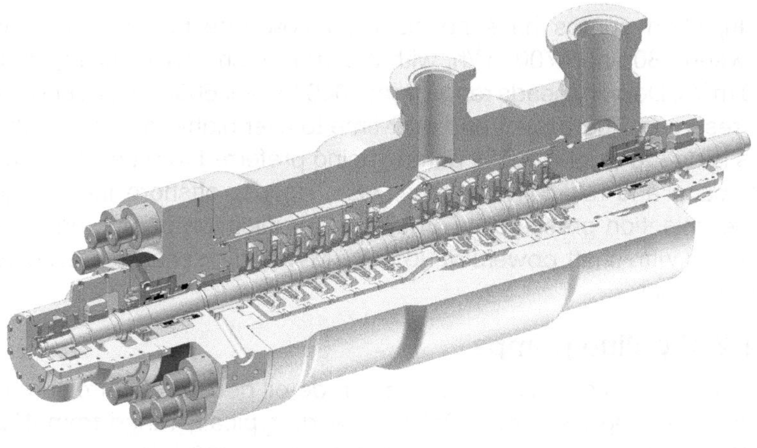

Figure 9.1 A Sulzer HPcp high pressure back-to-back injection pump

This is easily achieved with horizontally split casing designs. For barrel pumps this is achieved by using a full pull-out cartridge design that allows the complete hydraulic assembly including seals and bearings to be removed as a single unit.

Due to the high chlorine content of produced water and high impeller peripheral speeds necessitated by the high heads, special attention must be given to the choice of materials. Duplex and super duplex steels are now the most commonly used.

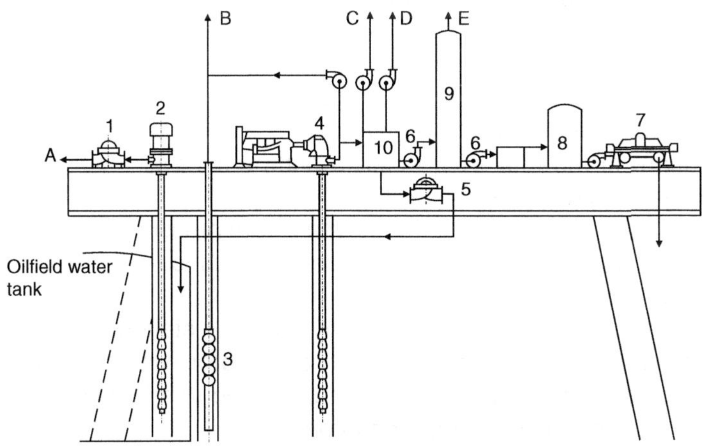

A Pipeline
B Fire fighting
C Wash water
D Cooling water
E Air venting

1 Pipeline pump
2 Booster pump (crude oil and ballast)
3 Fire fighting pump
4 Seawater lift pump
5 Charge pump (ballast and crude oil)
6 Auxiliary pumps
7 Injection pump
8 Seawater treatment module
9 Degassing module
10 Seawater system

Figure 9.2 Simplified schematic of typical processes on an oil platform

Injection pumps have no clear-cut flow rate ranges. Flows vary between 180 and 4700 m^3/h, with the current emphasis mainly around 720 m^3/h. Delivery heads range from 1000 to over 6000 m, depending on the reservoir. There is a trend over time to ever higher heads and flows. This is due to single, 100% pumps being preferred to parallel operation for space and reliability reasons. Also, deeper offshore fields require higher injection pressures. To date Sulzer injection pumps have been supplied with shaft powers of up to 27 MW and discharge pressures of over 600 bar.

9.1.2 Pipeline pumps

Pipeline pumps are used to convey crude oil or refined petroleum products over long distances. Delivery heads typically range from 400 to 1000 m. Flow rates vary between 290 and 5400 m^3/h. Single-stage double-entry horizontally split pumps with double volute are chiefly used for crude oil due to higher viscosity and higher pipeline friction. For refined products such as NGL and motor fuels, horizontally split multistage pumps are commonplace. For offshore crude oil shipping pumps where the pipeline friction head must be overcome by one pump, or terrains that require high static lift between pumping stations, barrel type casings may be required.

The design must be very rugged and service friendly as pumping stations are often located in remote areas without easy access to service facilities. In general the requirements of ISO 13709 (API 610) must be met. Pumps used to pump over mountain ranges are run in parallel since the head changes little with flow.

For long pipelines with little elevation change, it is mainly pipeline friction losses that need to be overcome. Since friction loss varies as the square of the flow rate, pipeline pumps for such applications are often arranged in series (Fig. 9.3).

The casings are designed for the maximum pressure encountered as specified in ISO 13709 (API 610). Tandem mechanical seals are common to reduce emissions.

Due to the remote location of many pumping stations, bearing RTD, seismic vibration, seal leak detection and motor winding temperature instruments are common. These are connected to a SCADA system for remote monitoring and pump operation.

Another pipeline application is the transport of seawater to inland injection pumping stations. Pumps capable of delivering 1800 to 14 400 m^3/h over long distances are needed. In Middle East locations, the delivery head consists almost entirely of line losses. Single- or multistage, single- or double-entry pumps are used.

Eighty-five to 95% of the cost of running a pipeline is energy cost. Efficiency is of crucial importance. Pumps are therefore chosen to closely match the duty requirements. Methods employed include

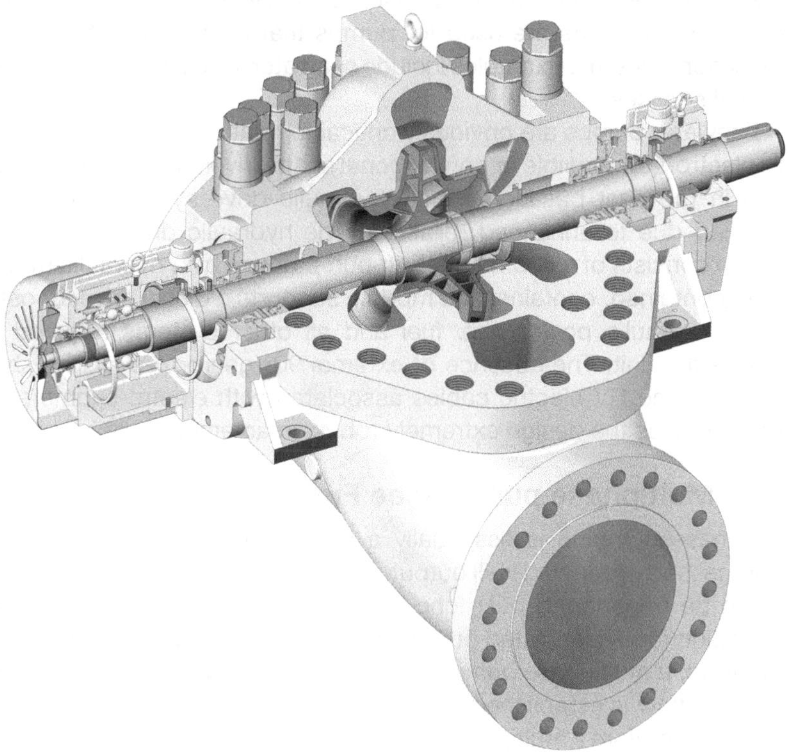

Figure 9.3 Sulzer HSB pumps are widely used in pipeline booster applications

designing hydraulics specifically for the duty, having the ability to change hydraulics to match projected future demand. Variable-speed drives are commonly used on product pipelines that pump diesel fuel in one batch and gasoline in the next. The advent of reliable VFDs has brought about substantial energy cost savings for pipeline operators.

9.1.3 Pumps for FPSO

Floating production storage offloading (FPSO) vessels are increasingly used to produce oil from offshore fields. Converted from conventional oil tankers or purpose built, they are able to operate in any depth of water and are relatively quick to put into service.

Pump applications on FPSO are similar to conventional platforms but with added demands in terms of footprint limitations, compressed delivery times and being able to accommodate deck roll. Footprint size is controlled by optimizing skid design to deliver the most compact layout. Deck movement and roll are compensated for using a rigid skid mounted on three pads. Oil return lines are inclined at an angle greater than the maximum specified roll angle to ensure positive oil flow under all conditions. Vertical shafts are used for pumps that do not need force feed lubrication systems. Examples include seawater lift, oil transfer and other general services.

Firewater pumps are obviously critical for platform safety. Traditionally electro-submersible or conventional shaft-driven designs have been used in the past but increasingly hydraulic drive systems are being specified due to enhanced reliability. The hydraulic drive fire fighting system consists of a submersible lift pump driven by a hydraulic motor. A self-contained, containerized module supports a diesel drive, booster pump, hydraulic power unit, fuel and all other systems required to operate the unit. The absence of external 90° gearboxes for line shaft pumps or the HV electric cables associated with electro-submersible pumps makes the design extremely robust in an emergency.

9.1.4 Multiphase pumps (see Fig. 9.4)

Multiphase pumping is essentially a means of adding energy to the unprocessed producing well output. This enables the water/oil/gas mixture to be transported over long distances without the need for prior separation (and associated pumps and compressors at each well pad). At startup, a multiphase pump may see high gas volume fraction (GVF). Once the flow is established, liquid fraction may increase and viscosity change. Therefore a multiphase pump must be designed for variable inlet conditions. It should also offer the possibility of a large operating envelope (in terms of flow and GVF variation). Operational flexibility has to be given priority over efficiency optimization for a narrow band of operation.

Sulzer Pumps' helico-axial multiphase pump range consists of 12 frame sizes covering total volumetric flow rates (oil, water and gas) at suction conditions (actual flow at pump inlet) from 22 000 BPD (146 m^3/h) up to 600 000 BPD (4000 m^3/h). It is based on the latest second generation helico-axial hydraulics (Poseidon license) developed by the Poseidon group (Institut Français du Pétrole, Total & Statoil).

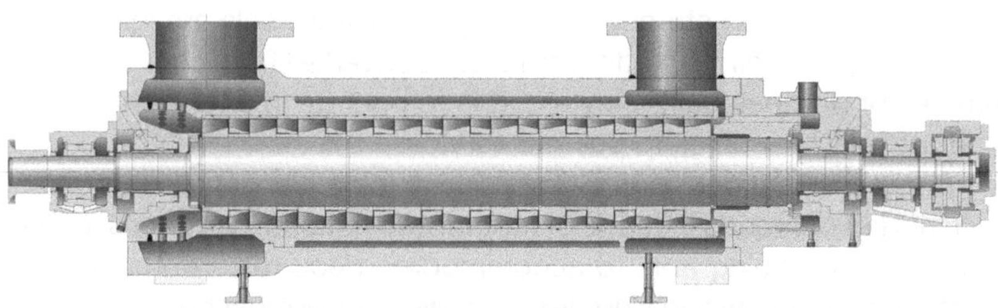

Figure 9.4 MPP multiphase pumps boost well output, handling both gas and liquids

Two mechanical designs are available: low speed to 4000 rpm, and high speed above 4000 rpm. The pump is an in-line multistage pump with a horizontally split inner case in a barrel (cartridge pull-out design). Each compression stage is comprised of a rotating helico-axial impeller and a stationary diffuser.

The main advantages of the helico-axial hydraulic are:

- Ability to pump any GVF from 0 (100% liquid) to 1.0 (100% gas).
- Mechanical simplicity and reliability (one single shaft, Rotodynamic principle).
- Compactness.
- Self-adaptation to flow changes.
- Very tolerant to solid particles (open type axial impeller, without tight clearances).
- Ability to achieve high differential pressure and flow rate.

9.1.4.1 MECHANICAL DESIGN

Based on Sulzer Pumps' experience with injection pumps and centrifugal compressors, a modular design for the multiphase pump range has been developed. The design is in general accordance with API 610. The selected layout is a multistage barrel in-line pump with axially split inner casing. The in-line layout is required to allow the transport of the solid particles contained in the pumped fluid. The impeller tip clearance is liberal and pressure rise is little affected by wear.

The diffusers are of split design and located in an axially split inner case. The axially split design allows an easy inspection of the rotating parts. In addition the rotor does not have to be dismantled during the pump assembly allowing low residual rotor unbalance to be achieved for a low level of vibration. For ease of maintenance, the hydraulic parts (complete rotor with impellers, diffusers, inner case and suction casing) are designed as a pull-out cartridge.

The shaft sealing is achieved with mechanical seals. In order to avoid gas or solid particles contaminating the faces and also to allow for dry running conditions, a pressurized external fluid is injected into the dual seals at about 5 bar over suction pressure. All wetted parts are typically duplex or super duplex SS for aggressive effluents with high chlorine content, H_2S, CO_2, sand and other contaminants. For wells with high sand load, Sulzer Metco's SUME coatings are used to dramatically extend service life.

9.1.4.2 HYBRID MULTIPHASE PUMPS

When the well head flowing pressure is suitable, the first few stages of a pump may be the helico-axial multiphase design, followed by radial flow liquid stages. The helico axial stages add enough pressure to get the vapor into the liquid state, and then the radial flow stages add much higher pressure rise per stage. Sulzer was one of the first to develop such a hybrid multiphase pump and has done extensive testing to prove the design. Sulzer's long experience with axial thrust balancing devices allows Sulzer MPPs and Hybrid MPPs to make much more pressure rise than other designs that only rely on thrust bearings.

9.2 PROCESS PUMPS FOR REFINERIES AND PETROCHEMICALS

Process pumps are applied in processes that move volatile and toxic liquids at high operating pressures and in temperatures that range from −162°C to +400°C. They are typically designed to ISO 13709 (API 610) standards. The pumped media can have densities that range from 0.25 to 1.4 kg/l.

Typically process pumps are designed as single-stage, dual volute, radially split, centerline mounted overhung pumps. For higher capacities and/or low $NPSH_A$ applications, the pumps are designed as single- or two-stage, between bearing, dual volute, centerline mounted pumps. In most cases, the process pumps are designed to be fitted with a wide variety of process appropriate mechanical seals and the seal chambers are designed to ISO 21049 (API 682) requirements. Some seal-less designs are also available.

For higher pressures and capacities multistage axially split or multistage barrel-casing pumps also designed to the requirements of ISO 13709 (API 610) are employed.

9.2.1 Refining (see Figs 9.5 and 9.6)

Refineries convert crude oil feedstock into products that are useful to mankind (gasoline, diesel, fuel oil, and chemical plant feedstock). Crude oil is separated into lighter fractions by heating it up to 400°C and allowing the "fractions" to boil off and condense in a process called fractional distillation. The various fractions are then routed to other processes within the refinery that clean and convert the hydrocarbon molecules into useful products. As an example, a modern full conversion refinery typically will yield as a volume percentage: 3% refinery gases; 50% gasoline; 33% distillates (diesel, jet fuel, heating fuel); 10% heavy fuels and other materials from a barrel of light sweet crude (i.e. WTI or Brent).

9.2.2 Synfuels

Synfuels are liquid fuels derived from gas (natural or liquefied natural gas – LNG) as well as from coal, heavy crude oils, oil sands or oil shale. The price of crude oil determines the economic attractiveness of synfuels. Typically the term "synfuels" is associated with fuels manufactured in a Fischer Tropsch conversion process. The conversion of oil sand and oil shale is particularly onerous on the process pumps employed in the plant due to high solids content in the early phases of the process stream. Pumps in oil sand or oil shale conversions need

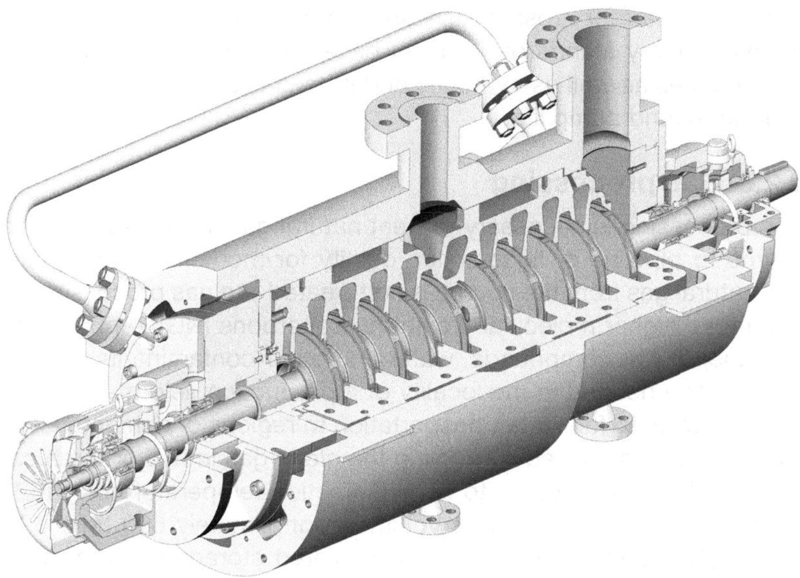

Figure 9.5 CP barrel pumps are used for high pressure/temperature applications in refineries

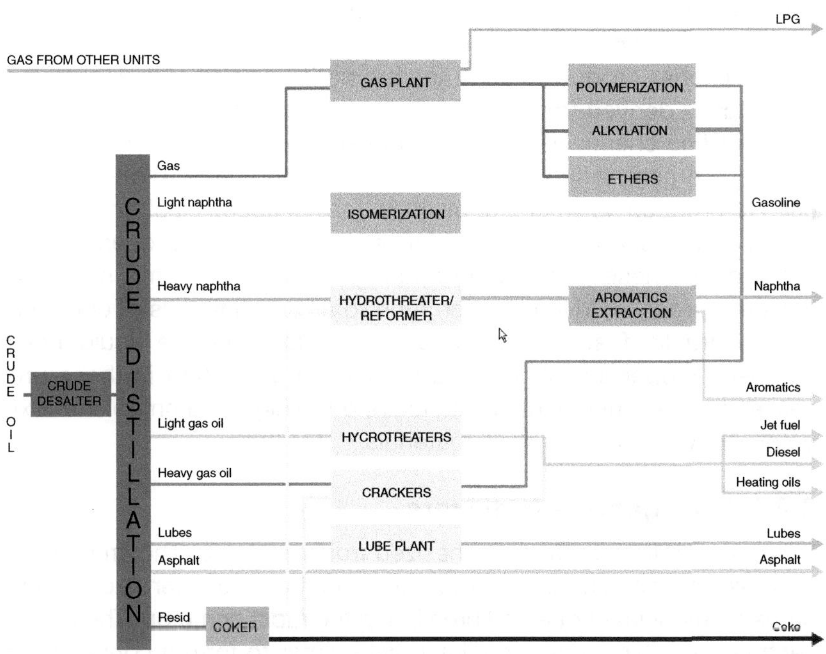

Figure 9.6 Typical refinery schematic

to be designed for erosion resistance in addition to compliance with normal refinery standards. Special base metallurgy, surface coating technologies as well as hydraulic and mechanical design features are used to extend pump life.

9.2.3 Gas processing

Gas processing is necessary to treat natural gas from the production source to pipeline or liquefaction quality for monetization purposes. The natural gas from the source is first treated in a gas processing unit to remove higher molecular weight hydrocarbons (NGL – butane and propane), sulfur compounds, water and other contaminants. Gas not immediately needed in the locality of a production field may be liquidified in order to facilitate transportation to regions of demand. Liquefaction is accomplished by cooling the clean gas in two or three cascade cooling cycles down to the liquefaction temperature of −162°C (−260°F) thereby reducing its volume approximately 600 times. The LNG is then transferred to heavily insulated storage tanks at atmospheric pressure, and from there it is loaded into LNG tankers for shipment.

9.2.4 Petrochemicals (see Fig. 9.7)

Petrochemicals are chemical products made from the output of petroleum refineries, gas plants or synfuel plants. The two main classes of primary petrochemicals requiring pumps manufactured in compliance with ISO 13709 (API 610) are olefins (including ethylene and propylene) and aromatics (including BTX – benzene, toluene and xylene). Petroleum refineries produce olefins and aromatics as part of their output of feed stocks derived from crude oil. From these basic materials a wide range of fine chemicals and other materials such as monomers, solvents, detergents and adhesives are produced. In turn monomers are used to produce polymers for plastics, resins, fibers, elastomers, lubricants, and gels. A wide range of refinery and chemical process pumps are used in these chemical-producing complexes. Pump specifications are process driven but common to all are materials of construction, hydraulic features and sealing solutions for the liquid being pumped. Proper specification is particularly important in these processes as the constituent products are often highly flammable or toxic before they become finished materials.

9.2.5 Nitrogenous fertilizers

Inorganic fertilizer is often synthesized from ammonia. The ammonia is used as a feedstock for the production of nitrogenous fertilizers, anhydrous ammonium nitrate and urea being the most common. These concentrated products may be diluted with water to form a concentrated

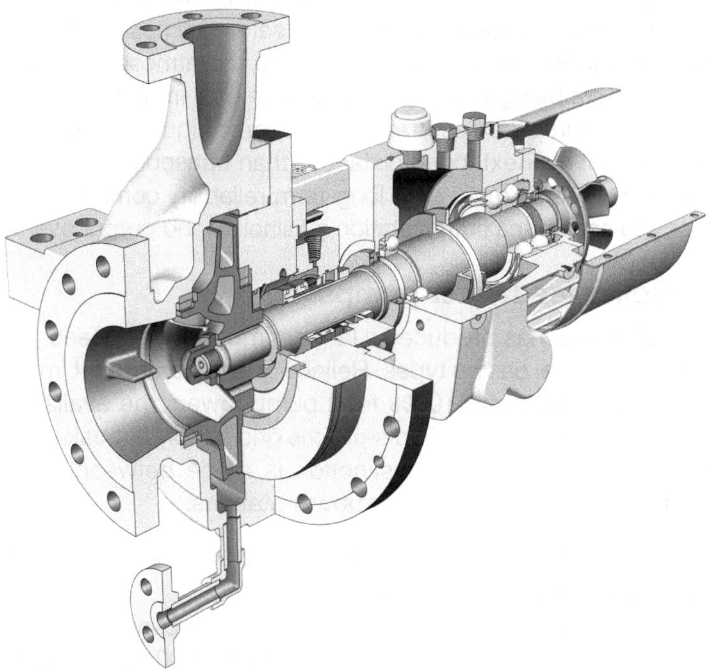

Figure 9.7 OHH ISO 13709 (API 610) process pumps are widely used in hydrocarbon processing applications

liquid fertilizer, UAN. Ammonia can also be used in combination with rock phosphate and potassium fertilizer to produce compound fertilizers.

Pumps and hydraulic power recovery turbines (HPRT) used in fertilizer production processes are handling liquid streams containing entrained gases and may be both corrosive and erosive as well as being highly toxic. Particular care must be taken for the selection of pumps with adequate NPSH margins, and the ability to handle entrained vapor. Design and control of the HPRT is also critical. Selection of the materials of construction is essential to ensure long and reliable life. In addition the sealing solution must ensure reliable sealing in adverse conditions. Reduced speeds and below normal impeller diameters are used to minimize the damaging effects of the entrained vapors.

9.3 POWER GENERATION

The most common forms of power generation are conventional fossil-fired power stations, nuclear, gas-fired combined cycle power plants and renewables. In nearly all cases (except hydro-electric generation and wind power) a heat source is used to produce steam that in turn drives a turbo-generator. The depleted steam is then cooled and condensed for returning to the boiler. Pumps form an essential part of the process, the key applications being boiler feed, condensate extraction, cooling water circulation and in the case of nuclear applications, safety pumps. With increasing concerns regarding the release of CO_2 into the atmosphere, scrubbing technologies are being introduced to capture CO_2 from the flue gases for safe disposal. Pumps are used in the transport of liquid phase CO_2 as this is more efficient over extended distances than transporting a gas. Pumps used in power plants must offer long-term reliability combined with high efficiency to ensure maximum station availability and output over time.

9.3.1 Boiler feedwater pumps

Since 1905 Sulzer has produced a large number of boiler feed pumps in segmental or barrel casing types. Reliability is of paramount importance for this duty. Failure of a 100% feed pump lowers the availability of a power station with potential to disrupt the grid supply at peak load times.

According to the duty a distinction is made between boiler feed pumps for fossil fired and nuclear power stations.

9.3.1.1 FEED PUMPS FOR FOSSIL-FIRED POWER STATIONS (SEE FIGS 9.8 AND 9.9)

Firing by coal, oil or natural gas

Depending on the boiler (subcritical $\leq$225 bar, supercritical $\geq$225 bar), feed pumps must provide heads from 1800 to 3800 m. The 100% flow rate in liters per second is 1 to 1.2 $\times$ the electrical output in MW. Depending on the availability demanded and the operating mode (base or peak load) the capacity is subdivided into 100%, 50% or $33\frac{1}{3}$%. The drive

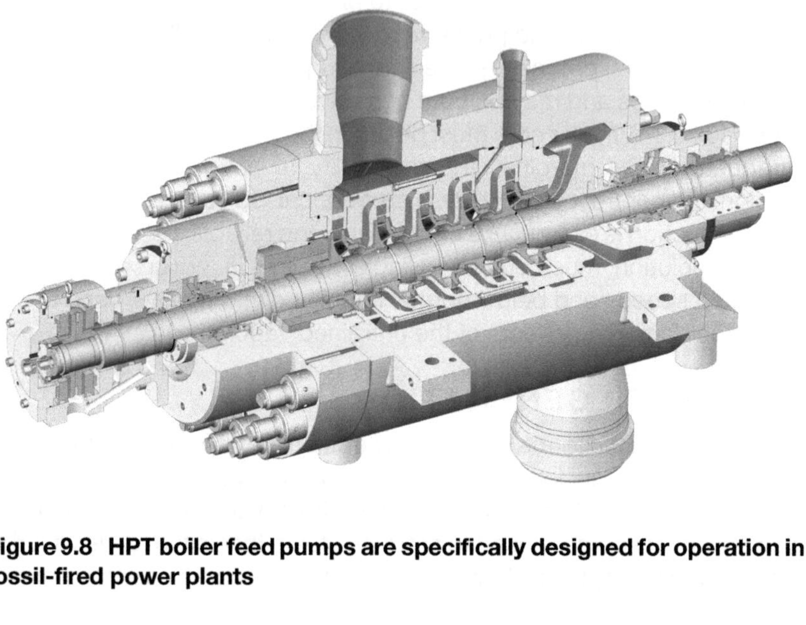

Figure 9.8 HPT boiler feed pumps are specifically designed for operation in fossil-fired power plants

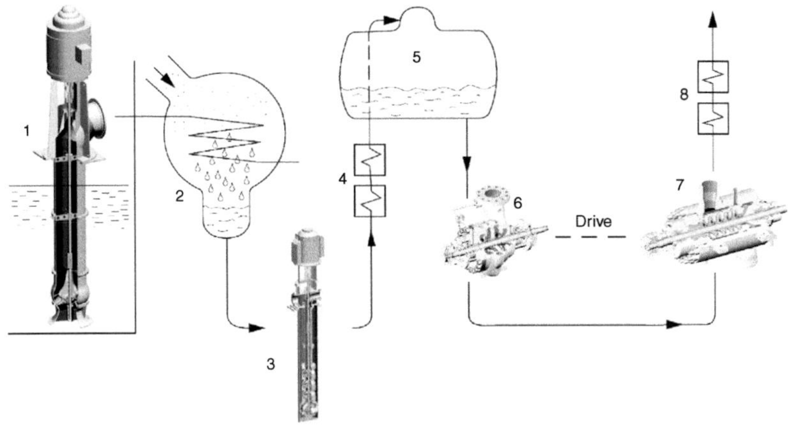

1 Cooling water pump
2 Condenser
3 Condensate pump
4 Low pressure feedwater heater
5 Feedwater tank
6 Booster pump
7 Feed pump
8 High pressure feedwater heater

Figure 9.9 Boiler feedwater system for conventional power stations

input of 100% feed pumps claims some 3–4% of the generation unit output. Segmental pumps are commonly employed for lower final pressures and flow rates (up to a station output around 360 MW for a full-load pump). For higher flow rates and pressures, barrel-casing pumps are almost universally used.

Depending on whether the high-pressure feed water heaters are arranged before or after the feed pump, the pump must be designed for a temperature of 220 to 240°C (resp. 170 to 190°C), though the latter is more usual.

9.3.1.2 FEED PUMPS FOR NUCLEAR POWER STATIONS (SEE FIG. 9.10)

Most commercial reactors today are either pressurized water reactors (PWR) or boiling water reactors (BWR) (Figs 9.11a and b).

Feed pumps for nuclear power stations have heads between 500 and 800 m. Feedwater temperatures range from 130 to 185°C and the 100% flow rate in liters per second is about 2 to 2.2 × the electrical station output in MW. Because nuclear stations operate with much lower live steam pressures (approx. 50 to 70 bar) than fossil-fired ones, the feedwater rate is much greater for any given station output. The drive input to the feed pumps claims about 1.5% of the unit output.

Due to the higher flow rate and lower head required, feed pumps for nuclear power stations are mainly single-stage double-entry designs.

The demand characteristic of the feedwater system for a PWR is practically a straight line; speed variation brings only slight economic gain. By contrast, feed pumps for the BWR may be speed regulated, because the system characteristic is somewhat steeper.

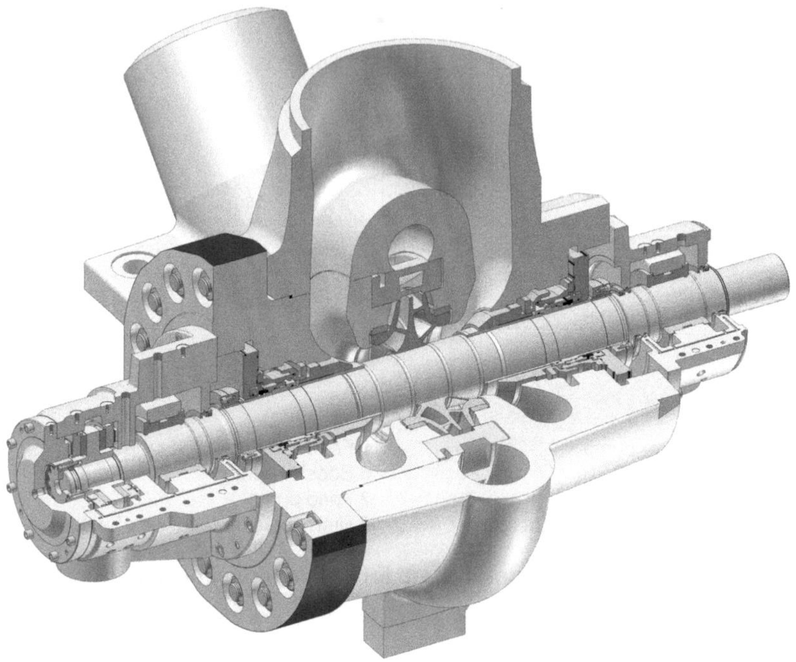

Figure 9.10 HPTd boiler feed pumps are specifically designed for operation in nuclear power plants

9.3.2 Booster pumps for feed pumps

Booster pumps are placed before the main pump. They provide the necessary suction head for the feed pumps, so that this does not cavitate within its specified operating range. For small installations single-stage overhung pumps designed for the high feedwater temperatures are used. Normally double-entry vertically split pumps are employed for high flow rates in order to cope with the small suction heads available (set by the height of the feedwater tank above the pump centerline).

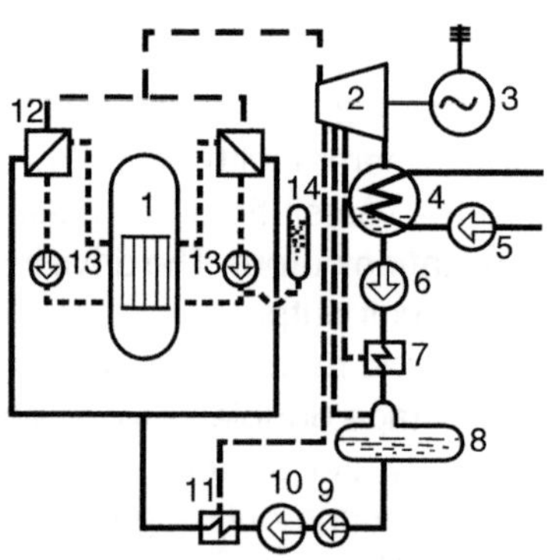

1 Reactor
2 Turbine
3 Generator
4 Condenser
5 Cooling water pump
6 Condensate pump
7 Low pressure feedwater heater
8 Feedwater tank
9 Booster pump
10 Feed pump
11 High pressure feedwater heater
12 Steam generator
13 Reactor recirculation pump
14 Pressurizing system

Figure 9.11a Schematic of a PWR

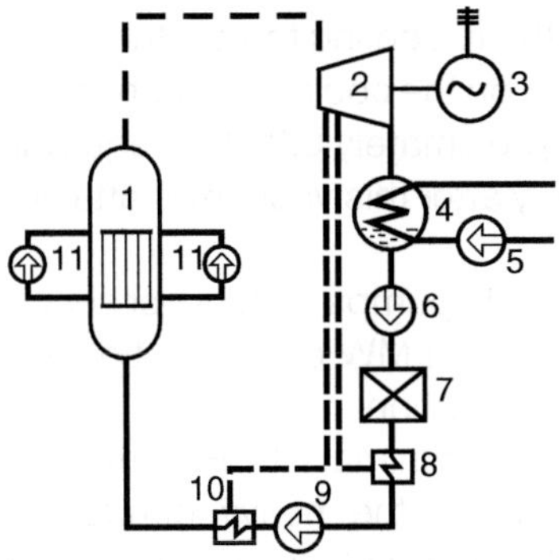

1 Reactor
2 Turbine
3 Generator
4 Condenser
5 Cooling water pump
6 Condensate pump
7 Condensate cleanup
8 Low pressure feedwater heater
9 Reactor feed pump
10 High pressure feedwater heater
11 Reactor recirculation pump

Figure 9.11b Schematic of a BWR

When designing booster pumps it must be borne in mind that turbo-set load variations result in pressure and temperature changes in the feedwater tank. Allowance must be made for these influences under transient operating conditions.

9.3.3 Condensate pumps (see Fig. 9.12)

From the collecting chamber of the condenser hotwell, condensate pumps deliver water at 25 to 35°C via the low-pressure heater into the

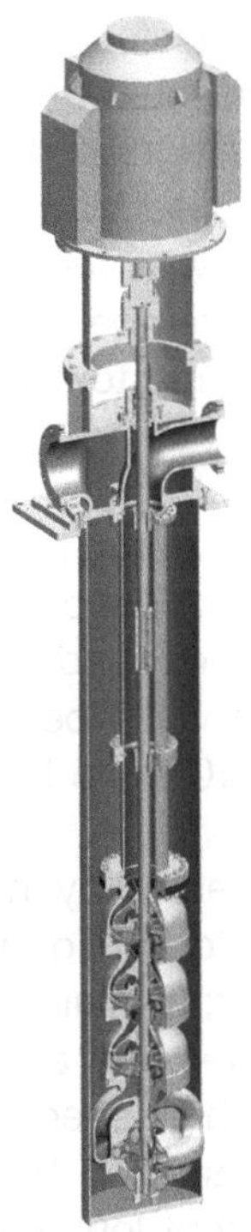

Figure 9.12 SJD (CEP) condensate extraction pump

feedwater tank. In thermal circuits with dry cooling towers the liquid may reach higher temperatures. Depending on the cooling water temperature the condenser pressure averages approximately 0.05 bar, equivalent to a vacuum of approximately 9.5 m. Only after this vacuum is attained can the main turbine deliver its full output.

The 100% flow rates of the condensate pumps in liters per second are about 1 to 1.2 times the electrical output in MW of conventional power stations and 2 to 2.2 times with nuclear stations.

The power station designer demands minimal NPSH from the pump maker, because raising the mounting of the heavy condenser (for 600 MW typically some 630 tons incl. water) entails costly construction work, while lowering the basement level of the building entails very high foundations costs. Owing to the small available NPSH, vertical slow-running pumps are used with special suction impellers or double-entry first stages. To maintain the system vacuum requirements, these pumps are supplied with a can.

9.3.4 Heater drain pumps (moisture separator drain pumps)

Arranged between condenser and feed pumps are the LP heaters, and between the feed pumps and steam generator (boiler) are the HP heaters. These produce heated steam bled from the turbine, which becomes partly condensed due to heat transfer into the feedwater. In smaller power stations the condensate is dewatered to the low pressure of the condenser in cascade circuitry. Here a loss of valuable energy is involved. In bigger plants this flow is recycled into the feed circuit by heater drain pumps. Such pumps operate with relatively high suction head (steam pressure corresponding to a temperature of 110 to 200°C). For this an NPSH of only 0 to 7 m (referred to the mounting level) is available.

9.3.5 Cooling water pumps

The purpose of a cooling water pump is to force water through the tube system of the condenser (Fig. 9.13 shows a compound cooling system). In conventional power stations, depending on the vacuum required, 20 to 35 liters per second of cooling water per MW electrical station output is needed, for nuclear stations 40 to 45 liters per second is needed, i.e. some 35% more.

The necessary head is governed by the kind of cooling system. With river water cooling it is about 8 to 15 m depending on the station location; with cooling towers 15 to 28 m; with air cooling 30 to 45 m.

Owing to the high flow rates and associated low heads of cooling water pumps, hydraulics with high specific speed are exclusively used.

On account of the small suction head (NPSH approx. 12 m) the speeds of these pumps are kept low. From about 600 min^{-1} down, a reduction gear could be justified by the motor costs.

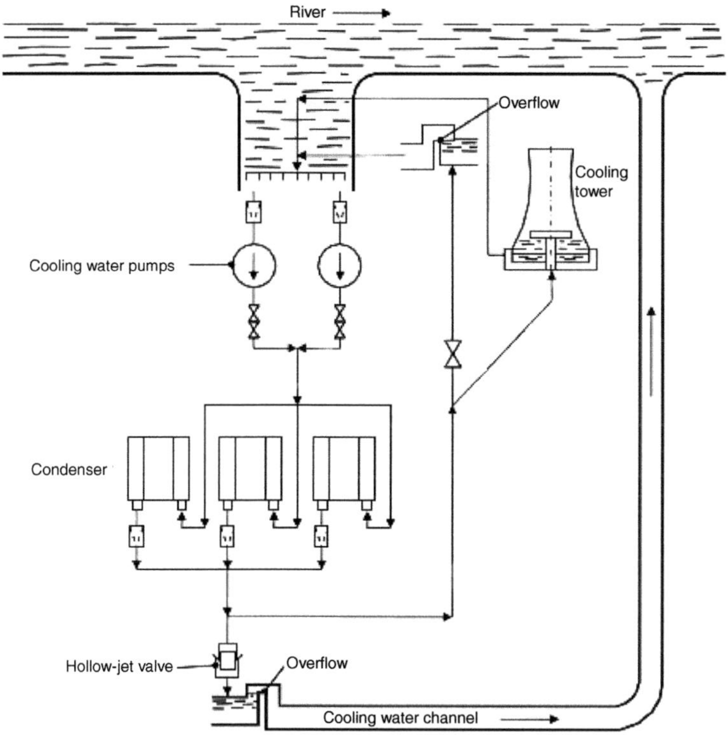

Figure 9.13 Typical power station cooling water schematic

Vertical turbine pumps with mixed-flow impellers or vertical single-stage concrete volute pumps are used as cooling water pumps for operation with river or seawater (Fig. 9.14). The concrete volute type is especially suitable for very large flows pumping seawater against low delivery heads. The hydraulic configuration of the inlet chambers should always be included in the design of a cooling water pump on account of the high specific speed. This enables all disturbing influences of the inflow to be limited and, especially in model tests, high reliability to be assured.

The cooling water pumps may be regulated by variable speed or rotor blade adjustment. Throttle control is almost never employed in modern systems. To assist maintenance, larger cooling water pumps are designed with a pull-out rotor assembly.

9.3.6 Main reactor coolant pumps and reactor circulating pumps for nuclear power stations

Reactor coolant pumps circulate the water past the fuel and out to the steam generator (BWR). The reactor circulating pump is circulating the primary cooling water within the reactor (PWR).

Figure 9.14 A Sulzer SJM vertical mixed-flow pump typically used in cooling water applications

9.3.7 Safety related auxiliary pumps for nuclear power stations

The safety related auxiliary pumps primarily assure reactor cooling at startup and shutdown and in cases of emergency. Depending on the reactor type, special pumps are needed. They are often specified to US codes (ASME code) and are subject to very exacting acceptance requirements with regard to quality assurance. To prevent radioactive leakage the seals are often filled with nitrogen as barrier gas.

If sleeve bearings are necessary, the oil lubrication system must enable the pump to start up within seconds without an additional electrical oil startup pump.

These pumps must be able to operate even during major earthquakes. This must be demonstrated mathematically or for preference on a special vibratory table producing stressing similar to that caused by an earthquake.

9.3.8 Fossil-fired flue gas and CO_2 scrubber pumps

Flue gas desulfurization scrubbers usually involve causing the flue gas to circulate through thin lime slurry in a contactor. The flow rates are large and the limestone slurry is abrasive. Depending upon customer preference, normal centrifugal or rubber-lined slurry pumps are commonly used.

CO_2 scrubbers are similar but may use one of a variety of different solvents. Ammonia and various recipes of amine are used. As above, pressures are relatively low and flow rates are high. CO_2 and water form carbonic acid so pump materials are typically 300 series SS as a minimum. The rich solvent returns from the contactor to a stripper where it is heated to drive off the gases and the solvent is recycled.

Compressors are then used to boost the CO_2 to supercritical pipeline pressures (>90 bar or 1300 psi). If the local seawater temperature is low enough, it may be more efficient to cool the CO_2 to a dense phase and pump it to the higher pressure needed for long pipelines and for injection.

CO_2 has a very low viscosity even at supercritical pressures so pipeline friction losses are relatively minimal. CO_2 has a surface tension less than 10% of that of propane, so careful attention is paid to welding and casting integrity.

The CO_2 can be injected into oilfields to enhance oil production. It can also be pumped into salt domes or hard rock for sequestration.

Understanding compressible supercritical fluid performance in pumps is important. Mechanical seal technology is also important.

9.4 WATER

Water is a natural resource essential for life. The collection, transfer, treatment for human consumption and subsequent wastewater disposal are, in most instances, completely dependent on pumps. Where freshwater is scarce or resources cannot support demand, desalination plants are used to convert seawater into freshwater thereby boosting the available supply. Many of the world's major food-producing regions rely on managed irrigation and land drainage systems to support crop production. Common to all these applications is the need for the highest possible efficiency as the cost of electricity to drive the pump over its lifetime is many times greater than the initial capital investment. Equally important is long-term reliability to ensure continuous dependable supplies with a minimum of maintenance.

9.4.1 Water pipeline pumps (see Fig. 9.15)

Water pipelines are employed to transport large flows of water over long distances (e.g. Riyadh in Saudi Arabia) and/or high heads (e.g. Caracas in Venezuela) with respect to the water sources.

Often it involves pumps with a combination of high flow (up to 8 m^3/s) and high head (up to 800 m) and as a consequence the pumps are of relatively large casing diameter and significant power input which requires a compromise between highest possible efficiency and smallest possible weight. Mostly axially split designs with single or multiple stages

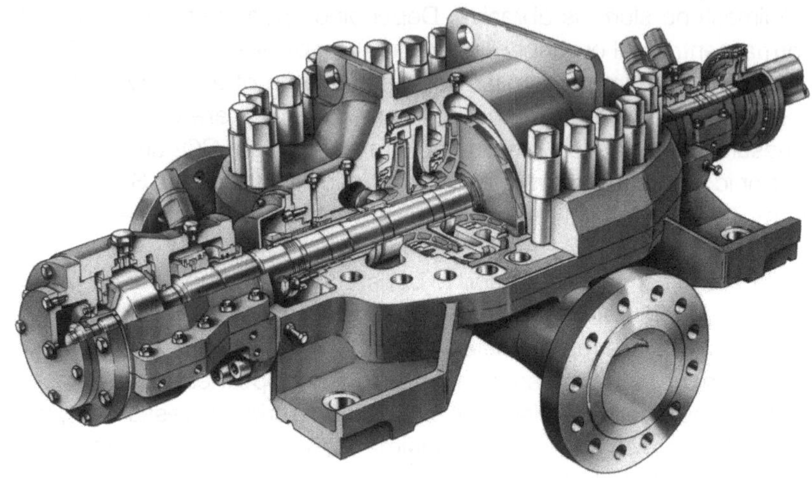

Figure 9.15 Sulzer HPDM pumps are individually tailored to suit specific project requirements for maximum efficiency

are employed to meet these requirements as well as ease of maintenance.

The pump speed is mostly dictated by the $NPSH_A$ and as such in certain cases it may be more economical to design the system with a booster pump to remove the $NPSH_A$ constraint and enable higher speed that will result in a more economical main pump.

9.4.2 Drinking-water supply pumps

Flow rates for drinking-water pumps are dictated by the number of consumers.

Distinctions are made between:

- river water pumping stations ($NPSH_{av}$ max. $\approx$13 m);
- lake water pumping stations ($NPSH_{av}$ max. $\approx$17 m); and
- booster pumping stations (large $NPSH_{av}$).

Submersible pumps, vertical pumps or horizontally split double-entry designs are used as feeder pumps. Where no feeder pump is fitted, single or multistage pumps, usually of the double-entry type, are used.

As a rough calculation, 200 to 500 liters per day per head are estimated for industrial cities, varying in a ratio of 1:5 according to the time of day and season.

The following possible forms of control exist for matching pump delivery to these fluctuations:

- combining similar pumps with constant speed in series and parallel;
- pumps operating in parallel to constant speed or with pole-reversing motors;
- speed control (good where the head consists primarily of line losses);

- variable-speed booster pumps followed by constant-speed main pumps;
- interposition of an elevated tank (the pumps are switched on and off by the level in it).

For parallel operation the pumps should have characteristics as steep as possible, so that the run-out point when they are operating singly is near 120% of the best point flow rate, at which the steep rise of the $NPSH_{req}$ has not yet begun.

9.4.3 Seawater desalination pumps

There are three common ways of producing desalinated water from seawater: reverse osmosis (RO), based on membranes technology, and multi-effect distillation (MED) and multistage flashing (MSF), both based in thermal evaporation technology.

Reverse osmosis works by using pressure to force seawater (or other raw feedwater) through a membrane, retaining the solute on one side and allowing the pure solvent to pass to the other side (Fig. 9.16). This is the reverse of the normal osmosis process, which is the natural movement of

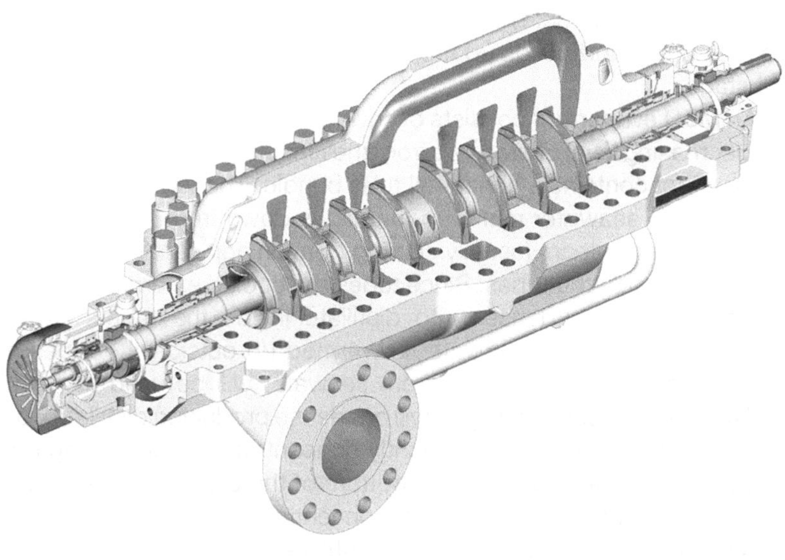

Figure 9.16 A high pressure Sulzer MSD multistage pump for reverse osmosis applications

solvent from an area of low solute concentration, through a membrane, to an area of high solute concentration when no external pressure is applied. For seawater a pressure of between 55 and 81 bar is required, depending on the seawater salinity and temperature, for maximum membrane effectiveness.

The pressure of the brine not passed through the membrane is used in energy recovery devices (pressure exchangers, turbochargers or Pelton wheel turbines) to improve efficiency. In addition to the high pressure pumps various other process pumps are required. All need to be manufactured in highly corrosion-resistant materials, super duplex grades being the most common.

An MSF evaporator consists of several consecutive stages maintained at decreasing pressures from the first stage (hot) to the last stage (cold). Seawater flows through the tubes of the heat exchangers where it is warmed by condensation of the vapor produced in each stage. The seawater then flows through the brine heater where it receives the heat necessary for the process (generally by steam coming from an adjacent power plant). At the outlet of the brine heater, when entering the first cell, seawater is overheated compared to the temperature and pressure of stage 1. Thus it will immediately "flash", i.e. release heat, and thus vapor, to reach equilibrium with stage conditions. The produced vapor is condensed into freshwater on the tubular exchanger at the top of the stage. The process is repeated in several stages until the last and coldest stage. The cumulated freshwater builds up the distillate production which is extracted from the coldest stage. Seawater slightly concentrates from stage to stage and builds up the brine flow which is extracted from the last stage.

An MED evaporator consists of several consecutive cells maintained at a decreasing level of pressure from the first (hot) to the last (cold). Each cell or effect contains a horizontal tube bundle. The top of the bundle is sprayed with seawater make-up that flows down from tube to tube by gravity. Heating steam (usually coming from an adjacent power plant) is introduced inside the tubes. Since tubes are cooled externally by make-up flow, steam condenses into distillate inside the tubes. By heat exchange (latent heat) seawater is warmed up outside the tubes and partly evaporates. That steam from seawater will be used as heating media in the next effect, where it will condense producing distillate water. Due to evaporation, seawater slightly concentrates when flowing down the bundle and gives brine at the bottom of the cell. In the last cell, the produced steam condenses in a conventional shell and tubes heat exchanger. This exchanger, called "distillate condenser" or "final condenser", is cooled by seawater. At the outlet of the final condenser, part of the warmed seawater is used as make-up of the unit; remaining flow is rejected to the sea. In the MSF process, due to higher temperatures and capacities, vertical can pumps are required.

In both MED and MSF processes, brine and distillate are collected from cell (or stage) to cell till the last one, wherefrom they are extracted by centrifugal pumps. A critical selection criterion for these pumps is a low $NPSH_R$. Since the evaporators are working at vapor equilibrium conditions, the only $NPSH_A$ is the evaporator geodetical head. In the MED process, horizontal pumps of end suction or double suction construction are typically used. Also high alloy material (duplex) in contact with sea-water and brine is required.

9.4.4 Irrigation pumps

Irrigation pumps are used in dry or arid regions to distribute surface water or water from subterranean sources. A distinction is made between pumps for sprinkler systems and for surface irrigation from storage basins with gravity feed to a system of channels. Sprinkler systems require delivery heads of around 100 m, surface irrigation reservoirs delivery heads of around 10 to 60 m.

For tapping water from depths down to about 40 m, multistage vertical pumps with above-ground motors are suitable. In special cases the limit of 40 m may be exceeded. The power limit for submersible pumps is now about 3500 kW, though this postulates very long submersible motors with lower overall efficiencies.

Figure 9.17 shows three typical irrigation pump configurations. For small delivery heads, axial or mixed-flow pumps are mostly employed. If the horizontal layout is adopted, a suction device must be provided for priming the pump, and reverse flow of water during shutdown must be

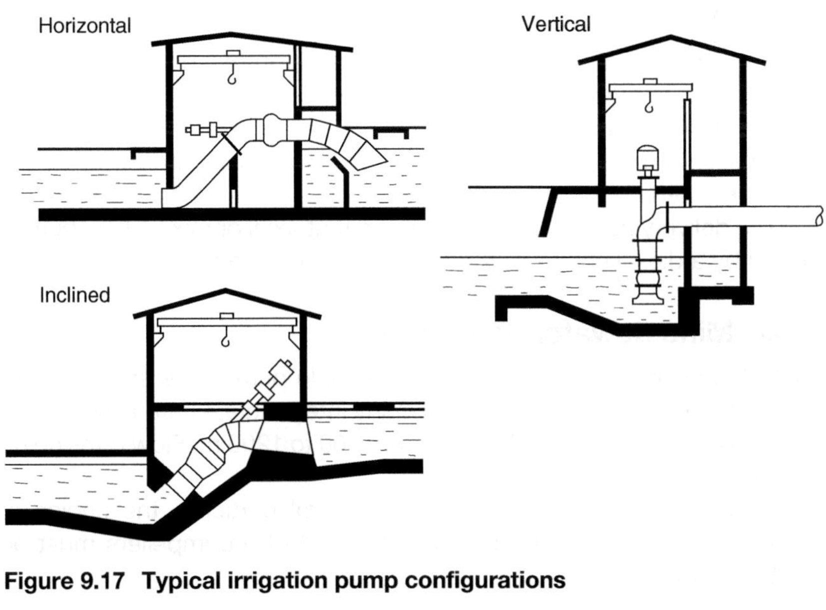

Figure 9.17 Typical irrigation pump configurations

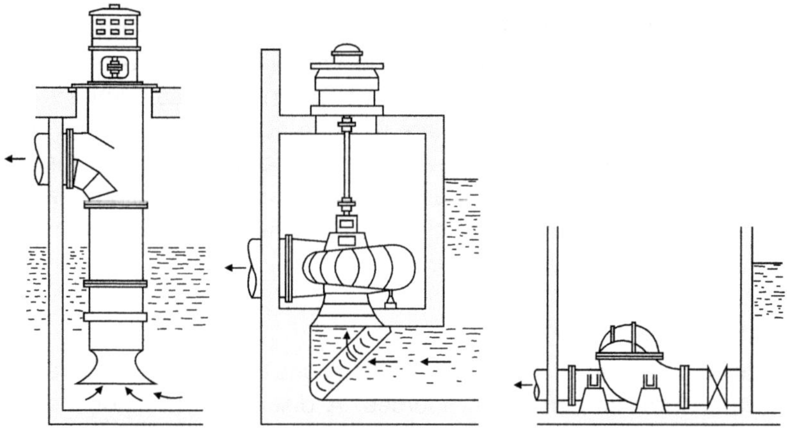

Figure 9.18 Wet and dry pump installations

prevented by opening an air inlet. With the vertical and inclined configurations the impellers are immersed in the tailwater, and the pumps are always ready for duty without venting. Reverse flow is then arrested by an automatic stop flap.

Figure 9.18 shows the difference between dry and wet installation. The advantage of dry installation is simplified dismantling of the pump internals for cleaning and inspection. Conventional, oil-lubricated bearings may be used too.

9.4.5 Drainage pumps (see Fig. 9.19)

These pumps are used to convey rainwater (stormwater) from low-lying areas to rivers after heavy showers. They are started automatically by float switches in accordance with the incidence of water. As the water level rises, further pumps are started. Simple design and reliability are decisive criteria.

The delivery head is usually small, so that axial or mixed-flow pumps driven by a diesel engine or electric motor are suitable.

9.4.6 Mine dewatering pumps

Multistage segmental pumps or submersible pumps are employed for mine drainage. Depending on how many intermediate natural reservoirs are available, delivery heads range from 500 to 1200 m. Flow rates go up to 300 l/s.

Since the water contains minute mineral particles, the choice of material is very important. Peripheral speed of the impellers must be limited to 45–50 m/s.

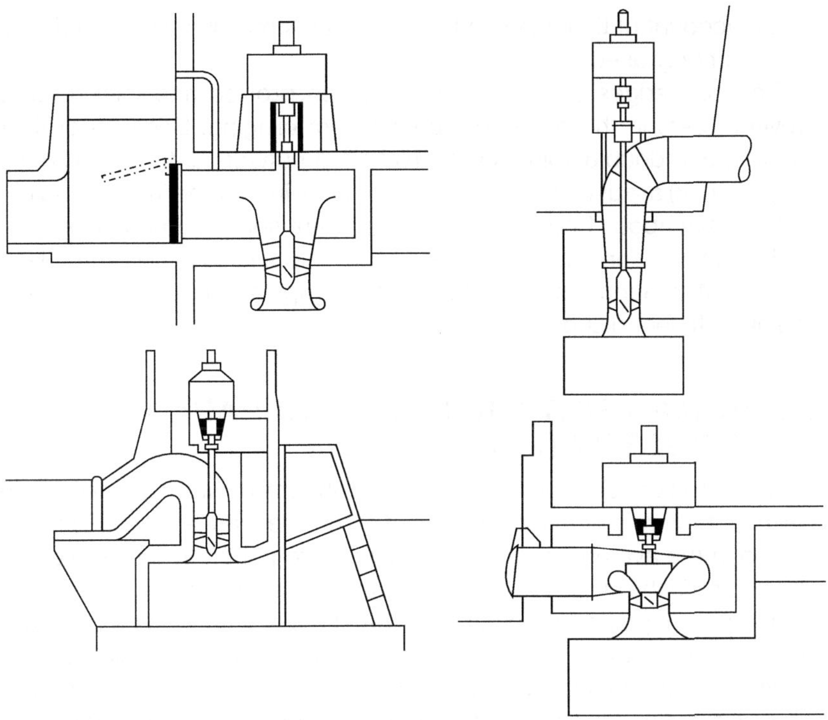

Figure 9.19 Drainage pump configurations

9.4.7 Sewage pumps

Approximately 200 to 750 liters of effluent per day are estimated to be produced per inhabitant though this may be trebled by rainfall.

The sewage pump is a centrifugal type for pumping water laden with solids and with varying chemical composition. Depending on the medium a distinction is drawn between pumps for the following:

- crude sewage (untreated sewage);
- wastewater (mechanically treated water from the sedimentation tank);
- sludge (activated, primary or digested sludge).

For crude sewage containing solid and long-fibered contaminants as well as sludge, the single-channel or spiral tube impeller is employed, with free passage equal to the nominal diameter of the suction branch. The free-flow impeller is used for contaminated liquids with an elevated content of gas and air.

Self-contained and multistage wide-channel impellers are employed with solids-laden liquids not prone to whirl formation. Self-contained wide-channel impellers are sized for defined maximum grain sizes determined by the number of vanes and the impeller inlet diameter. The minimum volute tongue clearance must also conform to these requirements.

On account of their direct intake, vertical pumps installed in a dry pit are commonly used.

The inlet edges of the impeller must be profiled and thickened to prevent fibers in the pumped medium from clinging. Because of their radial thrust, double volutes with thickened cutwaters (to minimize the clogging risk) are used for preference. Velocity in the pipelines should not be less than about 1 m/s, however, to prevent solid matter settling.

The conventional pumps have mechanical seals flushed with clean barrier water. Submersible motor-driven pumps are often used for sewage and effluent disposal.

9.5 PUMPS FOR THE PULP AND PAPER INDUSTRY (SEE FIG. 9.20)

For every ton of paper produced, about 150 tons of water are needed in the manufacturing process. As a result of this there are about 200 pumps utilized in the average mill converting wood to paper. When converting wood to pulp the stock is handled in a range of suspensions. In the initial stage after chipping and digestion, the stock still contains sand, hard knots, resin and bark. After bleaching it is then purified and thickened to higher concentrations. To cope with this, pumps with wide hydraulic passages are used; wear-proof internals are of particular importance.

Paper stock pumps must operate with the minimum possible pulsation to avoid varying paper thickness in the final product.

9.5.1 Pulp production

Paper and paperboard are mainly made from wood pulp. The wood chips are transformed into pulp either chemically or mechanically.

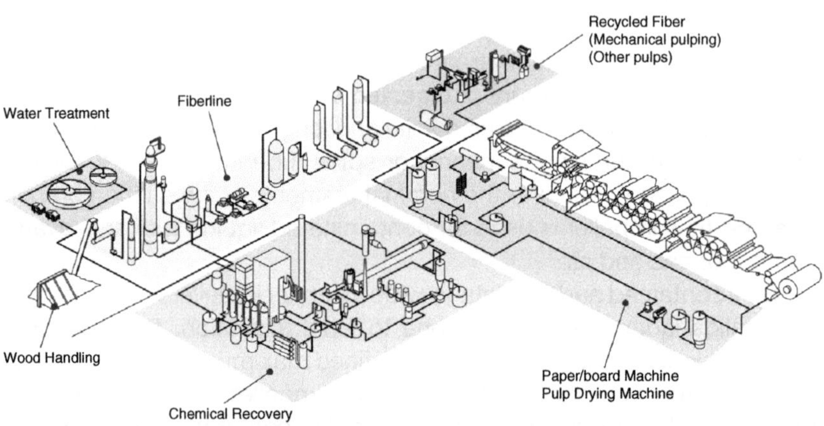

Figure 9.20 Typical pulp and paper mill process with recycled fiber plant.

In the chemical pulp manufacturing cooking process, fibers are separated from each other with the aid of chemicals and heat. After the cooking process, the pulp is bleached, because the lignin remaining causes brown staining. After bleaching the pulp will be pumped to the paper production plant using stock pumps. Pulp and paper mills are usually located within the same complex, enabling pulp to be piped directly into the paper machine. Otherwise the pulp is dried into large sheets (with a pulp drying machine), which are then trimmed to reduce their size and baled for storage or transport.

Paper made from chemical pulp is not only stronger and more durable, but also more transparent than paper made from mechanical pulp.

During the chemical pulp-making process only about half of the wood raw material is recovered. The cooking solution contains an abundance of wood-derived substances. After removing the water content the remaining fluid is burned in a recovery boiler to generate energy. This enables the bio-energy that the solution contains to be usefully exploited and chemical content to be recovered for reuse. Thus chemical pulp mills are more than self-sufficient in energy.

The mechanical pulp manufacturing process involves separating the wood fibers from each other either through grinding or refining. The mechanical stress separates the wood fibers from each other, but it also breaks them into shorter lengths. Paper made from mechanical pulp is opaque, making it especially suitable for printed products.

An important part of the paper manufacturing process is reuse of recovered paper. In the recycled fiber plant the pre-sorted paper is broken up in water. The remaining impurities and printing ink are then removed from the stock to produce recycled fiber for use as raw paper production material. Recycled fibers are mainly used to make newsprint, hygiene papers, board and chipboard.

9.5.2 Paper production

In the stock preparation plant the pulp/stock is screened and cleaned through a series of different stages and consistencies. A combination of stocks with additives is diluted with water to form a thin broth with a consistency level of only 1%. From the headbox at the wet end of the paper machine the mixture is spread onto a moving belt of plastic mesh wire, which allows the water to drain through. In the paper machine the dry-matter content is increased in stages. By the time the paper web leaves the drying section the web's dry-matter content is 95%.

By combining different types of pulp and using a variety of production methods, manufacturers can make papers with a wide range of properties each suiting specific needs. The quality and properties of paper can be further tailored by coating or glazing it.

Paperboard is made following the same general principles as in the case of paper, but is commonly multi-ply, whereas paper has only a single layer.

Manufacturing pulp and paper requires the use of large volumes of water, but most of it is recycled over and over again with the aim being to use as little water as possible.

9.5.2.1 PUMPS AND MIXERS FOR TYPICAL APPLICATIONS (SEE FIG. 9.21)

- *Stock/process pumps* (AHLSTAR^{UP} A). For normal pumping applications in all process stages.
- *Non-clogging pumps* (AHLSTAR^{UP} N). For applications where normal stock pumps cannot handle liquids due to plugging or abrasive wear. Applications include wood handling, recycled fiber plant, mechanical

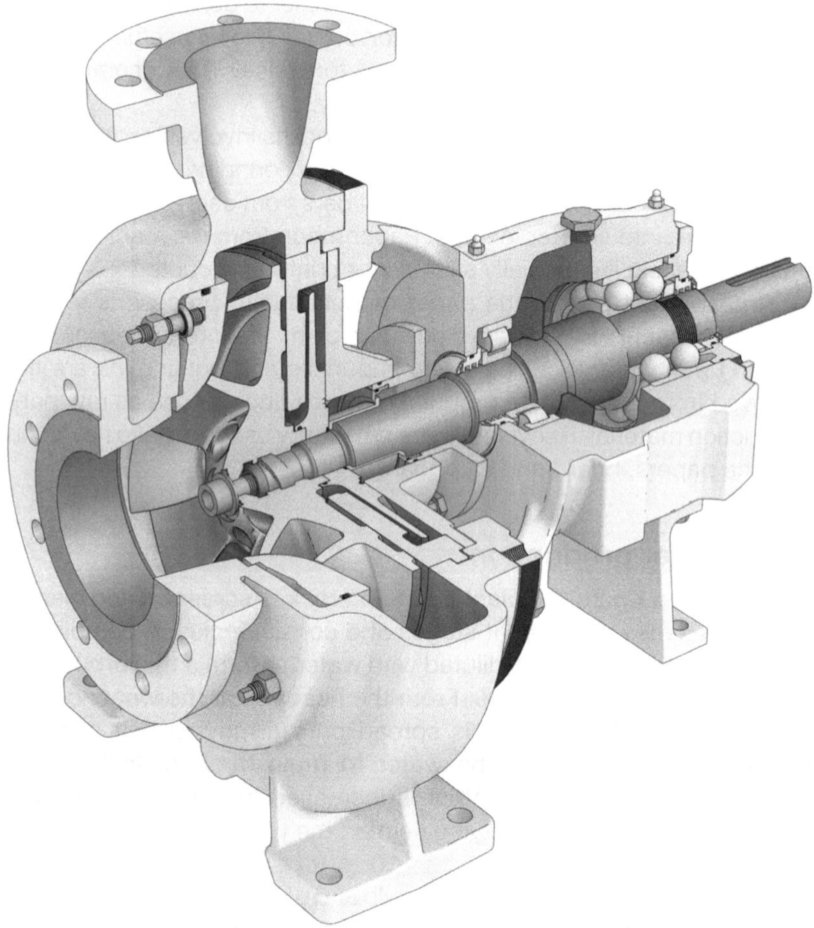

Figure 9.21 AHLSTAR^{UP} process pumps are specifically designed for pulp and paper applications

pulping, reject departments in paper and board machines, and wastewater treatment.

- *Wear-resistant pumps* (AHLSTAR^{UP} W). For pumping abrasive and erosive liquids, used in the chemical recovery plant, coating and finishing in stock preparation plus paper and board machines.
- *Digester pumps* (AHLSTAR^{UP} E). For pumping hot liquors in continuous and batch digester applications in the cooking plant.
- *Headbox feed pumps* (ZPP). To feed stock to the paper, board or tissue machine as well as to the pulp drying machine in chemical pulping.
- *High pressure multistage pumps* (MBN). For pumping boiler feedwater, shower water and sealing water. Used in mechanical pulping, paper, board, tissue, and pulp drying machines, and pumping boiler feedwater in recovery boiler power generation plants.
- *Non-clogging vertical pumps* (NVP, NKP). For all kinds of severe applications in seal pits and floor channels.
- *Medium Consistency (MC®) pumping systems* (MCE™, MC® Discharge Scrapers). For pumping of stock in the consistency range 6–18%. Applications: oxygen delignification and bleaching in chemical pumping, mechanical pulping and in the recycled fiber plant.
- *Chemical mixer* (SX mixer). For mixing both gaseous and liquid bleaching chemicals as well as steam into the paper stock. Applications: oxygen delignification and bleaching in chemical and mechanical pulping.
- *Horizontal and vertical agitators* (SALOMIX® SL,L). For mixing of stock, liquors and other liquids in all process stages where agitation is needed. Applications: chemical pulping, chemical recovery, mechanical pulping, recycled fiber plant, paper, board, tissue and pulp drying machine, coating and finishing, wastewater treatment.
- *Tower management systems* (SALOMIX® SL,L, TES, GLI, VULCA). For feeding and discharging from storage and bleaching towers, tanks, broke towers. Used in chemical pulping, chemical recovery, mechanical pulping, recycled fiber plant, paper, board, tissue and pulp drying machine, coating and finishing, wastewater treatment.

9.6 PUMPS FOR GENERAL INDUSTRY

The general industrial sector covers an extremely wide range of applications ranging from heavy steel manufacture through to the production of fine chemicals and the refining of foodstuffs. Many industrial processes require the reliable pumping of liquids that may be erosive, corrosive, hot, flammable or toxic. It is essential therefore that industrial process pumps are available as standard with a wide range of options in terms of materials of construction, hydraulics and shaft seals. As large numbers of process pumps are often required on a single production site interchangeability of subassemblies such as bearing housings can significantly reduce spare part stock holding requirements. Good knowledge of the individual processes is essential for the pump manufacturer to optimize pump selection to achieve the lowest total cost of ownership for the operator. The examples below illustrate some typical industrial processes.

9.6.1 Pumps for biofuels (see Fig. 9.22)

The most commonly used liquid biofuels are bioethanol and biodiesel. The main stages of the bioethanol production process are conversion of

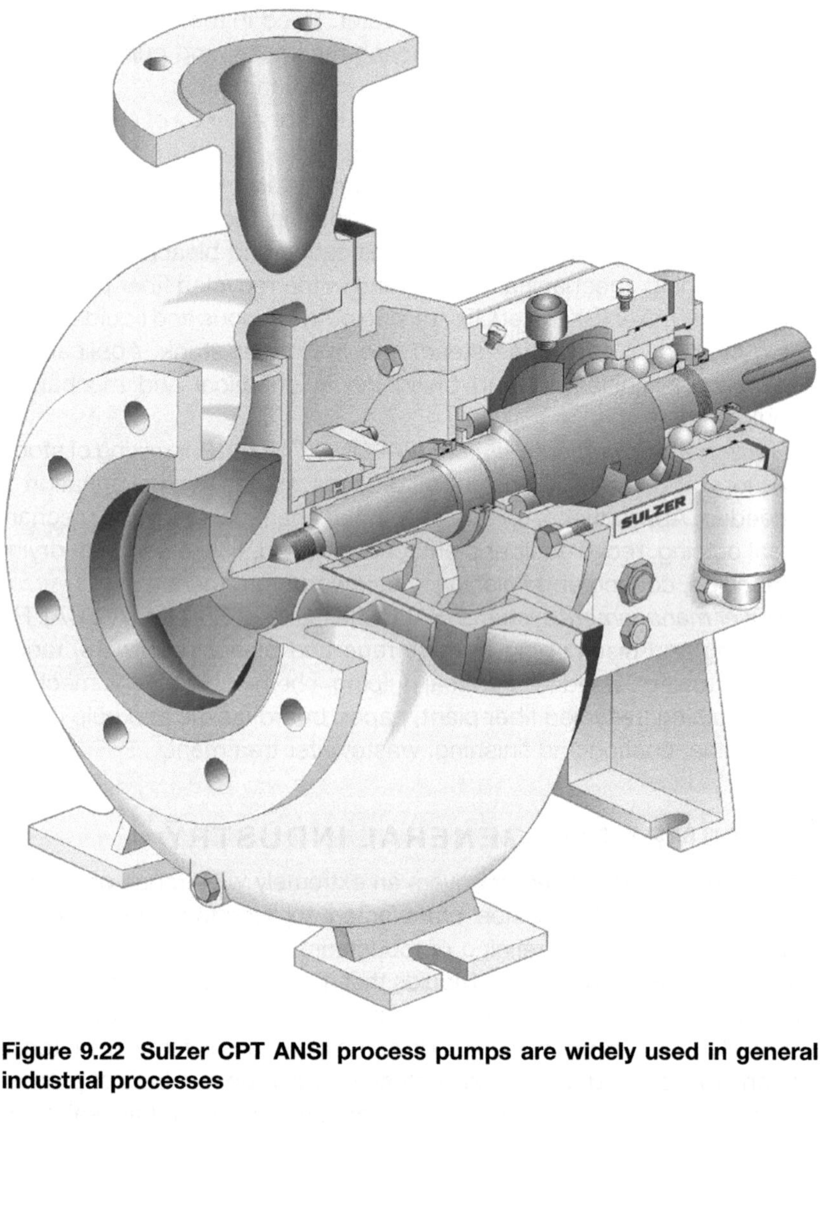

Figure 9.22 Sulzer CPT ANSI process pumps are widely used in general industrial processes

starch or sugar-rich biomass into simple sugars, fermentation, distillation and stillage handling. A more advanced second generation bioethanol production process consists of hydrolysis of ligno-cellulosic biomass (a wide variety of non-food feed stocks can be used), fermentation and distillation. First generation biodiesel is produced by extraction and transesterification of vegetable oil or animal fats using alcohols (typically methanol). Second generation biomass to liquid (BTL) processes are based on gasification of biomass and catalytic conversion into liquid fuels.

Single-stage industrial process pumps can be applied to most duties in the main production flow stream, fuel transportation, loading and storage tank areas. Duplex stainless steel is the recommended material for most applications with a closed seal water circuit being commonly required. Wear-resistant pumps may be needed in the early process stages depending on the contents of abrasive solids in the feedstock. The biomass consistency and its effect on viscosity must be considered when selecting the pumps. In applications with high fuel concentration (or methanol in biodiesel process), all the relevant local safety regulations must be followed. Vertical pumps and horizontally split casing pumps are used in cooling towers. Other pumps in the processes are mostly seal-less acid pumps and dosing pumps.

9.6.2 Pumps for the metal refining industry

Generally metal refining processes can be divided into two main categories: pyrometallurgical and hydrometallurgical processes. In pumping terms, although these processes are based on different technologies they present similar challenges for the pump designer.

A typical pumping environment would be sulfuric acid-based solutions and slurries. Low pH values due to the sulfuric acid, impurities from the ore and mineral solids in the solution or slurry combine to create demanding pumping applications in terms of corrosion and erosion. Applications become more onerous with increasing temperature and the addition of impurities such as chlorides and fluorides. Solids carried in the pumped fluid break down the passive layer on metallic surfaces, accelerating corrosive attack. To resist corrosive attack the pump's metallic parts in contact with pumped media should be selected for the specific conditions the individual pump will operate under. This applies equally to major components as well as fasteners, etc. To avoid erosion the materials of construction must be sufficiently hard to provide acceptable life. An alternative to hard metals is to use elastomer-lined pumps although these designs have low maximum temperature limits. Whatever the solution adopted, the pumps should run at slow speed and operate as closely as possible to the point of maximum efficiency to minimize internal recirculation.

An application particularly associated with metal production is electrolyte pumping where the pumped liquid is electrically charged. Galvanic corrosion can aggressively attack the pump, especially if the system piping is lined with plastic. In such applications the pump must be electrically isolated from the ground to avoid galvanic corrosion.

Approximately 85% of metal refining industry applications can be covered with carefully specified horizontal or vertical process pumps. The remaining 15% require slurry pumps (both normal and heavy duty designs). Shaft sealing may be single or double mechanical seal, with dynamic seals a popular option.

9.6.3 Pumps for the fertilizer industry

In fertilizer manufacture pumping demand varies with the individual manufacturing process. Common impurities such as chlorides, fluorides and silicon make the risk of corrosion and erosion much higher than with pure liquids. Material issues are important in phosphoric acid applications due to high process temperatures, impurities and the presence of solids. In many applications concentrations of ammonium nitrate create special considerations for pump and seal selection as it constitutes a risk of explosion. Most applications can be safely covered by appropriately specified and selected industrial process pumps.

9.6.4 Pumps for the sugar industry

Sugar is produced in two principal ways, from beet or cane.

In beet production sugar is extracted or leached from sliced sugar beets (cossettes), after which juice is purified through a series of milk of lime and CO_2 process steps. The filtered juice is evaporated (thickened) and sugar is crystallized from it. The juice is conveyed from one process stage to another by process pumps (Fig. 9.23).

Cane sugar is produced by extracting sugar from crushed cane. Depleted cane (bagasse) can be used as fuel in plant power generation; paper or cardboard can also be manufactured from surplus bagasse. After extraction raw cane sugar is refined into granulated white sugar and other sugar products.

Depending on the capacity of the sugar factory (typically 4000 to 12 000 tons of beet processed daily), some 50 to 100 centrifugal pumps, mostly single-stage, are employed; typically they run at four pole speeds (1500 or 1800 rpm). For media containing abrasive matter, pumps made of wear-resistant materials are used. If there is a risk of clogging, pumps

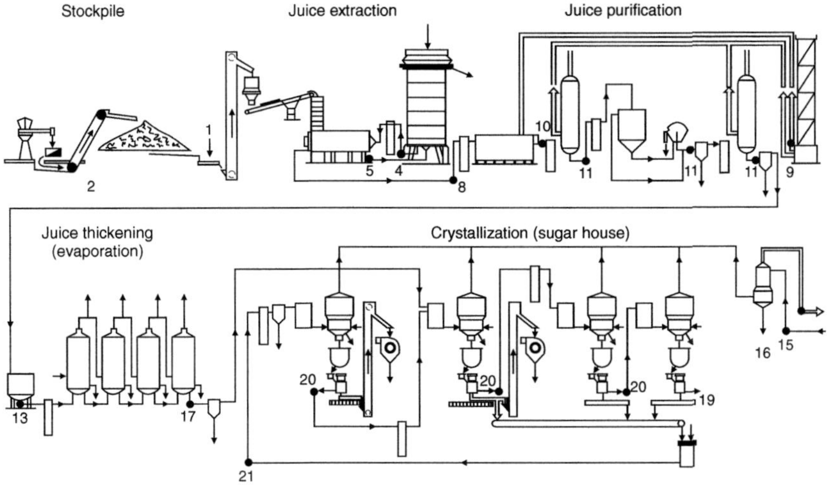

Process Flow Sheet

1 Wash water pump
2 Beet pump
4 Heating juice recirculation pump
5 Cossette-juice pump
8 Raw juice pump
9 Milk of lime pump
10 Prelimed juice pump
11 Carbonated juice pump
13 Thin juice pump
15 Condenser water pump
16 Cooling tower pump
17 Thick juice pump
19 Molasses pump
20 Run-off pump
21 Standard liquor pump

Figure 9.23 Sugar production from beet

designed with wider clearances are specified. In sugar crystallization degassing pumps have found a wide range of applications from pumping run-off liquids containing air/gas to replacing progressive cavity pumps in applications not previously suitable for centrifugal designs due to gas content in the fluid stream.

Index

Centrifugal Pump Handbook. DOI: 10.1016/B978-0-7506-8612-9.00014-0

E

F

G

H

I

9780750686129